Firefighters' Clothing and Equipment

Performance, Protection, and Comfort

Firefighters' Clothing and Equipment

Performance, Protection, and Comfort

Edited by
Guowen Song
Faming Wang

CRC Press
Taylor & Francis Group
Boca Raton London New York

CRC Press is an imprint of the
Taylor & Francis Group, an **informa** business

CRC Press
Taylor & Francis Group
6000 Broken Sound Parkway NW, Suite 300
Boca Raton, FL 33487-2742

First issued in paperback 2020

© 2019 by Taylor & Francis Group, LLC
CRC Press is an imprint of Taylor & Francis Group, an Informa business

No claim to original U.S. Government works

ISBN 13: 978-0-367-57068-2 (pbk)
ISBN 13: 978-1-4987-4273-3 (hbk)

Library of Congress Cataloging-in-Publication Data

Names: Song, Guowen, author. | Wang, Faming, author.
Title: Firefighters' clothing and equipment : performance, protection, and comfort / Guowen Song and Faming Wang.
Description: Boca Raton : Taylor & Francis, a CRC title, part of the Taylor & Francis imprint, a member of the Taylor & Francis Group, the academic division of T&F Informa, plc, 2019. | Includes bibliographical references and index.
Identifiers: LCCN 2018020266| ISBN 9781498742733 (hardback : alk. paper) | ISBN 9780429444876 (ebook)
Subjects: LCSH: Firefighters--Uniforms. | Firefighters--Equipment and supplies. | Fire extinction--Equipment and supplies.
Classification: LCC TH9395 .S66 2019 | DDC 628.9/2--dc23
LC record available at https://lccn.loc.gov/2018020266

Visit the Taylor & Francis Web site at
http://www.taylorandfrancis.com

and the CRC Press Web site at
http://www.crcpress.com

Contents

Preface

Personal protective equipment (PPE) and textile-based equipment are critical for firefighters to ensure their safety and health. Ineffective protection at a fire scenario with multiple hazards can cause injury and fatality as well as potentially increase property damage and loss. Fire reports [1] confirm that in the past four decades in the United States about 68,000 firefighters received burn injuries with more than 60 fatalities each year. The best approach for firefighters to mitigate burn injuries and reduce risk of death from unpredictable hazards is to apply high-performance PPE.

Firefighters encounter complex environments and conditions while performing their duties within a wide range of possible hazards. Thermal exposure, which may result from radiation, convection, hot liquid, steam, and/or hot solids, is the primary possible hazard exposure for firefighters. During combustion of structural materials, firefighters can encounter thermal hazards including collapsing fireground debris, hot liquid, and molten materials. In a fire scene, cool water from a hose can quickly become hot water, and then steam. Steam and wet air cause more serious burns because more heat energy can be stored in water vapor than in dry air. On the other hand, in some geographic regions, severe winter weather with sub-zero temperatures poses cold injury threats such as frostbite to firefighters, especially when the PPE system gets wet by sweat or hose water.

Current firefighter protective ensembles are heavy and stiff and suffer from reduced vapor permeability, all of which increase physiological strain. Statistics showed that overexertion, physical and thermal stresses, and medical issues account for 42% of the main causes of deaths. PPE are engineered with not only increased thermal protection but also increased bulk and weight. This affects efficiency and mobility, increasing the metabolic cost of work by up to 50% [2]. Furthermore, the fabric thickness and moisture barrier layers restrict body heat dissipation and create additional undue heat stress. The heat generated from working muscles, as well as the heat transferred from the local environment, generate increased thermoregulatory strain, putting more demand on the cardiovascular system. Uncompensated heat strain will greatly affect the performance, function, and health of the firefighters.

The current system tends to store large amounts of thermal energy during exposure to fire hazards, and this amount of stored energy can be discharged to the skin. Studies [3–5] have demonstrated that stored thermal energy contributes significantly to skin burn injuries, specifically compression burns. As a result, firefighters' arms and legs, knees, elbows, and shoulders, where SCBA (self-contained breathing apparatus) straps press the surrounding fabric against the skin, are vulnerable to burn injuries from stored energy discharge by compression. This compress burn actually relates to another major issue of concern for firefighters on their ability to sense the heat of the fire.

Additionally, existing product standards and testing protocols are not adequately developed to evaluate the risks caused by those hazards when combined with moisture on performance. During firefighting, protective clothing becomes wet from internal and external sources. At a high ambient temperature or during strenuous activity, the wearer perspires profusely, so clothing next to the skin becomes saturated with perspiration. Studies [6,7] show that the presence and distribution of moisture have a complex effect on heat transfer through insulating materials. In some occasions, if water vapor transfers to the human skin and condenses, steam burns may occur, as the water vapor also transfers the heat it absorbed to evaporate.

The core challenges for current PPE used for firefighters are the engineering and design of multifunctional performance for the high-level protection with minimum physiological burdens. The solution for this relies on the next-generation new textile materials, new discoveries on functional design and novel technology, as well as the understanding on mechanisms associated with heat and mass transfer in the human-clothing-environment system.

With this goal, we have developed this volume that presents an overview of the current state of understanding and knowledge for protective clothing and equipment, as well as issues and challenges associated with firefighter and other emergency first responders. This book includes 12 chapters and covers discussion on textile materials, clothing comfort and protective performance, human thermoregulation system, relevant methods and standards, instrumentation technologies for comfort and protection, 3D body scanning application, and human trials. In addition, future trends on smart firefighting clothing/equipment and numerical modeling and human skin burn are also presented.

We sincerely hope our efforts on this book will provide useful information and the present knowledge regarding hazards, firefighters, and protective clothing and equipment. This book may serve as a useful tool and technical source for scientists in textiles and clothing, mechanical engineering, and occupational safety and health. It is also our expectation that the book will provide a fundamental guide to educators, engineers, ergonomists, industrial hygienists, and designers in universities, research institutes, fire stations, and industrial companies.

We would like to express our sincere appreciation to all the authors and guest reviewers who devoted considerable effort to this book. We would also like to extend our thanks to the production team at Taylor & Francis for their patience with the editors and authors.

Guowen Song
Faming Wang
April 2018

References

1. Fahy, R. F., LeBlanc, P. R., Molis, J. L. (2017) Firefighter Fatalities in the United States-2016. Report, National Fire Protection Association, 2017. Available from https://www.nfpa.org/-/media/Files/News-and-Research/Fire-statistics/Fire-service/osFFF.pdf

2. Selkirk, G., McLellan, T. M. (2004) Physical work limits for Toronto firefighters in warm environments. *Journal of Occupational and Environmental Hygiene*, 1(4):199–212.

3. Song, G. et al. (2011) Thermal protective performance of protective clothing used for low radiant heat protection. *Textile Research Journal*, **81**(3):311–323.

4. Song, G., Chitrphiromsri, P., Ding, D. (2008) Numerical simulations of heat and moisture transport in thermal protective clothing under flash fire conditions. *International Journal of Occupational Safety and Ergonomics*, **14**(1):89–106.

5. Song, G., Barker, R. (2005) Effect of thermal stored energy on prediction of skin burn injury. In: *Proceedings of the 84th Textile Institute World Conference*, Raleigh, North Carolina, 2005.

6. Zhang, H., Song, G., Ren, H., Cao, J. (2018) The effects of moisture on the thermal protective performance of firefighter protective clothing under medium intensity radiant exposure. *Textile Research Journal*, **88**(8):847–862.

7. Barker, R. L., Guerth-Schacher, C., Grimes, R. V., Hamouda, H. (2006) Effect of moisture on the thermal protective performance of firefighter protective clothing in low-level radiant heat exposures. *Textile Research Journal*, **76**(1):27–31.

Editors

Guowen Song received his PhD degree in textile engineering, chemistry, and science at North Carolina State University's College of Textiles in Raleigh, North Carolina. Currently he is the Noma Scott Lloyd Chair in the Department of Apparel, Event and Hospitality Management (AESHM) at Iowa State University's College of Human Sciences. Song's academic interest is in functional textile materials, protective clothing, and systems to improve human health and safety. His work involves modeling studies of human physiology, textile materials, and protective clothing, development of devices and test protocols, and analysis of textile and clothing performance. His current focus is to establish a unique, multifaceted, cross-disciplinary research and education program that can integrate theoretical study, new technology discovery, and engineering with the aim of revolutionizing clothing system function and performance. These studies include lab simulations, application of instrumented manikin technology, and specially designed human trials, including 3D body scanning, and a human motion analysis approach. Dr. Song has published more than 100 scientific papers in peer-reviewed journals and conference proceedings. He authored several books and contributed a dozen chapters to books in his field of study.

Faming Wang is currently an assistant professor in clothing technology at the Institute of Textiles and Clothing (ITC) of The Hong Kong Polytechnic University. He earned his LicPhil and PhD degrees from Lund University (Sweden) under the supervision of world-renowned physiologist Professor Ingvar Holmér. After his PhD training, Dr. Wang joined Eidgössische Materialprüfungs-und Forschungsanstalt (EMPA Swiss Federal Laboratories for Materials Science and Technology, the ETH Domain, Switzerland) as a Marie-Curie Fellow. He later became a full professor (the youngest full professor in textiles and clothing in the history of Mainland China) in apparel design and engineering at Soochow University (Suzhou, China) in October 2013, and there he established the Laboratory for Clothing Physiology and Ergonomics (LCPE), a multidisciplinary research group for the study of the thermal interaction of the human body-clothing-environment system. To date, he has authored or coauthored more than 200 journal publications, conference papers/presentations, technical reports, books/chapters, and patents. His published work has been cited more than 1,000 times, and this gave him an h-index of 19 (Google Scholar, as of April 2018). In addition, he serves as an editorial member for several journals including *Journal of Thermal Biology* (a JCR-Q1 journal). He is also a founding member of the Asian Society of Protective Clothing (ASPC).

Contributors

Simon Annaheim
Laboratory for Biomimetic
    Membranes and Textiles
Empa-Swiss Federal Laboratories
    for Materials Science
    and Technology
St. Gallen, Switzerland

Martin Camenzind
Laboratory for Biomimetic
    Membranes and Textiles
Empa-Swiss Federal Laboratories for
    Materials Science and Technology
St. Gallen, Switzerland

Anna Dąbrowska
Central Institute for Labour
    Protection - National Research
    Institute (CIOP-PIB)
Department of Personal
    Protective Equipment
Lodz, Poland

Chuansi Gao
Thermal Environment Laboratory
Division of Ergonomics and Aerosol
    Technology
Department of Design Sciences
Faculty of Engineering
Lund University
Lund, Sweden

Ying Ke
School of Textiles and Clothing
Jiangnan University
Wuxi, China

Ziqi Li
Institute of Textiles and Clothing
    (ITC)
The Hong Kong Polytechnic
    University
Hung Hom, Kowloon, Hong Kong,
    China

Xiao Liao
The Hong Kong Research Institute of
    Textiles and Apparel (HKRITA)
Hong Kong SAR

Sumit Mandal
Laboratory for Biomimetic
    Membranes and Textiles
Empa-Swiss Federal Laboratories for
    Materials Science and Technology
St. Gallen, Switzerland

Emel Mert
Laboratory for Biomimetic
    Membranes and Textiles
Empa-Swiss Federal Laboratories for
    Materials Science and Technology
St. Gallen, Switzerland

Nazia Nawaz
Human Ecology and Clothing Science
School of Fashion and Textiles
RMIT University
Melbourne, Australia

Rajiv Padhye
School of Fashion and Textiles
RMIT University
Melbourne, Australia

Agnes Psikuta
Laboratory for Biomimetic
    Membranes and Textiles
Empa-Swiss Federal Laboratories for
    Materials Science and Technology
St. Gallen, Switzerland

René M. Rossi
Laboratory for Biomimetic
    Membranes and Textiles
Empa-Swiss Federal Laboratories for
    Materials Science and Technology
St. Gallen, Switzerland

Abu Shaid
School of Fashion and Textiles
RMIT University
Melbourne, Victoria, Australia

Yun Su
Department of Apparel, Events and
    Hospitality Management (AESHM)
College of Human Sciences
Iowa State University
Ames, Iowa

Olga Troynikov
Human Ecology and Clothing Science
School of Fashion and Textiles
RMIT University
Melbourne, Australia

Udayraj
Institute of Textiles and Clothing
    (ITC)
The Hong Kong Polytechnic
    University
Hung Hom, Kowloon, Hong Kong,
    China

Lijing Wang
School of Fashion and Textiles
RMIT University
Melbourne, Australia

Mengying Zhang
Department of Apparel, Events and
    Hospitality Management (AESHM)
College of Human Sciences
Iowa State University
Ames, Iowa

Textiles for Firefighting Protective Clothing

Abu Shaid, Lijing Wang, and Rajiv Padhye

1.1 Introduction

Firefighters' protective clothing (FPC) is the part of the safety outfit of firefighters on duty that protects them from dangers associated with heat, flame, hot or toxic liquid contact, abrasion, cuts, etc. No clothing material can withstand continuous exposure to flame or can provide comfort for an infinite time in hot environments. Hence, fire protective clothing does not necessarily mean the fabric is completely resistant to fire and heat. FPC is designed to save the firefighters from excessive heat and flash fire conditions by allowing them a time gap for a rescue mission, fighting fire, or withdrawing from direct flame contact. FPC fabrics are required not only to ensure that the clothing does not become a means of secondary ignition and spreading of fire and causing injury, but also to provide a certain degree of comfort from hot and humid situations both externally and internally, while still maintaining acceptable working efficiency through easy and quick movement. Hence, the fiber characteristic, heat source, intensity, time of exposure, and many other variables affect the protection performance. FPC typically consists of an overcoat, trouser, hood, and gloves. Other non-clothing assemblies may include self-contained breathing apparatus (SCBA), hand tools, ropes, etc. In this chapter, fibers and fabrics used in FPC will be primarily discussed, the fire retardant finishing on textiles will be noted, and finally the protection and comfort properties of FPC will be summarized.

1.2 Fibers Used in Firefighters' Protective Clothing

In general, FPC is a multilayer assembly composed of various types of woven and nonwoven fabrics. Conventional fibers such as cotton, wool, and viscose and high-performance fibers such as aramid, polybenzimidazole (PBI®), and polybenzoxazole (PBO) are used in FPC. Each type of fiber has its own advantages and disadvantages. One fiber may be effective in protecting from heat but may not be comfortable enough to wear; one may be comfortable but can be very expensive. In protective concern, one particular fiber may possess high tensile strength but may lack heat resistance, while another may have exceptionally good heat resistance but may lack tensile strength. From comfort perspective, one fiber may have good feel with smooth and soft handle but may lack moisture absorption, while another may have good moisture absorption properties but may lack of handle. The choice of fiber or fiber blend for FPC is challenging and requires deep understanding on the requirement of fiber properties. A balance between protection and comfort properties is essential for FPC.

Desired performance for FPC often needs to compromise with cost, availability, and processing limitation. To simplify the discussion, the fiber choice for FPC can be summarized in two general groups.

> **Group A:** High-performance fibers that are inherently flame-retardant (FR), such as polyamide (Kevlar®, Nomex®), polyimide fiber (P84®), PBI®, polybenzoxazole (Zylon®), modacrylic or oxidized acrylic (semicarbon), etc. Flame retardancy is inhabited into their structure at their synthesis stage (Holmes, 2000a), and these fibers have a limited oxygen index (LOI) value above 21% as shown in Figure 1.1 and Table 1.1. LOI indicates the percentage of oxygen that has to be present to support combustion after ignition (ASTM D2863-00—minimum oxygen concentration to support candle-like combustion of a polymer).
>
> **Group B:** Conventional fibers from both natural and synthetic origin that are not normally FR, but modified to do so after their natural production or synthesis, such as FR cotton, FR viscose, FR wool, etc.

According to Horrocks (2016), the conventional fibers treated with FR are usable up to 100°C continuous use, and those high-performance fibers are suitable for continuous use above 150°C. In this way, Group B fibers are more useful in flame-resistant bedding, upholstery, curtains, etc. other than firefighting application. Hence, some key high-performance inherently FR fibers from Group A are discussed here, and only polyester and FR viscose are noted in this chapter.

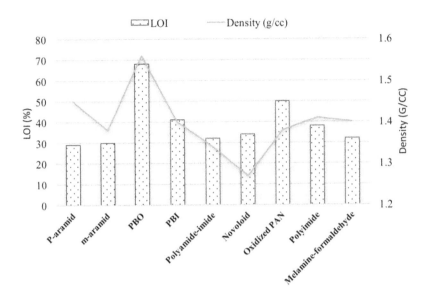

Figure 1.1

The limiting oxygen index and density of some prospective fibers for FPC.

1.2.1 Aromatic Polyamide

Aromatic polyamides (Figure 1.2) or aramid fibers are the most common fibers used in firefighters' garments. In aramid fibers, at least 85% of the amide linkages are attached directly to two aromatic rings; this distinguishes them from conventional polyamide fibers like nylon, which contains mostly aliphatic or cyclo-aliphatic units (Yang, 1993).

Among both types of aramid in Figure 1.2, poly-*p*-phenylene terephthalamide or *p*-aramid fiber is a liquid crystal polymer that is spun as a bundle of oriented polymer chain (crystal) and does not require any additional drawing step after immersion through the spinneret.

Liquid crystal polymer (LCP) fibers are either lyotropic or thermotropic. Lyotropic LCP, such as aromatic polyamide (*p*-aramid), have a high melting point that is close to their decomposition temperature. Hence, these polymers are not melt-spun but wet-spun. On the other hand, thermotropic LCP fibers such as aromatic polyester (Vectran®) are spun in the melt spinning process as these polymers melt at high temperature. Vectran will melt before decomposition when it is heated while *p*-aramid will produce char. The molten polymer will cause extra hazards while the char will provide an extra layer of heat and flame protection. Hence, Vectran is not used in FPC, but *p*-aramid is used. The prime attractions for the *p*-aramid fiber are its very high tensile strength, tear resistance, flame retardancy (see Table 1.1), and high stability to chemicals. *P*-aramid fiber is self-extinguishing and also possesses low electric conductivity. At the same time, a drawback of aromatic polyamide is their poor light stability together with loss of tear resistance (Lempa, 2009). *P*-aramid is available under different trade names from different manufacturers worldwide such as the most versatile Kevlar® fiber by DuPont, Technora® by Teijin, and Twaron® by Acordis.

Table 1.1 Summary of Some Prospective Fibers Which Have Favorable Characteristic to be Used in FPC in One or More Aspect

Fiber	T_m, °C (Melt)	T_p, °C (Pyrolysis)	T_i, °C (Ignition)	Continuous Operating Temp	LOI vol%	Density, g/cm³	Moisture Regain, %	Strength (cN/dtex)	Price* (US$/kg)
m-Aramid (Nomex)	375–430	425	> 500	200	29–30	1.38	4.5	4.8	20
p-Aramid (Kevlar)	560	> 590	> 550	190	29	1.45	4.5	20.3	25
Polyimide (P84®)	** (315, Tg)	**	**	260	38	1.41	3	3.5–3.8	**
Polyamide-imide (kermel®)	** (> 315, Tg)		**	200	30–32	1.34	~4	2.45–5.88	**
PBO (Zylon®)	650, decompose	**	**	310	68	1.56	0.6–2	37	130
PBI	**	> 500	**	250	41	1.43	15	2.4	180
Polytetrafluoroethylene (PTFE)	> 327	400	** (560, combustion)	275	95	2.2	0	1.4	50
Melamine fiber (Basofil®)	**	**	**	190–200	30–32	1.4	5	2–4	16
Polyester	255–260	420–447	480		17–20	1.38	0.4	8	3

(Continued)

Table 1.1 (Continued) Summary of Some Prospective Fibers Which Have Favorable Characteristic to be Used in FPC in One or More Aspect

Fiber	T_m, °C (Melt)	T_p, °C (Pyrolysis)	T_i, °C (Ignition)	Continuous Operating Temp	LOI vol%	Density, g/cm³	Moisture Regain, %	Strength (cN/dtex)	Price* (US$/kg)
Oxidized PAN fiber/semi-carbon fiber (Panox®)	**	**	**	200	55	1.38	10	1.6–1.7	10
Novoloid/cured phenol-aldehyde fiber (Kynol®)	**	>180	>2500	200	30–35	1.27	6	1.2–1.6	15–18
PVC	>180	>180	450	**	37–39	1.4	0	2.4–2.7	**
Cotton	**	350	350	**	16–18.4	1.52	7–8	1.5–4	**
Wool	**	245	570–600	**	25	1.31	14–18	1.1–1.4	**

Source: Bajaj, P., & Sengupta, A. K., 1992, Protective Clothing. *Textile Progress*, 22(2–4), 1–110; Bourbigot, S., & Flambard, X., 2002, *Fire and Materials*, 26(4–5), 155–168; Reprinted from R. A. B. L. Deopura, M. Joshi, & B. Gupta (eds.), *Polyesters and Polyamides*, Butola, B. S., Advances in Functional Finishes for Polyester and Polyamide-Based Textiles, pp. 306–325, Copyright 2008, with permission from Elsevier; Evonik, 2017b, P84® Fibre Characteristics. Retrieved 16 Feb. 2017 from http://www.p84.com/sites/lists/RE /DocumentsHP/p84-fibre-technical-brochure.pdf; Hearle, J. W., 2001, High-Performance Fibres: Elsevier; Horrocks, A., 2005, Thermal (Heat and Fire) Protection. *Textiles for protection*. Woodhead Publ. Ltd, Cambridge, 398–440; Horrocks, A., 2016, Technical Fibres for Heat and Flame Protection. *Handbook of Technical Textiles: Technical Textile Applications*, 2, 243; Kermel, 2013, Kermel®. High Performance Fibre. Retrieved 20 Feb. 2017 from http://www.kermel.com/fr/Production-of-High-Tech-non-flammables -Fibres-640.html; Kynol, 2012, Kynol Novoloid Fibers. Retrieved 20 Feb. 2017 from http://www.kynol.de/products.html; SGL Group, 2017, Panox® Thermally Stabilized Textile Fiber. Retrieved 20 Feb. 2017 from http://www.sglgroup.com/cms/international/products/product-groups/cf/oxidized-fiber/index.html?_locale=en; Swicofil, 2015, Polyamide-Imide PAI. Retrieved 20 Feb. 2017 from http://www.swicofil.com/products/223polyamideimide.html; Toyobo, 2015, The Strongest Fiber with Amazing Flame Resistance. Retrieved 14 Feb. 2017 from http://www.toyobo-global.com/seihin/kc/pbo/zylon_features.html.

* Price shown is for comparison only which is derived from ref (Hearle, 2001). Price does not reflect current market price.

** Not applicable or data not available.

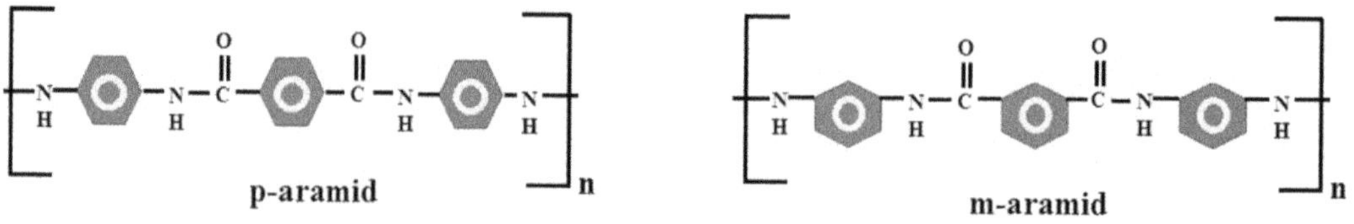

Figure 1.2

Chemical structure of aramid fiber.

Poly-*m*-phenylene isophthalamide or *m*-aramid shown in Figure 1.2 is another aromatic polyamide (Cook, 1980; Gohl, 1985; Moncrieff, 1975) commonly available in FPC. *M*-aramid fiber is also available under various trade names such as Nomex® and Corex®. *M*-aramid fibers are not as strong as the *p*-aramid but have superior heat resistance properties. This particular characteristic makes them a prime choice for the thermal liner in FPC.

1.2.2 Polybenzimidazole Fiber (PBI and PBI Gold)

Polybenzimidazole or PBI is nonflammable in air under normal conditions (Jackson, 1978). Though it is a costly fiber, it offers improved thermal protection and flame resistance property. The Federal Trade Commission (FTC) defined PBI as a manufactured fiber in which the fiber-forming substance is a long-chain aromatic polymer having recurring imidazole groups as an integral part of the polymer chain (Gooch, 2011). Imidazole derivatives are well known for their stability to high temperature (645°C) and resistance to the most drastic treatment with acids and bases (Hofmann, 1953). PBI is the condensation polymer from imidazole, which contains the repeating cores of benzimidazole (Vogel & Marvel, 1961) and also exhibits extraordinary heat stability. It is polymerized from the condensation reaction of tetra-aminobiphenyl and diphenyl isophthalate (Hagborg et al., 1968; Jackson, 1978). Though the PBI polymer development was started around the mid-1950s (Jackson, 1978), they became widely recognized in 1961 when Vogel and Marvel (1961) developed PBI with wholly aromatic structure, as shown in Figure 1.3. Due to the exceptional thermal stability of PBI, NASA started to use PBI in their space suits to protect the astronauts from fire in 1969. PBI was introduced for FPC in the 1980s to replace the old-fashioned Kevlar-leather-Nomex turnout gear and PBI continues its success for FPC. Many of the firefighters who died in the 9/11 terrorist attack were identified only by their PBI turnout gear (Pearson, 2007).

PBI can be processed in conventional textile machinery and can be blended with other similar fibers. Hence, to optimize the cost in relation to the desired

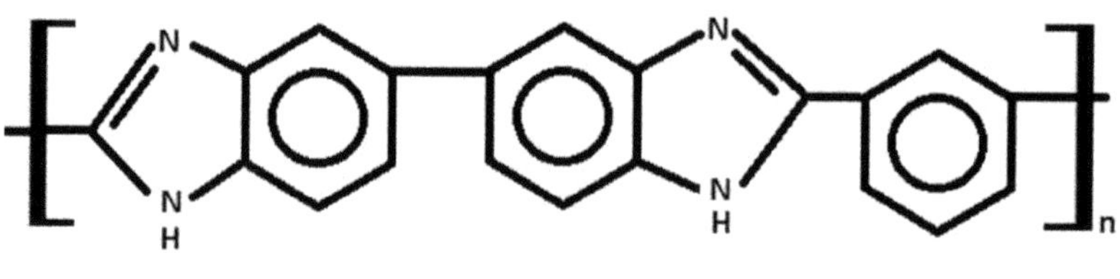

Figure 1.3

PBI structure.

1. Textiles for Firefighting Protective Clothing

performance, PBI is often blended with aramid. Researchers found that a 60/40 blend of aramid (Kevlar) and PBI can offer optimal overall fabric performance (Bajaj & Sengupta, 1992; Barker & Coletta, 1986). This blend ratio became well known by the name of PBI Gold. As a single fiber, PBI is much weaker than Kevlar as shown in Table 1.1. However, PBI Gold is reported for its high tear strength and high heat resistance in the FIRES (Firefighters Integrated Response Equipment System) project report by the International Association of Firefighters (IAFF) (Gore, 2010).

1.2.3 Polyimides

P84® is an inherently nonflammable, non-melting polyimide fiber manufactured by Evonik Fibers (Evonik, 2017a). Its structure is shown in Figure 1.4. The LOI of P84 is 38% and glass transition temperature (Tg) is 315°C, which are much better than aramids (Kevlar, Nomex, and PBI), as seen in Table 1.1. The fiber starts to carbonize above 370°C and is suitable for continuous use at temperature up to 260°C. P84 has suitable feel, comfort, and performance properties to be used on their own or in blend with FR viscose for protective clothing (Bajaj & Sengupta, 1992).

1.2.4 Polyamide-imide

Kermel® is a polyamide-imide fiber under the classification of meta-aramid. Its structure is shown in Figure 1.5. In France, Kermel is especially used for FPC (Bajaj & Sengupta, 1992). The fiber can withstand very high temperature up to 1000°C for a short time exposure such as flash fire condition. The fiber itself does not melt or decompose, but it slowly chars when exposed to very high temperature. Hence it is suitable as the outer layer of firefighter protective clothing. Kermel fiber is soft to handle due to its smooth circular surface and can be solution dyed. Kermel can be found in turnout gear as a blend with other high-performance fiber.

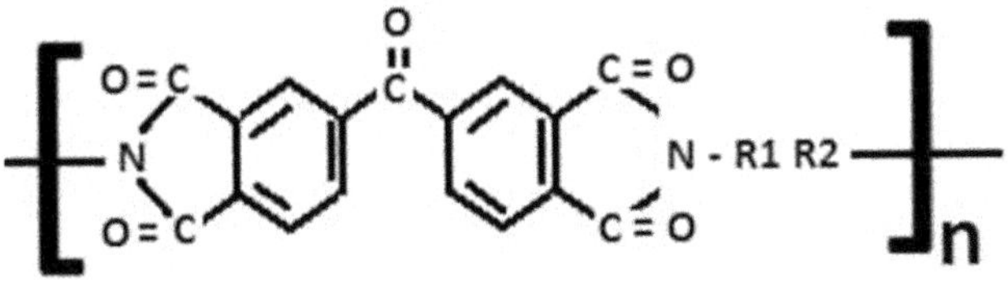

Figure 1.4

P84® chemical structure.

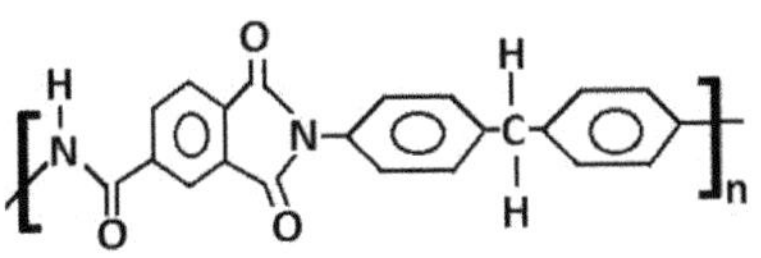

Figure 1.5

Kermel fiber structure.

1.2.5 PBO (Polybenzoxazole)

Zylon is another liquid-crystalline PBO (poly(p-phenylene-2,6-benzobisoxazole)) fiber, which is around 1.6 times stronger than Kevlar and 10 times stronger than PBI fiber. Its decomposition temperature is 650°C and LOI is 68% (Toyobo, 2015). Thus, the flame resistance and heat resistance of Zylon are significantly higher than those of PBI and aramid. However, the PBO fiber is not only heavier than aramid and PBI but also more costly. Hence, PBO often blends with aramid or PBI to provide a balance performance in FPC.

1.2.6 Novoloid (Cured phenol-aldehyde) Fibers

Novoloid is a highly flame-resistant fiber with soft handle and comparatively better moisture regain. Kynol® is a commercially available Novoloid fiber manufactured by GunEi Chemical, Japan (GCI, 2017). The fiber contains at least 85% crosslinked novolak, and is infusible and insoluble. In flame exposure, Kynol does not melt but gradually chars to complete carbonization. A Kynol fabric of 290 g/m² (gsm) can withstand 2500°C for 12 without breakthrough (Kynol, 2012), while the practical temperature to use is only around 150°C for long-term application. However, though Kynol has soft handle and comparatively higher moisture regain, it is comparatively a weak fiber with low tenacity. It can be blended with other high-performance fibers such as Nomex or FR viscose for improved fabric physical properties (Bajaj & Sengupta, 1992). Thus, novoloid fibers may not be usable as an outer layer, but they are ideal candidates for nonwoven felt structure for batting material in the thermal liner of FPC.

1.2.7 Oxidized PAN

The partial carbonization of polyacrylonitrile fiber results in oxidized PAN or semi-carbon fiber. Unlike carbon fiber, semi-carbon fiber retains acceptable textile properties after high-temperature oxidation. Semi-carbon fibers were commercially available from different manufacturers such as Celanese, Courtaulds, Asahi, SGL, and so on, but many of them are now obsolete. Panox® from SGL is an example of oxidized PAN fiber with 62% carbon content, which is produced by thermal stabilization of PAN in 300°C (SGL Group, 2017). The fiber does not burn, melt, or soften to drip. Panotex fabric made from Panox can withstand direct flame contact temperature in excess of 1000°C and is resistive to most common bases and acids (Bajaj & Sengupta, 1992). However, it has low strength. Hence, it is used in thermal liners of FPC as a blend with other fibers.

1.2.8 Melamine Fiber

BASF developed a comparatively cheaper and inherently flame-retardant melamine fiber called Basofil® in the 1980s (Murugesan & Gowda, 2012). It is a melamine-based staple fiber with high-temperature resistance properties. The main drawback of this fiber is its low strength. Hence, it can be blended with other high-performance fibers such as aramid to improve the strength and abrasion resistance (Horrocks, 2016). The prospect of its use in FPC is in the form of felt or nonwoven as a thermal barrier material.

1.2.9 Inorganic and Ceramic Fiber

Glass and ceramic fibers may be used in protective apparel primarily as nonwoven felt by blending with other fibers. Detailed properties of these fibers are extensively discussed in research literature (Bourbigot & Flambard, 2002; Moncrieff, 1975).

1.2.10 Conventional Fibers with Added Flame Retardancy

Some conventional textile fibers are used in FPC in blend or in rare case as a single internal layer. FR finishing of cellulose fiber is discussed in Section 1.4. In this section, FR polyester and viscose will be discussed in brief. FR additives are added during their synthesis process instead of FR chemical finishing.

a. Flame-Retardant Polyester

The linear polyester is normally highly flammable as it produces a variety of volatile and flammable products upon heating. Among the varieties of polyester fibers, poly(ethylene terephthalate) (PET) is the most common and widely used polyester fiber. By employing suitable flame-retardant strategies, the thermoplastic polyester can be made resistive to flame. The use of reactive flame-retardant comonomers, such as 2-carboxyethyl(methyl)phosphinic acid or 2-caboxyethyl(phenyl) phosphinic acid, is an effective strategy to achieve long-term fire-retardant effect. HEIM fiber (Fukui et al., 1973) commercialized by Toyobo is an example of FR polyester fiber.

b. FR Viscose

In 1974, researchers from Sandoz Ltd added phosphorus containing flame retardant (dioxaphosphorinane derivatives) in cellulose xanthate dope during the production of viscose rayon to impart flame retardancy (Mauric & Wolf, 1980). More recently, a patent was filed in 2015 and published in January 2017 where it is claimed to produce molded cellulose bodies (fiber, filament, nonwoven) with flame-resistant properties from cellulose and melamine cyanurate solution in an organic solvent (Niemz et al., 2017).

1.3 Membranes Developed for Protective Clothing

Membranes used in FPC are extremely thin (about 10 µm) microporous or hydrophilic polymeric film (Holmes, 2000c). Membranes used in protective clothing can be permeable, semipermeable, selectively permeable, or completely impermeable depending on the specific need of protection. Toxicological agents protective suit used by military is made of an impermeable film that does not allow liquid to pass through, not even water vapor. However, the standard-issue military chemical protection suits allow vapor exchange through a semipermeable absorptive carbon liner where the membrane blocks the liquid passing through but carbon absorbs the chemicals while the vapor passes through the fabric (Schreuder-Gibson et al., 2003). In this section, the membranes suitable for FPC will be discussed under two categories, vapor transportation through pores and vapor transportation by absorption.

1.3.1 Porous Membrane

Porous "perm-selective" membrane (Zhou et al., 2005) is a common element in various protective apparel like firefighters' and chemical protective clothing. It shows selectivity with respect to molecular solubility and diffusion through the polymer structure (Schreuder-Gibson et al., 2003). Water vapor passes through while the organic molecules are blocked. The selective membrane can be made of cellulose acetate, poly(vinyl alcohol), cellulosic cotton, and poly(allylamine) (Schreuder-Gibson et al., 2003). It is used in FPC as a moisture barrier.

Porous Gore-Tex® expanded PTFE (ePTFE) membrane is most widely used as a moisture barrier. Traditional Teflon (PTFE) seal tape for pipe joint breaks when stretched beyond a certain limit. However, a sudden hot stretch can allow it to expand about 800%, which creates thousands of tiny pores that allow the moisture vapor to pass through. In fact, Gore-Tex membrane contains more than 9 billion pores per square inch (Gore, 2017). It acts as a barrier to liquid moisture (water droplet) but allows water vapor (sweat) to escape to the environment. In 1969, Gore realized the potential of the pores and patented ePTFE membrane/film. Gore-Tex membrane acts as the moisture management medium. In FPC, various Gore-Tex moisture barrier layer fabrics are used such as CROSSTECH®, GORE® RT7100, GORE® PARALLON™, etc.

Alternatives of commonly used Gore-Tex membrane are also slowly emerging. Apart from ePTFE membrane, microporous membranes from PU (polyurethane) are also manufactured and marketed for FPC (Mukhopadhyay & Midha, 2016). PVDF (polyvinylidene fluoride) membrane is also available from other manufacturers for the same purpose (Holmes, 2000c) of Gore-Tex membrane. Nanofiber web or foam is another alternative to ePTFE membrane. Casting of an electrospun nanofiber layer on base fabric is another mechanism of pore formation and imparting breathability in protective clothing (Raza et al., 2014). Bagherzadeh et al. (2012) sandwiched electrospun nanofiber web between woven fabric layers to prepare a breathable textile barrier and compared its performance with Gore-Tex membrane. Serbezeanu et al. (2015) electrospun polyimide membranes on Kevlar base fabric to prepare a barrier material. Open cell foam is another porous vapor transmission technique that is used in protective clothing (Holmes, 2000a).

1.3.2 Nonporous Membrane

Unlike semipermeable and perm-selective membranes, impermeable types of membranes do not contain any pores. These can be polyester or polyurethane film, chemically modified to gain hydrophilic nature through the amorphous region of their polymeric system (Holmes, 2000c). Thus, the solid film prevents the liquid drops whereas the water vapor is attracted by the hydrophilic group of the chain modifying agent. Driven by the vapor density and heat, water vapor diffuses in the polymer system and finally escapes to the environment from the opposite surface.

1.3.3 Combined Porous–Nonporous Membrane

This is a combination of microporous and hydrophilic membrane. In this case, a nonporous hydrophilic coating is applied to a microporous membrane. In traditional microporous membrane, pinholes or oversized pores may cause water leakage. On the other hand, the pores can also be blocked by contamination (e.g., body oil, dirt, or other foreign materials). However, in this type of combined membrane, the nonporous hydrophilic layer seals the microporous membrane and offers better performance.

1.4 Fabrics and Nonwovens Used in FPC

Firefighters' protective gear is a system or assembly combining both textiles and nontextiles to serve the sole purpose of keeping firefighters safe and functional in various hazardous situations. The turnout coat worn by the firefighters has traditionally been a multilayer structure containing an outer shell, moisture barrier, and thermal barrier. The outer layer and the face cloth of the thermal barrier of FPC are normally plain or twill-woven fabric, whereas the moisture barrier and batting of the thermal barrier are nonwoven. In this section, the textile component of FPC (turnout coat and trouser) will be discussed. The following sections briefly highlight these three layers of FPC.

1.4.1 Outer Shell Fabric

The outer shell is the first line of defense for firefighters. The protection required has various aspects and no single fabric can meet all of those. The outer shell fabric should protect the firefighters from fire in flash-fire conditions or when entering a burning building. It needs to have sufficient tensile strength with acceptable abrasion and cut resistance to support crawling or climbing in rescue missions. FPC also needs to be lighter, flexible, and breathable to avoid heat stress and hindrance in movement. Hence, the fiber choice for outer layer is important and needs to consider price, performance, and comfort. The fiber for the outer layer is traditionally selected from any of the high-performance inherently flame-retardant fiber from Group A as discussed in Section 1.2. As approximately 21% oxygen is present in air, any fiber with a LOI value over 21% will not support combustion in air. Thus, the higher the LOI value is, the lower the flammability risk will be. In general, a fabric to be defined as flame retardant should have a minimum LOI value of 26–28% (Horrocks, 2016). Aramid, PBI, and their blends are commonly found in the outer layer fabric of current FPC due to their price and favorable properties. However, due to the emergence of new technologies and novelty fibers, manufacturers of FPC are continuously developing favorable blends with various fiber alternatives to bring optimum balance in price, comfort, and protection (Figure 1.6).

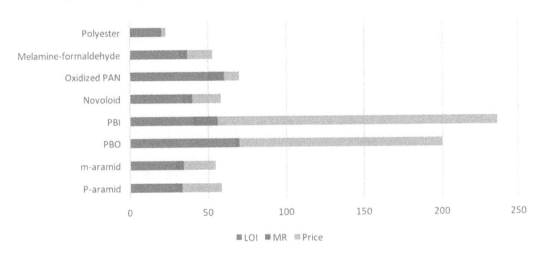

Figure 1.6

Price-Peace-Protection trilogy in terms of fire protection and moisture absorption.

It is a reality that no single fiber can meet all the requirements as an outer layer fiber. As an example, the widely used *p*-aramid is famous for its strength and abrasion resistance, but it suffers from degradation and strength loss when exposed to sunlight. *M*-aramid and PBI are comparatively more stable to ultraviolet degradation and have better heat resistance, but they lack strength. One fiber may have a high LOI value but lower melting temperature (such as PVC), whereas some non-melting fiber (such as Novoloid, wool) may have a comparatively lower LOI value.

The selection of a suitable fiber type and fabric structure is another important consideration. McQuerry et al. (2015) collected more than 250 FPC from various fire departments and evaluated their performance on ageing. Diversified material choice was seen in this study. The outer shell fabrics included Nomex, Kevlar, and PBI. Moisture barriers included Aquatech, Crosstech, Goretex, PTFE membranes, and RT71000. The thermal liners were composed of TenCate Caldura SL Quilt, aramid fibers, and E-89. Lee and Barker (1987) evaluated the thermal protective performance of 21 heat-resistant fabrics of knit, woven, and nonwoven structure. The fabrics tested were either aramid or PBI, and PBI blends with aramid or FR rayon. The woven fabrics were on twill, sateen, or plain structure with fabric weight ranging from 139 to 295 gsm. Fabric weight of tested knit structure ranged from 153 to 298 gsm and felt structure ranged from 180 to 295 gsm. Wang et al. (2011) evaluated the moisture transfer performance of several FPC assemblies that include nine outer layer fabrics composed of blends of aramid, PBI, Kermel, Tanlon, and carbon Antistatic. The fabric weight ranges from 150 to 260 gsm. Krasny et al. (1982) evaluated eight commonly used outer shell fabrics composed of aramid, novoloid, cotton, and their blends where the fabric weight ranges from 205 to 440 gsm. Hence, it is worthwhile to consider fabric construction to choose a suitable outer layer fabric from the wide range of commercial products. Table 1.2 presents some examples of commercial outer shell fabric from the recent market trend of fiber blend.

FPC is mostly constructed on basic weave structures. These include mainly plain or twill construction and their special derivatives such as ripstop construction, comfort twill, etc. Satin weave is not a practical option for FPC.

> **Plain weave** is the simplest, shortest, and most important woven fabric construction produced by alternative lifting and lowering of one warp yarn across one weft yarn. It has the maximum amount of interlacing possible in a woven fabric. Plain weave produces a tight cloth with firm structure, which is stronger than any other weave structure (Redmore, 2011; Wilson, 2001). Hence, at least 90% of the two-dimensional woven technical fabrics are constructed on plain weave (Sondhelm, 2000).
>
> **Twill weave** is another basic weave structure where diagonal rib lines become visible on the fabric surface. Rib lines usually run from the lower left to the upper right of the fabric where each end floats over or under at least two consecutive picks (Rouette & Schwager, 2001). Thus, the twill weave has more open construction with longer floats and fewer intersections.
>
> **Derived weave** structure is derived from the three basic types (plain, twill, and satin) or their combination. For example, leno, repp, and panama are derived from plain weave; honeycomb and herringbone are derived from twill; and crepe, shadow repp, etc. are derived from satin (Rouette & Schwager, 2001).

Table 1.2 Outer-Shell Fabrics from Various Manufacturers

Trade Name	Manufacturer	Description	Structure
Advance™	Southern Mills, Inc, doing business as TenCate Protective Fabrics USA Inc	60/40 blend of KEVLAR® and NOMEX® used for outer shell fabric, piece dyed	Ripstop
Nomex® IIIA	DuPont	93/5/2 Nomex-Kevlar-Carbon	Plain
Indura®	Westex by Millimen & Co	100% FR cotton	Rugged Twill
Fusion™	Safety components	60/40 or 50/50 Kevlar-Nomex blend	Ripstop
Armor AP™	Safety components	80/20 or 60/40 Nomex-Kevlar blend	Twill
PBI Max™	Safety components	70/30 PBI-Kevlar blend	Comfort-Twill
PBI Matrix®	Safety components	60/40 Kevlar®-PBI Gold Plus®	Plain
Armor 7.0 ™	Safety components	75/25 or 50/50 DuPont Kevlar-Nomex	Comfort-Twill
Millenia™	Southern Mills, Inc/TenCate Protective Fabrics USA Inc	60/40 Tejin Twaron or Technora-Zylon PBO	Ripstop
Advance Ultra	Southern Mills, Inc/TenCate Protective Fabrics USA Inc	60/20/20 Kevlar-Nomex - Zylon PBO	Ripstop
Omni Vantage™	Norfab corporation	40/30/30 Kevlar-Basofil-Nomex	Ripstop
Gemini™ XT	Southern Mills, Inc/TenCate Protective Fabrics USA Inc	55/37/8 Kevlar-PBI-Vectran or Technora	Plain
Pioneer™	Southern Mills, Inc/TenCate Protective Fabrics USA Inc	60/40 or 50/50 Nomex/Kevlar	Twill
Ultra®	Southern Mills, Inc/TenCate Protective Fabrics USA Inc	Kevlar/Nomex/PBO	Ripstop
Kombat™ Flex	Southern Mills, Inc/TenCate Protective Fabrics USA Inc	PBI/Kevlar	Twill
Dual Mirror®	Gentex corporation	Aluminized PBI/Kevlar	Ripstop
Tecasafe® Plus	Southern Mills, Inc/TenCate Protective Fabrics USA Inc	Modacrylic/cellulose/aramid	Twill
Defender M®	Southern Mills, Inc/TenCate Protective Fabrics USA Inc	Lenzing FR® Rayon-para aramid-Nylon	Ripstop
Brigade™	Southern Mills, Inc/TenCate Protective Fabrics USA Inc	Nomex®	Plain

Source: Component, S., 2017, Outer Shells. Retrieved 22 Feb. 2017 from http://www.safetycomponents.com/Fire/Outershells/; FireDex, 2017. Fxr Custom Turnouts-Materials. Retrieved 22 Feb. 2017 from http://www.firedex.com/product/fxr-custom-turnouts/; Honeywell, 2013, Honeywell First Responder Products. Retrieved 22 Feb. 2017 from http://www.honeywellfirstresponder.com/~/media/epresence/firstresponder/literature/pdf/selector%20guides/fabric%20selector.ashx?la=en; Topps, 2016, Flame-Resistant Fabric Information. Retrieved 22 Feb. 2017 from http://www.toppssafetyapparel.com/fabrics.html; Westex, 2017, Westex Fabric Brands. Retrieved 2 March 2017 from http://www.westex.com/fr-fabric-brands/; Veridian, 2017, Materials. Retrieved 22 Feb. 2017 from http://veridian.net/materials.php.

Ripstop weave is a widely used woven fabric structure in technical textile industry. It is designed to stop the ripping. Extra high strength yarn is weaved in regular interval within the normal base fabric to provide resistance to spreading of tear. The definition of ripstop weave is given as "very fine woven fabric, often nylon, with coarse, strong warp and filling yarns spaced at intervals so that tears will not spread. The same effect can be achieved by weaving two or three of the fine yarns together at intervals" (Wingate, 1979, p. 513). It is suitable for technical textiles that required resistance to tear. In 2005, Du Pont patented weave fabric structure specially designed for FPC with ripstop yarn component where the ripstop yarn has at least 20% more tensile strength than the body yarn (Zhu & Young, 2005). Nowadays, many manufacturers use ripstop construction for fabric intended as an outer layer fabric of FPC.

Miscellaneous: Manufacturers of technical textile fabrics developed innovative weave construction based on any of those basic structures for fulfilling specific requirements. *Rain drop* and *Comfort twill* are recent examples of such development. *Comfort twill* is based on patented Filament twill™ technology by *Safety Component*. This weave structure can be found in the outer layer fabrics such as PBI Max™, Armor AP™, or in thermal liner Glide™. It is claimed that this weave structure from Kevlar filament is over two times stronger than weave structure from Kevlar spun yarn, and also the fabric is more flexible, lightweight, and comfortable (Safety Components, 2017). Another such special construction is "Channeling raindrop" weave, which is being used in the Prism™ thermal liner. It is woven in a technical pattern, which positions a higher ratio of viscose. Since viscose has excellent wicking capability, such concentrated placement inside raindrop pattern absorbs and allows moisture to be channelled away from the wearer for enhanced comfort. Many other special weave structures are available for technical textiles, but those are not discussed here. Here the two special weave constructions highlight the idea of continuous weave modification trend in today's competitive market culture to meet specific requirements.

1.4.2 Moisture Barrier

The purpose of the moisture barrier is to impart breathability in FPC. This barrier layer makes the FPC impermeable to water while it allows the moisture vapor to pass through. In this way, firefighters remain protected against hot water and toxic liquids while the sweat can be vaporized and dissipated to the environment.

Moisture barriers may not be mandatory. In some countries, firefighters like to have their fire suits without a moisture barrier, whereas in other countries it is obligatory (Mäkinen, 2005). Microporous perm-selective membrane, hydrophilic nonporous membrane, or their combination can be used. In most cases, ePTFE membrane is used as a moisture barrier in a bicomponent structure with PU foam that creates an air cushion. Sometimes an additional hydrophilic coating is also applied on top of the ePTFE membrane. Table 1.3 shows some commercially available moisture barriers from various manufacturers.

1.4.3 Thermal Liner Fabric and Nonwoven

Thermal liner of FPC usually contains a nonwoven batting attached to a face cloth, as shown in Figure 1.7. The thermal liner is the most critical component in turnout gear as it has the biggest impact on thermal protection and heat stress reduction. Air is trapped in or between the nonwoven material of the thermal liner, and thus together with the moisture barrier, thermal liner provides 75% of the total thermal protection of a turnout garment (Young, 2010). Figure 1.7 shows a complete fabric assembly in an FPC where a nonwoven batting material is sewed to a facecloth. The face cloth of the thermal liner is conventionally a thin woven fabric from Group A as discussed in Section 1.2.

Table 1.3 Examples of Commercially Available Moisture Barriers

Trade Name	Manufacturer	Composition	Fabric
STEDAIR®3000	Stedfast	Bi-component ePTFE/FR PU	33/67 blend of Kevlar-Nomex® E89™. Nonwoven Spunlace
STEDAIR®4000	Stedfast	Bi-component ePTFE/FR PU	100% Nomex IIIA®. Woven Pajama-check
STEDAIR®Gold	Stedfast	Bi-component ePTFE/FR PU	80/20 blend of Nomex IIIA®-PBI. Woven Pajama-check
CROSSTECH® 3-layer	W.L Gore	Bi-component ePTFE/FR PU	Nomex, Woven Pajama-check
CROSSTECH® black	W.L Gore	Bi-component ePTFE/FR PU	100% Nomex IIIA®. Woven Pajama-check
GORE® RT7100	W.L Gore	Bi-component ePTFE/FR PU	15/85 blend of Kevlar®-Nomex®. Needle punched Nonwoven
Porelle® Membrane	Porelle	Bi-component ePTFE/FR PU or hydrophilic	
Entrant®	Toray Industries	Microporous hydrophobic PU film	

Source: FireDex, 2017, Fxr Custom Turnouts-Materials. Retrieved 22 Feb. 2017 from http://www.firedex.com/product /fxr-custom-turnouts/; Honeywell, 2013, Honeywell First Responder Products. Retrieved 22 Feb. 2017 from http://www.honeywellfirstresponder.com/~/media/epresence/firstresponder/literature/pdf/selector%20guides /fabric%20selector.ashx?la=en; Mukhopadhyay, A., & Midha, V., 2016, Waterproof Breathable Fabrics. *Handbook of Technical Textiles: Technical Textile Applications*, 2, 27; Porelle, 2017, PTFE Membranes- Waterproof, Windproof and Heat Resistant. Retrieved 24 Feb. 2017 from http://www.porellemembranes.co.uk/en/membranes/ptfe -membranes/; Toray, 2017, Toray- All About Sports Fabrics. Retrieved 24 Feb. 2017 from http://www.torayen trant.com/en/about/.

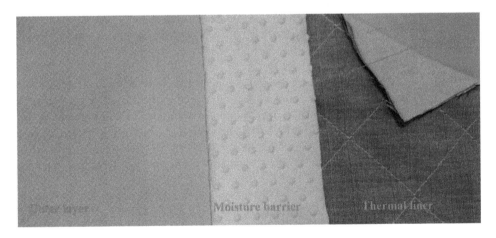

Figure 1.7

Fabric assembly in FPC: from left, Nomex outer layer, STEDAIR moisture barrier, and thermal liner showing nonwoven batting and woven Nomex face cloth.

Nonwoven is a fabric structure built without twisting of the fiber (as in yarn) or interlacing of yarn (as in fabric), resulting in a lofty textile construction unless pressed. Hence, nonwoven textiles can accommodate high volume of air pocket in much lower weight of textile, which can be used as a barrier to heat transmission. In knitted or woven fabrics, fibers are held together by frictional force.

However, in nonwovens, fibers are held together either by frictional force (fiber entanglement) or adhesive system (thermal or chemical bonding). Initially a layer of fiber web (batt) is laid by a suitable method like carding machine, air laying, water laying, spun laying, etc., and then the laid fibers are bonded by means of mechanical entanglement such as needle punching, hydroentanglement, stitch binding, etc. or by means of a thermal/chemical agent (Smith, 2000). Two commonly available nonwoven batting materials of thermal liner are "needle punched" and "spunlace." Table 1.4 shows some examples of commercial thermal liner with their construction, combination, and manufacturer.

Needle punched: Needle punching is a simple stitching of fiber batt with entangled fiber. In this case, a set of needles is punched through the laid fiber batt and then withdrawn. During the needle descending, a strand of fiber is caught by the barbs of the needle and forms loops. On upward withdrawal of the needle, the fiber strand is released from the needle barbs, forming an entangled stitching fiber strand in the laid fiber batt. This is a "down-punched" method. Similarly it can be "up-punch" or a combination of both (a "double-punch").

Spunlace nonwoven: The term "spunlace" was derived from its laying mechanism, not the bonding technique as it was in needle-punched nonwoven. In "spun laying" mechanism, as soon as the filaments are extruded from the spinnerets, they are directly drawn (usually by air) downward to lay into a batt. Then the batt is bonded (typically hydroentanglement) to nonwoven. As the filaments are directly laid to make fabric just after their extrusion from raw material, spunlace nonwoven production is called the shortest possible textile route from polymer to fabric (Smith, 2000). The production process is so fast that a commercially spunlace production line with production speed up to 400 m/min (Andritz, 2016) is available.

Needle punched battings are normally bulkier than spunlace batting. Normally two-layer spunlace batting is used in thermal liner while one layer of needle punched batting does the job in identical condition. However, needle-punched battings usually offer higher thermal protection performance ratings, while spunlace battings offer better total heat loss performance (TenCate, 2016). Needle-punched battings are normally less flexible and less durable than spunlace battings.

1.5 Chemical Finishing

Ignition and burning is a gas phase reaction (Bourbigot et al., 2003) where a mixture of combustible gases with ambient oxygen yields a flame and generates heat (Duquesne & Bourbigot, 2010). The combustion cycle of a solid starts by absorbing heat, which decomposes the solid and creates many smaller components. These smaller molecules diffuse to the surface and escape to the environment, creating a gloomy flame. The pyrolytic decompositions produce carbonaceous char with combustible and noncombustible gases (Duquesne & Bourbigot, 2010). In their way to combustion, combustible gases are oxidized by the oxygen present in the environment and produce more heat, which is absorbed by the remaining parts of the solid, and the combustion cycle continues. Reducing the flammability of textile

Table 1.4 Thermal Barrier Combination in Commercially Available Thermal Liner

Trade Name	Manufacturer	Composition Face Cloth	Weave Face Cloth	Batting
Glide Ice™	Safety Components	60% Kevlar® with 40% Nomex/Lenzing FR®	Twill	2 Layers Nomex spunlace
Prism™	Safety Components	68/21/11 blend of Aramid, FR viscose, and Polyamide	Channeling Raindrop	Nomex/Kevlar needle punch
Bravo™	Safety Components	51/32/17 blend of aramid, FR viscose, and polyamide	*NK	2 layer, 50/50 blend of aramid and FR viscose
XLT-Lite™	Starfield	100% Nomex®	Plain weave	Reprocessed Nomex, Needle punch
Aralite® NP	TenCate Protective Fabric	100% Nomex®	Plain weave	Kevlar®/Nomex®, Needle punch
Aralite® SL3	TenCate Protective Fabric	100% Nomex®	Plain weave	3 Layers Kevlar®/Nomex® E-89™, Spunlace
Defender® M	TenCate Protective Fabric	85-25-10 blend of Lenzing FR®-Twaron®-Nylon	Plain weave	2 Layers Kevlar®/Nomex®, Spunlace
PBI Thermal GUARD™	PBI performance products, Inc	Proprietary PBI Blend	Plain weave	80% Aramid, 20% PBI, Spunlace
Caldura® NPi	TenCate Protective Fabric	61/34/5 blend of Kevlar®-Lenzing FR®-Nylon	Twill	Aramid, Needle punched

(Continued)

Table 1.4 (Continued) Thermal Barrier Combination in Commercially Available Thermal Liner

Trade Name	Manufacturer	Composition Face Cloth	Weave Face Cloth	Batting
Caldura® SL2i	TenCate Protective Fabric	61/34/5 blend of Kevlar®-Lenzing FR®-Nylon	Twill	2 Layers Kevlar®/Nomex® E-89™, Spunlace
Glide™ Pure	Safety Component	60/26/14 blend of Kevlar®-Nomex®-Lenzing FR®	Filament Twill	50/50 Kevlar®-Nomex®, Needlepunch
Glide™ 2 layer	Safety Component	60/26/14 blend of Kevlar®-Nomex®-Lenzing FR®	Filament Twill	2 Layers Kevlar®/Nomex® E-89™, Spunlace
Quantum3D® SL2i	TenCate	61/34/5 blend of Kevlar®-Lenzing FR®-Nylon	Twill	2 Layers Kevlar®/Nomex®, Spunlace
Glide ICE™ PBI G2	Safety Component	61/26/14 blend of Kevlar®-Nomex®-Lenzing FR®	Twill	2 Layers 80/20 Aramid-PBI, Spunlace
Q-8™	TenCate Protective Fabric	Meta aramid- FR modacrylic	Woven	Aramid-FR Rayon. Needle punched
Defender M® SL2	TenCate Protective Fabric	Lenzing FR® Rayon-para aramid-Nylon	Twill or plain	2 Layers Kevlar-Nomex. Spunlace
Quantum4™	TenCate Protective Fabric	Aramid, FR Rayon, FR Nylon	Twill	80% Nomex, 20% PBI, Spunlace
Ultraflex®	DuPont	100% Nomex®	Plain weave	Aramid 50/50 Kevlar-Nomex

Source: Component, S., 2017, Outer Shells. Retrieved 22 Feb. 2017 from http://www.safetycomponents.com/Fire/Outershells/; Globe, 2016, Globe Turnout Gear-Materials. Retrieved 5 March 2017 from http://globeturnoutgear.com/education/materials; Honeywell, 2013, Honeywell First Responder Products. Retrieved 22 Feb. 2017 from http://www.honeywellfirstresponder.com/~/media/epresence/firstresponder/literature/pdf/selector%20guides/fabric%20selector.ashx?la=en; TenCate, 2016, Firefighter Protection Lives in TenCate Thermal Barriers. Retrieved 5 March 2017 from http://tencatefrfabrics.com/fire-service/firefighter-thermal-barriers/; Veridian, 2017, Materials. Retrieved 22 Feb. 2017 from http://veridian.net/materials.php.

Note: Nomex® 89™ is a nonwoven fabric produced by DuPont and used in moisture barrier with ePTFE membrane. *NK: Not known.

fibers is basically the interruption of this complex combustion process at one or several points of their combustion cycle, which can be achieved either by chemical incorporation or mechanical addition of fire retardant into the fiber polymer molecules (Joseph & Ebdon, 2008). In general, textile flame can be retarded by using inherently flame-retardant polymer, by chemically modifying the conventional polymer to gain flame retardancy, or by applying flame-retardant chemicals/particles to the material by coating the surface or incorporating them into the structure (Bourbigot & Duquesne, 2007; Duquesne & Bourbigot, 2010). High-performance flame-retardant fibers and modified flame-retardant fibers are discussed in Section 1.2. In this section, the principle of flame retardancy is briefly discussed, and then the mode of chemical flame-retardant finishing is noted.

1.5.1 Three Simplified Principles of Flame Retardancy

To retard the flame, it is necessary to interfere with the combustion process in one or several stages. The reactions interfering with the combustion process may act either in condensed or gas phase. Breakdown of polymer to withdraw from flame and formation of char or ceramic-like structure on material surface is the type of retarding effect that takes place in the condensed phase. In the gas phase, the retarders interfere the exothermic reactions and cool down the system to reduce the emission of pyrolysis gases (Duquesne & Bourbigot, 2010). The flame retardancy action can be simplified in three basic principles: (1) the halogen system that interferes in oxidation reaction, (2) the char formation by phosphorus and boron compounds where the retarding action is done by forming an incombustible char on the material surface, and (3) in the endothermic principle where the decomposition of additives absorbs heat to cool down the system and delay the reaching of ignition temperature. Chlorinated paraffins or polybrominated diphenyl ethers are halogenated compounds that liberate halogen atoms that interfere with the oxidation process (Joseph & Ebdon, 2008) and retard the flame. Phosphorus and boric acid based additives work on intumescence phenomenon principle (Bourbigot et al., 2004). They create a low thermal conductive shield that protects the material by reducing the degradation rate and thus lowering the emission of pyrolysis gases (Duquesne & Bourbigot, 2010). For example, phosphorus additives form an incombustible char that protects the residual polymer under char from flame (Joseph & Ebdon, 2008). Di- or tri-alkyl phosphates, tri-aryl phosphates, etc. (Duquesne & Bourbigot, 2010) are typical examples of organic phosphates. A high amount of char means less combustible product for the flame to continue. On the other hand, the retardant effect of hydroxide is threefold. It goes through endothermic decomposition that cools the material, forms inert gases that dilute the flammable gases mixture, and also forms an oxide protective barrier (Duquesne & Bourbigot, 2010). The use of aluminium trihydroxide is an example of such type of retardancy.

1.5.2 The Mode of Flame Retardant Finishing of Textiles

Flame retardant finish can be durable, semidurable, and nondurable. A flame-retardant finish will be durable if the retardant can trap itself in the polymeric structure of the fiber by creating a chemical bond or can be on the substrate surface by durable coating (Horrocks, 2013). Durable reactive FR can be applied to a natural fiber that contains reactive sites for bonding, whereas surface coating is the most common technique to impart flame-retardant finish for synthetic fiber with little reactivity (Opwis et al., 2011). Nondurable flame-retardant finishing can be applied for disposable textiles such as wall covering, party costumes, or medical gowns (Weil & Levchik, 2008a) and hence less relevant to the current discussion. As the present topic is FPC, only durable and semidurable flame-retardant finishing on cellulosic materials is noted here.

a. **THP Salts or Proban Process:** In this case, a phosphorus-containing material, based on tetrakis(hydroxymethyl)phosphonium (THP) salt (THPC, THPS, THPOH), is reacted with urea to form Proban, which is then padded onto cellulosic fabric (Horrocks, 1986). The bulk of FR finished cellulosic fibers are based on the THP derivatives. It can be applied through a pad–dry–cure–oxidation process (Charles, 1992). After padding and drying, the insoluble polymer formed by the Proban is trapped inside the fiber void and interstices of yarn by mechanical means. The detailed chemistry of the Proban process is discussed by the inventor Robert Cole (1978) and also by Frank et al. (1982). Indura® fabric manufactured by Westex is an example of Proban-treated cotton fabric (blended) that is used in FPC.

b. **Pyrovatex Process:** This is a chemical bonding process where a flame-retardant chemically reacts with cellulose and forms a durable bond. In 1969, Ciba authors (Aenishänslin et al., 1969) demonstrated this process of cross-linking cotton by treating with methylol carbamate agent that formed *in situ* in the fiber. The commercial product is known as Pyrovatex® CP where the main molecule is $(CH_3O)_2P(=O)CH_2CH_2C(=O)$ $NHCH_2OH$ (generally called N-methylol dimethyl phosphonopropionamide) (Weil & Levchik, 2008a). The process uses reactivity of cellulose with N-methylol (Charles, 1992). One disadvantage of Pyrovatex is the slow release of loosely held formaldehyde when stored. Hall, Horrocks, and Roberts (1998) demonstrated the process of lowering formaldehyde release, and together with Ciba, Horrocks (Weil & Levchik, 2008a) developed low formaldehyde grade Pyrovatex. Details on both Proban and Pyrovatex process are commonly available in many other related literatures (Horrocks, 1986; Weil & Levchik, 2008b; Yang, 2013).

c. **Semidurable FR Finish of Cellulose:** Semidurable finish can survive water soaking but generally cannot withstand alkaline laundering conditions. One approach is to heat cellulose with phosphoric acid or ammonium phosphates to produce cellulose phosphate. However, as the phosphorylation occurs in glucose unit, it degrades the polymer chain and causes yellowing (Weil & Levchik, 2008a). Use of urea or dicyandiamide as a coreactant along with phosphoric acid or ammonium salt minimizes the damage. Today dicyandiamide salt of phosphoric acid is commercially available for FR finish of cellulose. Again, the use of organic phosphonic acid instead of inorganic phosphate salts/acid and the combination of boric acid and urea has been proven as a more effective semidurable FR finish on cellulose (Dermeik et al., 2006). Flammentin® FMB produced by Thor Specialties and Pyrovatim® PBS produced by Ciba are examples of semi-durable FR finishes.

1.6 Thermal Protective Properties

Protection expected through clothing is multidimensional. For traditional clothing, the expectation is basically social and determines its effectiveness from modesty and cultural value to fashion discerning consumers. However, for protective clothing, values are determined on specific protection performance, limiting cultural or fashion perspective, and even compromising comfort. For firefighting garments, the main concern is protection against heat, which can be experienced from any uncomfortably hot object or direct/indirect contact with flame. The human body can control its internal temperature at a certain level when an external or internal condition changes. Specific central and peripheral nervous systems continuously sense the temperature instability in the body and try to keep a balance through biological action (Li & Wong, 2006). However, under extreme weather conditions, the body needs protection for survival. One of the most important functions of clothing is to protect a wearer against extremes of environmental temperature—either heat or cold (Slater, 1977; Ukponmwan, 1993). The measure of the insulation of a material is its thermal resistance. It is defined as the temperature difference between the two faces divided by the heat flux and has the unit of Km^2W^{-1}. The heat flow can be any form such as conductive, convective, or radiative. Textiles are generally low heat conductive. Hence, convective and radiative heat resistance are primary concerns for designing firefighting protective clothing.

1.7 Thermal Comfort Properties

British Standard BS EN ISO 7730:2005 defines "thermal comfort" as a mental condition of an individual that expresses satisfaction in a thermal condition. Designing protective clothing for firefighters is a challenging task as it requires making a compromise between two crucial but conflicting factors, that is, maximizing thermal protection and minimizing heat stress (Holmér, 1995; Wang et al., 2013). Thermal protection is undoubtedly the primary concern for FPC; however, its effect on metabolic heat stress is also an important consideration (Fanglong et al., 2007). Hence, the FPC needs to be built with a balance of these two factors. The key to thermal comfort is the condition of skin-clothing microclimate, which depends on two vital factors, humidity and temperature. The level of thermal comfort of human body is determined by the heat and moisture balance between body and environment. In brief, this type of comfort can be termed as "thermophysiological wear comfort," which refers to the heat and moisture transport properties of clothing and the way the clothing helps to maintain the heat balance of the body (Song, 2009; Tashkandi et al., 2013). It does not simply depend on one or two key ingredients such as thermal conductivity or insulative behavior of the clothing, but many other minor details, such as environment and wearer's physiomental condition, need to be considered as well. Air velocity affects comfort; the thermal insulation reduces with increased wind velocity as compared with insulation in still air (Ukponmwan, 1993, p. 19). Moisture content of textiles increases the thermal transmissivity (Ukponmwan, 1993, p. 19). Thus, heat, air, and moisture transport properties should be taken into consideration to predict wearer thermal comfort (Barker, 2002; Yoo & Barker, 2005).

1.7.1 Moisture Management

How well a specific fabric type can manage the moisture plays a significant role in wearer comfort. Wearer's perception of moisture comfort sensations and clothing comfort is directly related to absorption of moisture or body sweat by the garment in the garment-skin microclimate, and moisture transportation through and across the fabric where it is evaporated (Benisek et al., 1987; D'Silva et al., 2000; Holme, 2002; Hu et al., 2005). The sweat produced during a firefighting activity should be absorbed from the skin by the surface of the next-to-skin wear and then gradually transferred to the layers further out in the clothing assembly. The use of multiple layers in FPC makes the liquid transfer difficult by reducing the moisture management capability of the garment, resulting in accumulation of sweat on the next-to-skin layer. As a consequence, the wearer suffers increased wet clinginess and thermal discomfort (Houshyar et al., 2017), which may restrict the work time of the firefighter. The fabric does not allow the passage of water vapor, which results in the condensation of water vapor and formation of liquid moisture that is a direct reason of the sensation of discomfort (Srdjak, 2009). Moisture management properties of a clothing are vital in this case. Unlike the simple determination of fabric absorbency and wicking properties, a moisture management test measures the behavior of dynamic liquid transfer in clothing materials. For example, the moisture transportation through FPC is a multidimensional process as fractional amount of moisture can be absorbed in first contact surface, some amount can go through the fabrics, and a small amount can be absorbed in the other surface. For a firefighter, the transfer of external liquid to the skin is not desirable, whereas it is highly expected that the clothing used in FPC should allow the sweat to escape to the environment. These two desirable properties are self-conflicting for any fabric type. How much moisture will be absorbed in the first surface and how much will go to the opposite of the fabric depends on many fabric attributes such as fabric construction, surface finish, etc. Therefore, it is important to analyze the moisture management property of fabrics intended for firefighting gear.

Another important parameter to look for in an FPC fabric is the permeability of moisture, which can be defined in two aspects, transfer of liquid and the permeability index. The former provides an idea of how much protection the fabric type used may offer against harmful liquids from external source, whereas the latter expresses the degree of evaporative cooling (of sweat) of the fabric. The permeability index considers the dry heat resistance of a textile fabric and relates it with the escape of water vapor (sweat evaporation) through the clothing.

1.7.2 Heat Transfer

The temperature between the thermal liner and the firefighter's undergarment can reach from 48°C to 62°C before receiving burn injuries (Rossi, 2003). It has been identified that the pain threshold of human skin is around 44°C (Hardy et al., 1952; Stoll & Greene, 1959). When the skin temperature exceeds this threshold, the absorbed energy determines if and how severe burns will be received (Stoll & Chianta, 1968). The skin receives second-degree burns when skin temperature approaches 55°C (2012). Thus, there is a time gap between starting to feel pain and receiving a second-degree burn. The time between the two points, that is, the time when the skin starts to feel pain and when it receives irreversible burns, is called the pain alarm time (Rossi & Bolli, 2005). This time frame is crucial for a firefighter as it indicates the time available for the wearer to respond to the warning and escape from a danger zone before receiving burns. A longer time gap offers a higher chance for the firefighter to avoid injury. Heat-absorbing materials such as phase change material and thermal insulation material, e.g., aerogel, may be used for thermal regulation management and increase the pain alarm time (Shaid et al., 2016a; Shaid et al., 2016b). Hence, in designing protective clothing, the temperature behind the thermal liner is decisive. The lower the temperature behind the thermal liner is, the cooler the skin temperature will be.

1.7.3 Air Permeability

Air permeability, being a biophysical feature of textiles, determines the ability of air to flow through the fabric. It can be greatly affected by the finishing treatment (Bivainytė & Mikučionienė, 2011). The clothing normally is a barrier to free air flow to the body. Easy air flow increases the rate of evaporation of body moisture and reduces the body temperature. However, increased wind velocity reduces fabric thermal insulation compared with the fabric insulation in still air. Different fabric types allow different amounts of air to pass through and consequently have different levels of thermal comfort. Air permeability of a fabric has two opposite sides. For the winter clothing, high air permeability fabric is not desirable. More windproof fabric will be a smart choice for winter garments. On the other hand, for summer clothing, more air permeable fabric may be a good choice. Hence, no universal conclusion can be derived whether more or less air permeable fabric is better. Opposite ideas are also true in FPC. In the view of moisture transport through air, where air acts as a carrier of moisture, the high air permeable fabric has an advantage. However, in respect of incoming hot air toward the body, less air permeable fabric has the benefit. Therefore, when the firefighter has to work for a long time in comparatively low heat conditions, more air permeable fabric will allow quick release of metabolic heat as the heat and moisture transfer is outward from the body in this scenario. On the other hand, when a firefighter has to enter into a flashover condition, such as a burning building, for a quick rescue mission, less air permeable garment could be a smarter choice as heat flow will be toward the body in this circumstance. In any case, protective clothing should be designed to cope with the worst conditions.

1.8 Summary

Multilayer FPC is a system or assembly where all the components play their role to fulfill the single objective of keeping the firefighter safe at work. However, each component of FPC has its own influence on the overall performance of the clothing system, and this influence solely depends on the properties of the respective element of the clothing. Hence, it is crucial to properly understand, consider, and then carefully select the components of a firefighter's clothing from a wide variety of textiles available in the current market. The chapter provides a general understanding on the textiles used in FPC, which may be a starting point to look for the FPC-related manufacturers, consumers, or any other parties of interest.

References

Aenishänslin, R., Guth, C., Hofmann, P., Maeder, A., & Nachbur, H. (1969). A New Chemical Approach to Durable Flame-Retardant Cotton Fabrics. *Textile Research Journal, 39*(4), 375–381.

Andritz. (2016). Spunlace Line Solutions. Retrieved 5 March 2017 from http://www.andritz.com/products-and-services/pf-detail.htm?productid=7407

Bagherzadeh, R., Latifi, M., Najar, S. S., Tehran, M. A., Gorji, M., & Kong, L. (2012). Transport Properties of Multi-Layer Fabric Based on Electrospun Nanofiber Mats as a Breathable Barrier Textile Material. *Textile Research Journal, 82*(1), 70–76.

Bajaj, P., & Sengupta, A. K. (1992). Protective Clothing. *Textile Progress, 22*(2–4), 1–110.

Barker, R. L. (2002). From Fabric Hand to Thermal Comfort: The Evolving Role of Objective Measurements in Explaining Human Comfort Response to Textiles. *International Journal of Clothing Science and Technology, 14*, 181–200.

Barker, R. L., & Coletta, G. C. (Eds.) (1986). *Performance of Protective Clothing: A symposium sponsored by ASTM Committee F-23 on Protective Clothing.* The First International Symposium on the Performance of Protective Clothing, Raleigh, N.C.

Benisek, L., Harnett, P. R., & Palin, M. J. (1987). Influence of Fibre and Fabric Type on Thermophysiological Comfort. *Melliand Textilber. Eng., 12*(68), 878.

Bivainytė, A., & Mikučionienė, D. (2011). Investigation on the Air and Water Vapour Permeability of Double-Layered Weft Knitted Fabrics. *Fibres & Textiles in Eastern Europe, 19*(3), 69–73.

Bourbigot, S., & Duquesne, S. (2007). Fire Retardant Polymers: Recent Developments and Opportunities. *Journal of Materials Chemistry, 17*(22), 2283–2300.

Bourbigot, S., & Flambard, X. (2002). Heat Resistance and Flammability of High Performance Fibres: A Review. *Fire and Materials, 26*(4–5), 155–168.

Bourbigot, S., Le Bras, M., Duquesne, S., & Rochery, M. (2004). Recent Advances for Intumescent Polymers. *Macromolecular Materials and Engineering, 289*(6), 499–511.

Bourbigot, S., LeBras, M., & Troitzsch, J. (2003). Fundamentals-Introduction. In *Flammability Handbook.* Munich.

Butola, B. S. (2008). Advances in Functional Finishes for Polyester and Polyamide-Based Textiles. In R. A. B. L. Deopura, M. Joshi, & B. Gupta (Ed.), *Polyesters and Polyamides* (pp. 306–325): Elsevier.

Charles, T. (1992). Chemistry and Technology of Fabric Preparation and Finishing. *Department of Textile Engineering, Chemistry and Science College of Textiles North Carolina State University Raleigh, North Carolina.*

Cole, R. (1978). Flameproofing of Textiles. United States patent US 4078101. https://www.google.com/patents/US4078101

Component, S. (2017). Outer Shells. Retrieved 22 Feb. 2017 from http://www.safetycomponents.com/Fire/Outershells/

Cook, J. G. (1980). *Handbook of Textile Fibres* (4th ed., vol. 1). Herts, UK: Merrow Publishing Co. Ltd.

Dermeik, S., Braun, R., & Lemmer, K. H. (2006). Flame-Retardant Compositions of Methanephosphonic Acid, Boric Acid and an Organic Base. United States patent US 6981998 B2, https://www.google.com/patents/US6981998

D'Silva, A. P., Greenwood, C., Anand, S. C., Holmes, D. H., & Whatmough, N. (2000). Concurrent Determination of Absorption and Wickability of Fabrics: A New Test Method. *Journal of the Textile Institute, 91*(3), 383–396.

Duquesne, S., & Bourbigot, S. (2010). Flame Retardant Nonwovens. In R. Chapman (Ed.), *Application of Nonwovens in Technical Textiles*. New York: Woodhead Publishing Limited.

Evonik. (2017a). P84® Polyimide Fibers. Retrieved 16 Feb. 2017 from http://www.p84.com/product/p84/en/products/polyimide-fibers/pages/properties.aspx

Evonik. (2017b). P84® Fibre Characteristics. Retrieved 16 Feb. 2017 from http://www.p84.com/sites/lists/RE/DocumentsHP/p84-fibre-technical-brochure.pdf

Fanglong, Z., Weiyuan, Z., & Minzhi, C. (2007). Investigation of Material Combinations for Fire-Fighter's Protective Clothing on Radiant Protective and Heat-Moisture Transfer Performance. *Fibres and Textiles in Eastern Europe, 15*(1), 72.

FireDex. (2017). Fxr Custom Turnouts-Materials. Retrieved 22 Feb. 2017 from http://www.firedex.com/product/fxr-custom-turnouts/

Frank, A. W., Daigle, D. J., & Vail, S. L. (1982). Chemistry of Hydroxymethyl Phosphorus Compounds: Part III. Phosphines, Phosphine Oxides, and Phosphonium Dydroxides. *Textile Research Journal, 52*(12), 738–750.

Fukui, N., Kato, Y., & Masai, Y. (1973). Fireproof, Thermoplastic Polyester-Polyaryl Phosphonate Composition. United States patent US3719727 A.

GCI. (2017). Kynol® Novoloid Fibers. Retrieved 21 Feb. 2017 from http://www.gunei-chemical.co.jp/eng/product/kynol.html

Globe. (2016). Globe Turnout Gear- Materials. Retrieved 5 March 2017 from http://globeturnoutgear.com/education/materials

Gohl, E. P. G. (1985). *Textile Science—an Explanation of Fiber Properties*. Melbourne: Longman Cheshire Pty Limited.

Gooch, J. W. (2011). *Polyimide Fiber*: Springer.

Gore, W. L. (2010). Fabric Performance-Technical Fabrics. Retrieved from http://www.gore-workwear.co.uk/remote/Satellite/2.1-Technical-Fabrics/2.1-Technical-Fabrics?sectorid=1173326935021

Gore, W. L. (2017). The Gore-Tex® Membrane. Retrieved 22 Feb. 2017 from http://www.gore-tex.com/en-us/technology/gore-tex-membrane

Hagborg, W. E., Bohrer, T. C., Chen, D. H., & Prince, A. E. (1968). *PBI Fiber Processes*: DTIC Document.

Hall, M., Horrocks, R., & Roberts, D. (1998). Minimisation of Formaldehyde Emissions. In R. Horrocks (Ed.), *Ecotextile '98: Sustainable Development* (pp. 63–70): Elsevier.

Hardy, J. D., Wolff, H. G., & Goodell, H. (1952). Pricking Pain Threshold in Different Body Areas. *Experimental Biology and Medicine, 80*(3), 425–427.

Hearle, J. W. (2001). *High-Performance Fibres*: Elsevier.

Hofmann, K. (1953). *Imidazole and Its Derivatives. 1 (1953)* (Vol. 6). Interscience Publishers.

Holme, I. (2002). Survival 2002: Performance Garments. *Textile Horizons, 5*(6), 7–8.

Holmér, I. (1995). Protective Clothing and Heat Stress. *Ergonomics, 38*(1), 166–182.

Holmes, D. A. (2000a). Textiles for Survival. In A. R. Horrocks & S. C. Anand (Eds.), *Handbook of Technical Textiles*, (Vol. 12, pp. 461–489): Woodhead Cambridge.

Holmes, D. A. (2000b). Textiles for Survival. *Handbook of Technical Textiles, 12*, 461.

Holmes, D. A. (2000c). Waterproof Breathable Fabrics. *Handbook of Technical Textiles*, 282–315.

Honeywell. (2013). Honeywell First Responder Products. Retrieved 22 Feb. 2017 from http://www.honeywellfirstresponder.com/~/media/epresence/firstre sponder/literature/pdf/selector%20guides/fabric%20selector.ashx?la=en

Horrocks, A. (1986). Flame-Retardant Finishing of Textiles. *Coloration Technology, 16*(1), 62–101.

Horrocks, A. (2005). Thermal (Heat and Fire) Protection. *Textiles for protection. Woodhead Publ. Ltd, Cambridge*, 398–440.

Horrocks, A. (2016). Technical Fibres for Heat and Flame Protection. *Handbook of Technical Textiles: Technical Textile Applications, 2*, 243.

Horrocks, A. R. (2013). Overview of Traditional Flame Retardant Solutions (Including Coating and Back-Coating Technologies). In A. R. Horrocks, F. Carosio, & G. Malucelli (Eds.), *Update on Flame Retardant Textiles: State of the Art, Environmental Issues and Innovative Solution* (pp. 123–178): Smithers Rapra Technology Ltd.

Houshyar, S., Padhye, R., & Nayak, R. (2017). Effect of Moisture-Wicking Materials on the Physical and Thermo-Physiological Comfort Properties of Firefighters' Protective Clothing. *Fibers and Polymers, 18*(2), 383–389.

Hu, J., Li, Y., Yeung, K.-W., Wong, A. S. W., & Xu, W. (2005). Moisture Management Tester: A Method to Characterize Fabric Liquid Moisture Management Properties. *Textile Research Journal, 75*(1), 57–62.

Jackson, R. H. (1978). PBI Fiber and Fabric—Properties and Performance. *Textile Research Journal, 48*(6), 314–319.

Joseph, P., & Ebdon, J. (2008). Flame-Retardant Polyester and Polyamide Textiles. *Polyesters and Polyamides*, 306.

Kermel. (2013). Kermel®: High Performance Fibre. Retrieved 20 Feb. 2017 from http://www.kermel.com/fr/Production-of-High-Tech-non-flammables -Fibres-640.html

Krasny, J., Singleton, R., & Pettengill, J. (1982). Performance Evaluation of Fabrics Used in Fire Fighters' Turnout Coats. *Fire Technology, 18*(4), 309–318.

Kynol. (2012). Kynol Novoloid Fibers. Retrieved 20 Feb. 2017 from http://www .kynol.de/products.html

Lee, Y. M., & Barker, R. L. (1987). Thermal Protective Performance of Heat-Resistant Fabrics in Various High Intensity Heat Exposures. *Textile Research Journal, 57*(3), 123–132.

Lempa. (2009). *Cht Fibers*. Paper presented at Training for the Teacher conference, National Institute of Textile Research and Development, Savar, Bangladesh.

Li, Y., & Wong, A. S. W. (2006). *Physiology of Thermal Comfort*. Cambridge: The Textile Institue, CRC Press, Woodhead Publishing Limited.

Mäkinen, H. (2005). Firefighters' Protective Clothing. *Textiles for Protection*, 622–647.

Mauric, C., & Wolf, R. (1980). Dioxaphosphorinane Derivatives as Flameproofing Agents. US Patent 4220472 A.

McCarthy, L., & Marzo, M. (2012). The Application of Phase Change Material in Fire Fighter Protective Clothing. *Fire Technology, 48*(4), 841–864.

McQuerry, M., Klausing, S., Cotterill, D., & Easter, E. (2015). A Post-Use Evaluation of Turnout Gear Using NFPA 1971 Standard on Protective Ensembles for Structural Fire Fighting and NFPA 1851 on Selection, Care and Maintenance. *Fire Technology, 51*(5), 1149–1166.

Moncrieff.W. (1975). *Man-Made Fibres* (5th ed.). London: Newnes-Butterworths.

Mukhopadhyay, A., & Midha, V. (2016). Waterproof Breathable Fabrics. *Handbook of Technical Textiles: Technical Textile Applications, 2,* 27.

Murugesan, H., & Gowda, R. V. M. (2012). Characterization of Flammability and Low Stress Mechanical Properties (Compression and Shear) of Basofil® Fibers and Its Blends. *Chemical Science Review and Letters, 1*(1), 35–44.

Niemz, F. G., Krieg, M., Mooz, M., Bauer, R. U., & Riede, S. (2017). Flame-Resistant Molded Cellulose Bodies Produced According to a Direct Dissolving Method. United States patent application US20170016148A1.

Opwis, K., Wego, A., Bahners, T., & Schollmeyer, E. (2011). Permanent Flame Retardant Finishing of Textile Materials by a Photochemical Immobilization of Vinyl Phosphonic Acid. *Polymer Degradation and Stability, 96*(3), 393–395.

Pearson, C. (2007). PBI History, *The Role of Chemistry in History* (Vol. 2016): Dickinson College.

Porelle. (2017). PTFE Membranes- Waterproof, Windproof and Heat Resistant. Retrieved 24 Feb. 2017 from http://www.porellemembranes.co.uk/en/mem branes/ptfe-membranes/

Raza, A., Li, Y., Sheng, J., Yu, J., & Ding, B. (2014). Protective Clothing Based on Electrospun Nanofibrous Membranes. In *Electrospun Nanofibers for Energy and Environmental Applications* (pp. 355–369): Springer.

Redmore, N. (2011). Woven Textile Design. In A. Briggs-Goode & K. Townsend (Eds.), *Textile Design-Principles, Advances and Applications* (pp. 31–54): Woodhead Publishing Limited.

Rossi, R. M. (2003). Fire Fighting and Its Influence on the Body. *Ergonomics, 46*(10), 1017–1033.

Rossi, R. M., & Bolli, W. P. (2005). Phase Change Materials for the Improvement of Heat Protection. *Advanced Engineering Materials, 7*(5), 368–373.

Rouette, H.-K., & Schwager, B. (2001). *Encyclopedia of Textile Finishing* (Vol. 23): Springer Berlin.

Safety Components. (2017). Filament Twill Technology. Retrieved 6 March 2017 from http://www.safetycomponents.com/FTT/

Schreuder-Gibson, H. L., Truong, Q., Walker, J. E., Owens, J. R., Wander, J. D., & Jones, W. E. (2003). Chemical and Biological Protection and Detection in Fabrics for Protective Clothing. *MRS Bulletin, 28*(8), 574–578.

Serbezeanu, D., Popa, A. M., Stelzig, T., Sava, I., Rossi, R. M., & Fortunato, G. (2015). Preparation and Characterization of Thermally Stable Polyimide Membranes by Electrospinning for Protective Clothing Applications. *Textile Research Journal, 85*(17), 1763–1775.

SGL Group. (2017). Panox® Thermally Stabilized Textile Fiber. Retrieved 20 Feb. 2017 from http://www.sglgroup.com/cms/international/products/product -groups/cf/oxidized-fiber/index.html?__locale=en

Shaid, A., Wang, L., Islam, S., Cai, J. Y., & Padhye, R. (2016a). Preparation of Aerogel-Eicosane Microparticles for Thermoregulatory Coating on Textile. *Applied Thermal Engineering, 107,* 602–611.

Shaid, A., Wang, L., & Padhye, R. (2016b). The Thermal Protection and Comfort Properties of Aerogel and PCM-Coated Fabric for Firefighter Garment. *Journal of Industrial Textiles, 45*(4), 611–625.

Shaid, A., Wang, L., Padhye, R., & Bhuyian, M. A. R. (2018). Aerogel Nonwoven as Reinforcement and Batting Material for Firefighter's Protective Clothing: A Comparative Study. *Journal of Sol-Gel Science and Technology*, DOI: 10.1007/s10971-10018-14689-10978.

Slater, K. (1977). *Textile Progress—Comfort Properties of Textiles* (Vol. 9). Mancheser: The Textile Institute.

Smith, P. A. (2000). Technical Fabric Structures-3. Nonwoven Fabrics. *Handbook of Technical Textiles, 12*, 130.

Sondhelm, W. S. (2000). Technical Fabric Structures–1. Woven Fabrics. *Handbook of Technical Textiles, 12*, 62.

Song, G. (2009). Thermal Insulation Properties of Textiles and Clothing. *Textiles for Cold Weather Apparel*, 19–32.

Srdjak, M. (2009). Water Vapour Resistance of Knitted Fabrics under Different Environmental Conditions. *FIBRES & TEXTILES in Eastern Europe, 17*(2), 73.

Stoll, A. M., & Chianta, M. A. (1968). *A Method and Rating System for Evaluation of Thermal Protection*: DTIC Document.

Stoll, A. M., & Greene, L. C. (1959). Relationship between Pain and Tissue Damage Due to Thermal Radiation. *Journal of Applied Physiology, 14*(3), 373–382.

Swicofil. (2015). Polyamide-Imide PAI. Retrieved 20 Feb. 2017 from http://www.swicofil.com/products/223polyamideimide.html

Tashkandi, S., Wang, L., & Kanesalingam, S. (2013). An Investigation of Thermal Comfort Properties of Abaya Woven Fabrics. *Journal of the Textile Institute, 104*(8), 830–837.

TenCate. (2016). Firefighter Protection Lives in TenCate Thermal Barriers. Retrieved 5 March 2017 from http://tencatefrfabrics.com/fire-service/firefighter-thermal-barriers/

Topps. (2016). Flame-Resistant Fabric Information. Retrieved 22 Feb. 2017 from http://www.toppssafetyapparel.com/fabrics.html

Toray. (2017). Toray- All About Sports Fabrics. Retrieved 24 Feb. 2017 from http://www.torayentrant.com/en/about/

Toyobo. (2015). The Strongest Fiber with Amazing Flame Resistance. Retrieved 14 Feb. 2017 from http://www.toyobo-global.com/seihin/kc/pbo/zylon_features.html

Ukponmwan, J. O. (1993). *The Thermal-Insulation Properties of Fabrics* (Vol. 24). Manchester: The Textile Institute.

Veridian. (2017). Materials. Retrieved 22 Feb. 2017 from http://veridian.net/materials.php

Vogel, H., & Marvel, C. (1961). Polybenzimidazoles, New Thermally Stable Polymers. *Journal of Polymer Science, 50*(154), 511–539.

Wang, Y., Zhang, Z., Li, J., & Zhu, G. (2013). Effects of Inner and Outer Clothing Combinations on Firefighter Ensembles' Thermal-and Moisture-Related Comfort Levels. *Journal of the Textile Institute, 104*(5), 530–540.

Wang, Y., Zong, Y., Li, J., & Zhao, M. (2011). Evaluating the Moisture Transfer Property of The Multi-layered Fabric System in Firefighter Turnout Clothing. *Fibres & Textiles in Eastern Europe, 19*(6), 101–105.

Weil, E. D., & Levchik, S. V. (2008a). Flame Retardants in Commercial Use or Development for Textiles. In *Flame Retardants for Plastics and Textiles - Practical Applications (2nd ed.)* (pp. 265–302): Hanser Publishers.

Weil, E. D., & Levchik, S. V. (2008b). Flame Retardants in Commercial Use or Development for Textiles. *Journal of Fire Sciences, 26*(3), 243–281.

Westex. (2017). *Westex Fabric Brands*. Retrieved 2 March 2017 from http://www
.westex.com/fr-fabric-brands/

Wilson, J. (2001). Weave and Woven Textile Design. In J. Wilson (Ed.), *Handbook
of Textile Design* (pp. 82–92): Elsevier Ltd.

Wingate, I. B. (1979). *Fairchild's Dictionary of Textiles*: Fairchild Publications.

Yang, C. (2013). Flame Resistant Cotton. *Handbook of Fire Resistant Textiles*, 177.

Yang, H. (1993). *Kevlar Aramid Fiber*: John Wiley & Sons Ltd.

Young, R. (2010). Understanding Turnout Gear. *Fire Apparatus and Emergency
Equipment Magazine*, 15(10).

Yoo, S., & Barker, R. L. (2005). Comfort Properties of Heat Resistant Protective
Workwear in Varying Conditions of Physical Activity and Environment.
Part I: Thermophysical and Sensorial Properties of Fabrics. *Textile Research
Journal, 75*, 523–530.

Zhou, W., Reddy, N., Yang, Y., & Scott, R. (2005). Overview of Protective Clothing.
Textiles for Protection, 3–30.

Zhu, R., & Young, R. H. (2005). Fire-Retardant Fabric with Improved Tear, Cut,
and Abrasion Resistance. United States patent US6840288 B2.

2

Firefighters' Protective Clothing and Equipment

Sumit Mandal, Martin Camenzind,
Simon Annaheim, and René M. Rossi

2.1 Introduction

Fire is an integral part of our daily lives; thus, the probability of occurrence of a fire incidence remains high. In a situation of fire incidence, efficient suppression of fires by firefighters is crucial to minimize human and economic loss (Kahn et al., 2012). To effectively suppress the fire, every firefighter needs to wear thermal protective clothing (Song, Mandal, & Rossi, 2016). Along with protective clothing, they also have to wear protective equipment such as helmet, gloves, boot, etc. (Ferguson & Janicak, 2015). The protective clothing and equipment help to mitigate firefighter injuries and fatalities by providing them proper protection and comfort (Rossi, 2003). Generally, the protective clothing and equipment certified from the International Organization for Standardization (ISO) or European Committee for Standardization (EN) or National Fire Protection Association of USA (NFPA) are widely used—ISO-certified products are mainly used in Latin America and the Asia/Pacific Region, EN-certified products are mainly used in the European Union, and NFPA-certified products are mainly used in the United States and Middle East. Although these three organizations are different, they work together to ensure that a certified product could be globally acceptable for providing protection and comfort to firefighters.

In the last few decades, consistent research and development has occurred in the field of firefighters' protective clothing and equipment (Ferguson & Janicak, 2015; Song et al., 2016). As a result of this effort, different types of firefighters' protective clothing (station uniform, turnout gear) and equipment (helmet, gloves, boot, breathing apparatus) have been developed. For providing protection to firefighters from various hazards, including thermal (e.g., flame, radiant heat, hot liquids), physical (e.g., sharp objects, slippery floor, falling debris), biological (e.g., blood-borne pathogens), and chemical (e.g., toxic liquids and air vapors), faced by them in a fire incidence, features of the clothing (a combination of various factors, namely fiber types and properties, yarn types and properties, fabric types and properties, and/or garment designs and properties) and equipment (a combination of various factors, namely related to designs and/or materials) have been constantly modified. These modifications help to increase the integrity, thermal insulation capacity, and/or liquid/air impermeability and thus increase the protective performance of the clothing and equipment. By maintaining the proper thermal insulation capacity and liquid/air impermeability, features of the protective clothing and equipment have also been modified to provide better thermophysiological (by properly regulating their metabolic heat and sweat vapor) and/or sensorial (by providing proper ergonomically fit sensation to their skin and body parts) comfort for firefighters. For example, moisture barrier and thermal liner were incorporated in the fabric system (used in a turnout gear) to increase its liquid/air impermeability and thermal insulation capacity so as to provide better protection to firefighters from hot liquids and high heat hazards (Mandal et al., 2013; Mandal et al., 2014); the collar of the protective clothing was designed with long zip to ensure the complete closure of the jacket up to the firefighters' necks to provide better protection for firefighters from various hazards (Uniforms and Equipment, n.d.); boots were designed with dual density cushioned rubber sole to provide better comfort (e.g., orthotic superiority, greater ankle support) for on-duty firefighters (Uniforms and Equipment, n.d.).

In this chapter, types and features of firefighters' protective clothing and equipment are thoroughly discussed to develop better understanding for achieving the best possible clothing performance for the protection and comfort of firefighters. The discussion on the clothing features elaborates several factors associated with firefighters' protection and comfort such as fiber types and properties, yarn types and properties, fabric types and properties, and garment designs and properties. The comfortable design aspects and protective materials used for the development of firefighters' protective equipment are also explained. Finally, the key issues related to firefighters' protective clothing and equipment are highlighted.

2.2 Types and Features of Firefighters' Protective Clothing

Firefighters' thermal protective clothing is a combination of a tunic and a trouser (Bennett, 2013). Over this trouser, another similar trouser is also worn sometimes by firefighters. In this case, the air layer trapped in between the trouser and over-trouser could enhance the thermal insulation capacity of the firefighters' protective clothing, and that could provide an extra protection to firefighters especially from thermal hazards like flame and radiant heat. From time to time, firefighters are also required to operate on or near roadways or in areas where they need to be clearly visible by others. For these working situations, firefighters wear a lime yellow vest with silver reflective tape. This vest ensures that a firefighter is in a non-firefighting situation, and can be adequately seen by passing motorists or other firefighters. Nevertheless, these vests are not used when firefighters are engaged in actual firefighting situations.

In general, firefighters wear two different types of thermal protective clothing, "station uniforms while working in fire stations" and "turnout gears while fighting fires such as structural building fires, wildland fires, and vehicle fires" (Ward, 2014). The necessity of a station uniform is to be a comfortable clothing that they can wear while working in the fire stations; this uniform should not become an obstruction when a firefighter is required to put on his/her turnout gear. Naturally, turnout gear is to be worn over the station uniform during any emergency when a firefighter is called upon (Figure 2.1). It is required that turnout gear should possess high protective and comfort performances to provide adequate safety to working firefighters in a fire incidence. At present, the commercially available firefighters' protective clothing should meet the design and performance requirements set by various standard organizations based on their respective test methods: ISO 11613:1999 or ISO 11999-3:2015, EN 469:2005, and NFPA 1971:2013. Nevertheless, these standards mainly set the protective performance requirement for firefighters' clothing. For example, as per ISO 11613:1999 standard, the protective performance of firefighters' turnout gear fabrics under flame hazards [measured based on ISO 9151:2016 Heat Transmission Index at 24°C (HTI_{24}), which is equivalent to increasing the skin temperature 24°C] can be categorized into two levels: Level 1: $HTI_{24} \geq 13$ s, and Level 2: $HTI_{24} \geq 17$ s. Based on Figure 2.1, firefighters' turnout gear is also encircled with a band of fluorescent

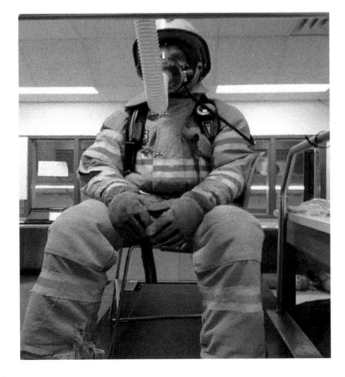

Figure 2.1

A firefighter in turnout gear over his station uniform.

or combined retroreflective/fluorescent material on each of the arms, legs, and torso regions of the garments. As per ISO 11613:1999 guideline, the minimum area of this band should not be less than 0.2 m² in firefighters' turnout gear.

For both types of firefighters' protective clothing, chemically modified fire-retardant [fire-retardant (FR) wool, FR cotton, FR polyester] and/or inherently fire-resistant [e.g., aramid, poly(aramid-imide), polyimide, polybenzimidazole, polybenzoxazoles] fibers are generally used as a raw material (Fan & Hunter, 2009; Horrocks, 1996). These fibers are converted into yarns through the spinning process (Klein, 1987); thereafter, these yarns are processed through the preparatory-weaving (starch sizing to enhance the smoothness and strength, heat setting of yarns) and then weaving (interlacing two set of yarns in power loom like Sulzer, rapier) to develop the shell fabrics (Marks & Robinson, 1976). As a station uniform is made up of single-layered fabric system (Song et al., 2011), only the shell fabric is processed through garmenting (fabric spreading, cutting, and sewing) for the manufacturing of the station uniform (Figure 2.2) (Glock & Kunz, 2005; Nayak & Padhye, 2015).

A turnout gear consists of a moisture barrier and a thermal liner along with the shell fabric (Keiser, Becker, & Rossi, 2008; Keiser & Rossi, 2008; Mandal et al., 2013). For the manufacturing of moisture barrier, a fabric is first constructed from the fibers through the weaving or nonwoven (bonding fibers through chemical, mechanical, heat, or solvent treatment) process. Next, this fabric is laminated or coated with chemical substances (polyurethane, polytetrafluoroethylene) or bonded with a membrane to convert it into the moisture barrier (Figure 2.3). Additionally, the thermal liner is manufactured by sewing or laminating a woven/nonwoven fabric with a nonwoven fabric (Figure 2.3). After developing the shell fabric, moisture barrier, and thermal liner, all these three fabric layers are assembled to make a composite fabric system (Figure 2.4). In this composite fabric system, shell fabric is used as an outer layer to directly face the hazards, moisture barrier is used as a middle layer, and thermal liner is used as an inner layer that is in contact with firefighters' skin. This composite fabric system is processed through the garmenting to manufacture the turnout gear (Figure 2.3). Rarely, the shell fabric of a composite fabric system could be leather based, but the fat content of the leather (tested in accordance with ISO 4048:2008) should not exceed 15% as per ISO 11612:2015 standard.

Figure 2.2

A schematic flowchart of manufacturing of a firefighter's station uniform.

2. Firefighters' Protective Clothing and Equipment

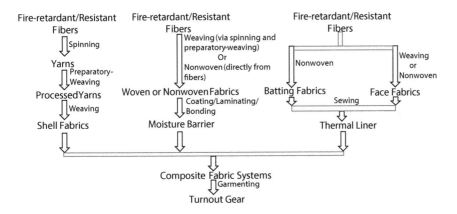

Figure 2.3

A schematic flowchart of manufacturing of a firefighter's turnout gear.

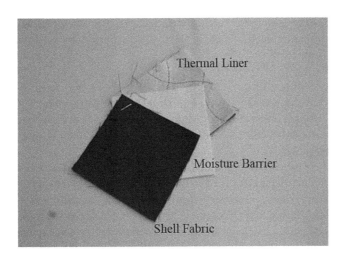

Figure 2.4

A composite fabric system for a firefighter's turnout gear.

Based on the above discussion, fire-retardant/resistant fibers are converted into thermal protective clothing to cover the complex geometry of firefighters' bodies. Thus, the features of a firefighter's protective clothing depend on various factors such as fiber types and properties, yarn types and properties, fabric types and properties, and garment designs and properties (Rezazadeh & Torvi, 2011; Song, Chitrphiromsri, & Ding, 2008; Ukponmwan, 1993). These factors mainly contribute to maintain the integrity, thermal insulation capacity, and/or liquid/air impermeability of the fabrics/clothing, which ultimately affect the protective and comfort performances of the clothing (Song et al., 2016; Song & Mandal, 2016). Most importantly, if a factor enhances the integrity, thermal insulation capacity, and/or liquid/air impermeability of the clothing, the transfer of thermal energy slows down through the clothing under different thermal hazards faced by firefighters in a fire incidence (e.g., flame, radiant heat, hot surfaces, molten metal substances, steam, and hot liquids); this situation enhances the thermal protective performance of firefighters' clothing. However, the transfer of metabolic heat and sweat vapor from firefighters' bodies to their ambient environment becomes disrupted (causing heat stress) due to the enhancement of thermal insulation capacity and liquid/air impermeability of the fabrics/clothing; this situation results in decreased thermophysiological comfort performance of firefighters' clothing.

To thoroughly understand the protective and comfort performances, many researchers have studied the impact of various factors (e.g., fiber/yarn/fabric types and properties, and garment designs and properties) on the integrity, thermal insulation capacity, and liquid/air impermeability of the clothing (Ghassemi, Mojtahedi, & Rahbar, 2011; Hatch, 1993; Hearle, 1963, 1991; Ho et al., 2011; Reischl & Stransky, 1980; Shalev & Barker, 1984; Sun et al., 2000; Varshney, Kothari, & Dhamija, 2011). The following section discusses their findings to develop better understanding for achieving the best possible clothing performance for the protection and comfort of firefighters. Contextually, it is notable that the protective and comfort performances of firefighters' clothing can be effectively measured by using various standard methods recommended by different organizations [ISO, NFPA, American Society for Testing Materials (ASTM)]; these standardized performance evaluation methods are discussed in a separate chapter of this book.

2.2.1 Fiber Types and Properties

As indicated earlier, the fibers used in firefighters' protective clothing are chemically modified fire-retardant and/or inherently fire-resistant fibers. As a result, the limited oxygen index (LOI; i.e., minimum percentage of oxygen required for combustion) of these fibers is always above 21% (Bajaj, 1992; Bajaj & Sengupta, 1992; Bourbigot, 2007).

The techniques for manufacturing the chemically modified fire-retardant fibers are mainly dependent on the nature of the combustible substrate fibers: synthetic fibers (namely polyester, nylon, and acrylic) or natural fibers (namely wool, cotton, and viscose) (Horrocks, 1986, 2003; Samanta, Baghchi, & Biswas, 2011; Tasukada et al., 2011). For the synthetic substrate fibers, the flame-retardant materials (halogen, nitrogen, silicon, and phosphorus) are incorporated into polymerization during the fiber production process through melt spinning or doped into the spinning bath during the fiber production through solution spinning. Both synthetic and natural substrate fibers can also be finished by flame-retardant materials to convert them into chemically modified fire-retardant fibers. When chemically modified fire-retardant fibers are exposed to high heat thermal hazards, these flame-retardant materials form the gaseous-vapor phase (nonvolatile ester compound) or solid-condensed phase (carbonaceous char compound) on the fibers or these fiber based fabrics (Figure 2.5). Formation of these phases insulates the underlying substrate fibers or fabrics and helps to maintain the integrity of firefighters' protective clothing.

In the case of inherently fire-resistant fibers, the use of flame-retardant materials as an additive or finishing agent on the combustible substrate fibers is not required (Bajaj, 1992; Bajaj & Sengupta, 1992; Bourbigot & Flambard, 2002). The chemical structures of the inherently fire-resistant fibers [aramid, poly(aramid-imide), polyimide, polybenzimidazole, polybenzoxazoles, melamine-formaldehyde, phenolic, chlorinated, fluorinated, polyphenylene sulphide, polyetheretherketones, polyether imide, polyacrylate, semi-carbon, glass, and ceramic] are such that they do not easily melt or catch the fire. As an example

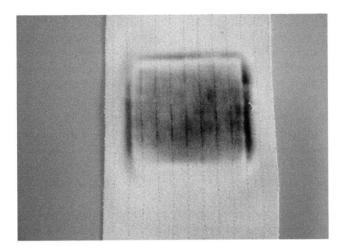

Figure 2.5

Carbonaceous char on flame-retardant (phosphine oxide) treated cotton fabric.

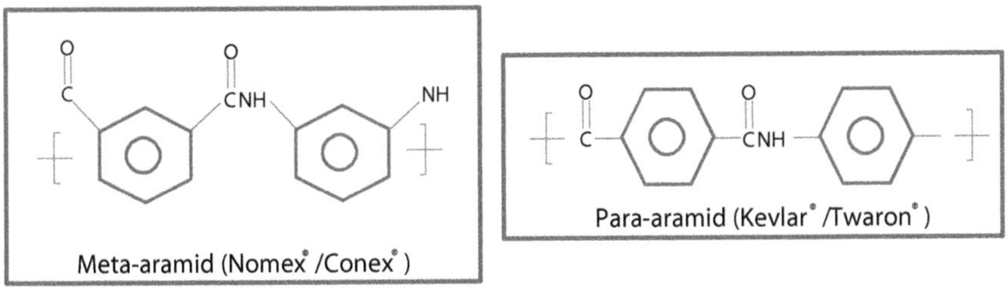

Meta-aramid (Nomex®/Conex®)

Para-aramid (Kevlar®/Twaron®)

Figure 2.6

Chemical structure of aramid fibers.

(in Figure 2.6), the linked and highly oriented aromatic groups in the backbone polymer chains of the aramid fibers do not easily break down into ignitable molecular fragments. The presence of hydrogen bonding in aramid fibers also affects the orientation of polymer chains and provides a basis for improving the orientation, crystallinity, and breaking strength of the fiber molecules under a thermal hazard. Additionally, the high dissociation energies of C–C and C–N bonds in the main polymer chain give high thermal stability to the aramid fibers (Tsvetkov & Shtennikova, 1978; Wang & Chen, 2005; Yoda, 1962). Depending on the position of the aromatic groups, the chemical structure of the aramid fibers could differ and increase the strength—namely, the rod-like structure of para-aramid fibers possesses more strength than meta-aramid fibers. Altogether, the chemical structure of the inherently fire-resistant fibers helps to maintain the integrity of firefighters' protective fabrics or clothing under various hazards, by increasing their strength. As per the ISO 11612:2013, at least, these fibers should help to maintain the tensile (tested in accordance with 13937-2) and tear (tested in accordance with ISO 3377-1) strength of the outer layer shell woven fabrics more than 300 and 10 N, respectively.

Based on the above discussion, chemically modified fire-retardant fibers and inherently fire-resistant fibers could help to maintain the integrity of the firefighters' protective clothing. At the same time, the properties of these fibers (fineness, cross section, length, and crimp) could also help increase the thermal insulation capacity of the firefighters' protective clothing by trapping more dead air within its structure. For example, the fiber surface-to-volume ratio is high in the case of fine fibers, and the surface of fine fibers can trap a lot of dead air in comparison to the coarse fibers; noncircular/profiled or hollow fibers could trap more dead air because these fibers have more loops or holes within their structures than circular fibers; also, the lengthy and crimped fibers could trap more air on their surfaces than the short and non crimped fibers (Hes, 2008; Marom & Weinberg, 1975; Rao & Gupta, 1992).

2.2.2 Yarn Types and Properties

Depending on the spinning processes (e.g., ring spinning, rotor spinning, air jet spinning), the types of yarns used in firefighters' protective clothing can be named as spun yarns, rotor yarns, texturized yarns, and filament yarns. Notably, texturized, spun, and rotor yarns possess more protruding fibers on their surface than filament yarns. As a result, the thermal insulation capacity of texturized/spun/rotor yarns–based firefighters' protective clothing is higher and provides better protection to firefighters (Ghassemi, Mojtahedi, & Rahbar, 2011; Hatch, 1993; Ramachandran, Manonmani, & Vigneswaran, 2010).

Along with the different types, the properties of the yarns, namely, yarn twist (twist can be measured by ASTM D 1422:2013 standard), are important to increase the thermal insulation capacity of the firefighters' protective clothing (Rengasamy & Kawabata, 2002). As found, the fibers in a yarn could be highly twisted during spinning; this compact yarn can trap less dead air and lower the thermal insulation capacity of the firefighters' clothing. Nevertheless, if the fibers are loosely twisted in a yarn, this yarn cannot trap the dead air due to high natural convection on its surface; this situation also lowers the thermal insulation capacity of firefighters' clothing. Moreover, the arrangement of the fibers in a yarn during twisting could affect the thermal insulation capacity of the clothing. If the fibers lie parallel to each other, the yarn can trap more dead air to enhance the thermal insulation capacity of the clothing (Hatch, 1993; Song, 2009).

2.2.3 Fabric Types and Properties

All the fabrics are generally porous and their structure comprises two phases: (1) a gaseous phase consisting of water vapor and/or dry air; and (2) a solid phase consisting of fibers/yarns (as well as bound water, if hygroscopic) (Figure 2.7)

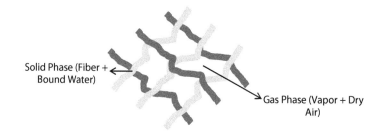

Solid Phase (Fiber + Bound Water)

Gas Phase (Vapor + Dry Air)

Figure 2.7

Hygroscopic porous nature of fabrics.

(Chitrphiromsri & Kuznetsov, 2005; Song, 2009; Song et al., 2008). Depending on the structure of these two phases, the types and properties of the fabric vary.

Fabric systems used for firefighters' protective clothing are generally a combination of three types of fabrics: woven, knitted, and/or nonwoven (Mandal & Song, 2014). Among all of these fabrics with same weight, the nonwoven fabrics can trap more dead air than other fabrics (Abdel-Rehim et al., 2006). As a result, the nonwoven fabrics–based firefighters' protective clothing displays high thermal insulation capacity, especially under some thermal hazards (flame, hot surface contact), where conductive and convective heat transfer predominates through the clothing. Nevertheless, the trapped air could be lower in a nonwoven fabric when it is compressed (especially when firefighters come in contact with hot surface) and that can lower the thermal insulation capacity of the clothing. Additionally, the compression recovery of the nonwoven fabrics is low, which can permanently lower the thermal insulation capacity of the clothing. Also, nonwoven fabrics cannot act as a good thermal insulator under radiant heat hazard because open fiber surfaces can easily radiate the heat toward firefighters' bodies. Altogether, the low compression recovery and high radiative heat transfer property can lower the thermal protective performance of the clothing (Mao & Russel, 2007; Qashou, Tafreshi, & Pourdeyhimi, 2009). To lower the radiative heat transfer through the nonwoven fabrics, many researchers recommend modifying the structure of the fabrics by increasing the fiber volume, using the finer fibers or tortuous fiber orientation (Barker & Heniford, 2011; Gibson & Lee, 2007). Furthermore, yarns used for the manufacturing of knitted fabrics comprise looped structure, and these loops can trap a lot of dead air and enhance the thermal insulation capacity of firefighters' protective clothing (Au, 2011; Spencer, 2001). The thermal insulation capacity can also be enhanced by modifying the structure of the knitted fabrics (e.g., developing three-dimensional spacer knitted fabrics, incorporating extra films or fiber surface on the regular knitted fabrics) (Mao & Russel, 2007). As the woven fabrics are generally manufactured by interlacing two different sets of yarns (warp and weft), these fabrics do not comprise any loop or trap air (Celcar, Meinander, & Gersak, 2008; Frydrych, Dziworska, & Bilska, 2002; Marks & Robinson 1976; Matusiak & Sikorski, 2011). Eventually, the thermal insulation capacities of the woven fabrics are the lowest. Nevertheless, due to the interlacement, the strength of the woven fabrics is high; this situation could contribute to maintaining the integrity of firefighters' protective clothing. Notably, by changing some parameters of the woven fabrics (warps/wefts per inch, weave design, finishing), the thermal insulation capacity can be changed. For example, the number of interlacements in a plain weave designed fabric is much higher than the satin weave designed fabrics, and that results in more trapped air or thermal insulation capacity for the plain weaved fabrics (Matusiak & Sikorski, 2011). It has also been found that the resin finishing could help to enhance the thermal insulation capacity of the woven fabrics (Frydrych et al., 2002).

Along with the types, the properties of the fabrics (emissivity, air permeability, weight, thickness, density, moisture content, thermal conductivity, and heat capacity) also substantially contribute to enhance their thermal insulation capacity (Barker & Heniford, 2011; Fan, Luo, & Li, 2000; Farnworth, 1983, 1986; Keiser & Rossi, 2008; Lawson et al., 2004; Lee & Barker, 1986; Matusiak & Sikorski, 2011; Qashou et al., 2009; Shabaridharan & Das, 2012; Shalev & Barker, 1983; Shekar et al., 1999; Song et al., 2011). A fabric with high liquid/

air permeability and emissivity [liquid permeability, air permeability, and emissivity can be measured by the American Association for Textile Chemists and Colorists (AATCC) 195, ASTM D 737, and Xenon reflectometer, respectively] reflects less and/or absorbs/transmits more thermal energy toward firefighters' bodies under thermal hazards such as heat and hot liquids; this situation results in lower thermal insulation capacity or thermal protective performance of the fabrics (Barker & Heniford, 2011; Fan et al., 2000; Keiser & Rossi, 2008). However, a fabric with high weight and thickness (weight and thickness can be measured by ASTM D 3776 and ASTM D 1777, respectively) could trap a lot of dead air within its structure and could enhance the thermal insulation capacity (Shalev & Barker, 1983; Song et al., 2011). Among two fabrics with similar thickness, the fabric with low density could show the high thermal insulation capacity (Mao & Russel, 2007; Shalev & Barker, 1983). It is further notable that the structural properties of two same density fabrics can be quite different; one fabric may be loosely woven from highly twisted yarn, and the other may be closely woven from loosely twisted yarns. So, these changes in structural properties could vary the thermal insulation capacity of the fabrics (Barker & Lee, 1987). Similar to the density, the moisture content of the fabrics have a complex effect on the thermal insulation capacity (Barker et al., 2006; Lee & Barker, 1986). If the moisture content in a fabric is low (≤ ~15%), the thermal conductivity of the fabric increases and could transmit the more thermal energy toward firefighters' bodies (i.e., lowering the thermal protective performance of the clothing). However, if the amount of moisture is high, the heat capacity of the fabric increases, and that could help to store more thermal energy inside the fabrics. This situation will contribute to transmitting less thermal energy toward firefighters' bodies and enhancing the thermal protective performance of the clothing. Notably, when firefighters get exposed to the high-intensity thermal hazards (~84 kW/m² flame or radiant heat), the moisture inside the fabric systems could convert into steam, which could generate burns on firefighters' bodies and lower the thermal protective performance of the clothing.

2.2.4 Garment Designs and Properties

It has been found that the garment designs and properties such as stitch and seam, size, closures (e.g., pockets cuffs, collar), and fasteners (e.g., hooks, loops, buttons) have an impact on the thermal insulation capacity and/or liquid/air permeability of the clothing (Berglund & Gonzalez, 2006; Crown et al., 1998; Holmer & Nilsson, 1995; Konarska et al., 2006; Reischl & Stransky, 1980). When different clothing parts (e.g., collar, sleeves) are sewn together in the garmenting process, different types of stitch (e.g., chain stich, lock stitch, zigzag stitch) and seam (e.g., plain seam, flat seam, lapped seam) are used (Glock & Kunz, 1999; Mandal & Abraham, 2010). These stitching and seaming processes could structurally damage the fabrics with the strikes of sewing needles. This structural damage could reduce the integrity of the firefighters' clothing by reducing the seam strength. To avoid the problem of reduced seam strength, as per ISO 11612:2013 guideline, the minimum requirement of the seam strength in woven fabric (tested in accordance with ISO 13935-2:2014) and burst strength in knitted fabric seam (tested in accordance with ISO 13938-1:1999) should be 225 N and 100 kPa/50cm², respectively. Depending on the stitch and seam, the thermal insulation capacity of the clothing could also vary. Notably, the thermal insulation capacity could be enhanced by covering the stitch and seam area with a special type of fabric or by introducing stitchless and seamless firefighters' protective clothing in the market.

Furthermore, the size of the garment with respect to wearers' (firefighters') bodies is important to provide protection and comfort (Crown et al., 1998; Holmer & Nilsson, 1995; Mah & Song, 2010; Song, 2007). If the garment size is too small, the size of the microclimate region (i.e., the air gap thickness in between the clothing and wearers depending on various aspects such as gender, body shape and size, draping of the fabrics used in the clothing) becomes low and vice versa. The too low size of the microclimate region [can be measured using 3D body scanning method as described by Psikuta et al. (2012) and Lu et al. (2013)] could promote the conductive (due to the contact between the hot garment and firefighters) and radiative (due to the radiative heat exchange between hot garment and skin) heat transfer from the thermal hazards to firefighters' bodies; the too high size of the microclimate region also promotes the natural convection (presence of environmental thermal energy in the region) and transmits the thermal energy toward firefighters. Both too low and high sizes of the microclimate region actually lower the thermal insulation capacity or thermal protective performance of firefighters' clothing. Eventually, it is required to decide a typical or optimal size of the microclimate region to provide better protection to firefighters; based on the earlier research work, the average microclimate size has been standardized at 6.35 mm (Crown et al., 2002; Kim et al., 2002; Torvi, 1997). Contextually, it is notable that the metabolic heat and sweat vapor generated from firefighters' bodies are also transmitted through the microclimate region toward the clothing via convection, conduction, radiation, evaporation, and/or moisture absorption/diffusion (Lu et al., 2013; Morozumi, Akaki, & Tanabe; 2012). As a result, the microclimate region may comprise the considerable amount of moisture. As the moisture has high thermal conductivity, it may help to transmit the thermal energy from the clothing to the firefighters' bodies and lower the thermal protective performance of the clothing.

Furthermore, various types of closures and fasteners attached with the firefighters' clothing could disturb the microclimate region and thus promote the transmission of thermal energy toward firefighters (Berglund & Gonzalez, 2006; Konarska et al., 2006). For example, if the collar of the firefighters' clothing is open, it may promote the transmission of the thermal energy from a fire hazard to the firefighters' bodies and generate burns. Considering this, it is necessary to properly design the closures and attach the fasteners. As an example in this regard, the collar of the protective clothing is now designed with long zip to ensure complete closure of the jacket up to the firefighters' neck.

2.3 Types and Features of Firefighters' Protective Equipment

To protect different body parts (head, eyes and face, hands and arms, feet and legs) from various hazards, firefighters need to wear different types of protective equipment. This includes helmet, flash hood, breathing apparatus, gloves, boots, and high visibility safety vest (Personal Protective Equipment, n.d.). Before making firefighters' protective equipment commercially available, it is necessary to remember that equipment such as helmets, flash hoods, gloves, and boots should generally meet the design and minimum performance requirement as set by ISO 11999:2015 (ISO 11999-5 for helmet, ISO 11999-9 for flash hoods, ISO 11999-4 for gloves, ISO 11999-6 for boots) and NFPA 1971:2013. The breathing apparatus should also meet the requirements set by NFPA 1981:2013.

2.3.1 Helmet

Protecting firefighters from potential head injuries by falling debris and heat is important because head injuries can impair them for life or can be fatal. It is recommended that every firefighter wear a proper helmet (Figure 2.8). The designs and materials of helmets used by firefighters should be durable, long-lasting, impact absorbent, hazards protective, and as lightweight as possible (Uniforms and Equipment, n.d.; Personal Protective Equipment, 2004).

The firefighters' helmet was first invented by Jacobus Turck in 1730 (Hasenmeier, 2008). This helmet was made of leather with a high crown and wide brim. Then, Henry T. Gratacap modified its design in 1836. This design was a reinforced dome-shaped leather helmet with a front shield and brim rolling to a long back tail. The Gratacap's helmet was suitable for providing the protection to firefighters from falling debris and water that could run off the back of the helmet.

Although the design aspects of the Gratacap's helmet have changed little over the years, the materials used for the developing the firefighters' helmet have drastically changed. Instead of using leather as was used in traditional helmets, modern helmets can be manufactured by using metals (e.g., nickel, brass, aluminum), polymers, plastics, or composite materials. Nevertheless, composite materials

Figure 2.8

A firefighter wearing helmet.

2. Firefighters' Protective Clothing and Equipment

(could be reinforced with the aramid fibers) are widely used in modern firefighting helmets (U.S. Fire Administration Emergency Incident Rehabilitation, 2008). One of these composite materials–based helmets is called Formula 1 (F1) helmet made by Gallet (presently a subsidiary of MSA Safety Incorporated, USA) in France. This handmade F1 helmet is developed using synthetic materials (carbon fiber, polyethylene, and aramid) covered with galvanized nickel. The F1 helmet is designed in such a way that it could cover the maximum head area of firefighters as well as give a wide field of vision to them. This also covers a number of ergonomics and safety-based requirements, such as clearance between the head and the shell of the helmet. Before selling in the commercial market, the F1 helmet has to pass through extreme deformation and fragmentation tests indicated in EN 443:2008 standard (penetration of 1 kg pointed blade, shock absorption while impacted using 5 kg falling striker). Currently, the F1 helmet is used in nearly 85 countries. Although the F1 helmet is widely used by firefighters while working in vehicle and wildland fires, this is not suitable for structural building fires. The helmet used for structural firefighting is generally made by thermoplastic glass fiber composite materials. This helmet is designed in such a way that a rear brim is longer than the front brim and a face shield is attached to the front of the helmet. This is designed to give more protection to firefighters' neck and face from thermal and physical hazards. Nevertheless, goggles could give better protection to the eyes of firefighters. Recently, a glass fiber, lightweight, smaller brim, and rounded edge–based helmet called "Metro helmet" has been developed by several leading helmet manufacturers. This helmet is ergonomically designed to provide better sensorial comfort (e.g., noise buffeted) to firefighters; the smaller brims of these sensors also put little obstruction in other body parts such as the neck, especially when firefighters look upward. Notably, in many countries (e.g., Germany, United States, Canada), the color of the helmet also denotes the official rank of a firefighter. A yellow helmet is usually worn by rank-and-file firefighters; red or orange helmets are usually worn by captains; and white helmets are typically worn by fire chiefs. In the commercial market, as per ISO 11999-5:2015 guideline, the helmet can be categorized as Type 1 that meets the performance requirement set by EN 443:2008 and Type 2 that meets the performance requirement set by NFPA 1971:2013. These Type 1 and Type 2 helmets should also meet the performance requirement set by ISO 11999-3:2015 standard under specific thermal hazards such as flame or radiant heat.

2.3.2 Flash Hood

Although helmets can protect the firefighters' head, it cannot provide protection to ears, neck, and part of the face. Therefore, every firefighter needs to wear a flash hood under their helmet (U.S Fire Administration Emergency Incident Rehabilitation, 2008; Components of Wildland Fire PPE, 1993), which is generally designed according to the ISO 11999-9:2015, NFPA 1971:2013, or NFPA 1975:2013 standard (Figure 2.9). This flash hood is a one-piece garment made by inherently fire-resistant fibers–based fabrics, namely meta-aramid fiber–based knitted fabric of 200 g/m^2 (Uniforms and Equipment, n.d.); this garment has a face opening for ventilation and bib (sides, front, and back) (U.S. Fire Administration Emergency Incident Rehabilitation, 2008). This fabric is most often double plied with only one seam running from the top center of the face opening, over the top and down the bottom of the bib. This garment is tightly placed around the face of a firefighter to protect his/her face, ears, and neck. When worn, the flash hood

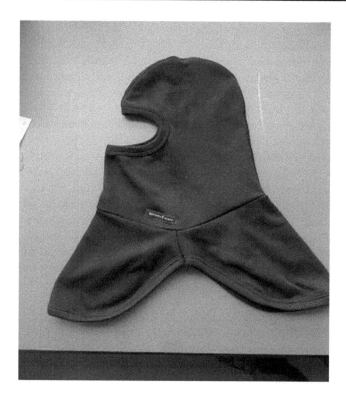

Figure 2.9

Firefighters' flash hood.

is first tucked into the collar of the firefighters' turnout gear. Then, the self-contained breathing apparatus (SCBA) mask is donned and the hood is pooled over the face seal to cover any exposed skin; some flash hoods are designed to accommodate a specific respirator mask. Many firefighters complain that the flash hood provides an encapsulation to their faces that restricts the ability of their ears to receive proper communication from their colleagues. Considering this, some hoods are composed of mesh materials in the ear region to facilitate communication. The flash hoods used by American firefighters may also have mesh materials on the top of the hood (sitting underneath the helmet) to provide a means for heat to escape the hood. As per ISO 11999-9:2015 standard, there are two types of commercially available firefighters' flash hoods (FH), FH1 and FH2. Here, the protective performance of FH2 is generally higher than the FH1; for example, HTI_{24} of FH1 and FH2 are ≥ 8 s and ≥ 11 s, respectively. ISO 11999-9:2015 also recommends that the fibers used for the manufacturing of flash hoods should not generate any molten debris, hole, and/or will not promote any after flame burning of ≤ 2 s.

As the flash hoods are manufactured by using the knitted fabrics, they are stretchable due to the looped form yarns in the fabrics. Eventually, the flash hoods are available only in one size that fits all. Although flash hoods are available in one size only, these hoods must be selected carefully by firefighters to fit properly with other protective equipment, including the SCBA mask (U.S. Fire Administration Emergency Incident Rehabilitation, 2008). Due to repeated stretching of the flash hoods over the firefighters' face and head, some hoods could also quickly lose their shape. This change in the shape of the flash hoods could lead to a reduction of protection for the firefighters. To overcome this problem, NFPA 1971:2013 recommended a test for evaluating face opening size of the flash hood after repeated doffing and donning of the hood on a head manikin.

2.3.3 SCBA

SCBA is a device worn by firefighters while carrying out interior offensive structural firefighting or working in areas where they may be exposed to high temperatures, oxygen deficiency, toxic substances, smoke concentration, dust, heat radiation, or burning embers (Park et al., 2014; Types of Breathing Apparatus, n.d.; Uniforms and Equipment, n.d.). As these fire environments are immediately dangerous to human life or health, the SCBA device helps to provide breathable air to firefighters. In fact, an SCBA allows a firefighter to safely explore the smoke-filled areas to search for unconscious victims (Browne, n.d.).

A SCBA typically has three main components: a 4–6 L air [a mixture of oxygen (21%) and nitrogen (79%) gasses] cylinder having an air pressure of 150 to 380 bars, a pressure regulator, and a face-piece assembly (face-piece lens, a hose, and an exhalation valve) (Figure 2.10). These three components are connected together and mounted to a carrying frame. During extended firefighting operations, empty air cylinders can be quickly replaced with fresh ones and then refilled from larger tanks in a cascade storage system or from an air compressor brought to the firefighting scenarios.

The air cylinders in SCBA are generally made up of aluminum, steel, or carbon fiber–based composite materials. As composite materials–based air cylinders are lightweight, these cylinders are gaining in popularity among firefighters; however, the life span of these composite cylinders is short. To check the quality (leaks, strength) of the air cylinder, the used cylinders are always hydrostatically tested (by filling the cylinder with colored liquid to check for leaks and by pressuring the cylinder at a specified pressure) every 5 years. It is also necessary to note that the amount of air in the cylinder can support the breathing of a working firefighter for certain duration, and this duration can be estimated by {Volume of the Cylinder (in liters) × Pressure of the Cylinder (in bars)}/40 – 10.

When a firefighter uses SCBA, the air from the cylinder travels through the high-pressure hose to the regulator (Types of Breathing Apparatus, n.d.). Depending on the SCBA model, the regulator consists of a mainline valve and a bypass valve for the normal and emergency operations, respectively. In the normal operation, the regulator reduces the pressure of the cylinder air to slightly above atmospheric pressure and controls the flow of air to meet the respiratory requirements of the firefighter. When the firefighter inhales, a pressure differential is created in the regulator. In this situation, the diaphragm of the regulator moves inward and tilts the mainline admission valve so that low-pressure air can flow into the inhalation connection; the diaphragm is then held open to create the positive pressure. During the exhalation, the diaphragm moves back to the closed position so that air cannot flow into the face-piece assembly. Notably, some SCBA units have regulators that fit into the face-piece assembly; on other units, the regulator is on the firefighter's chest or waist strap. Also, firefighters should keep an eye on the pressure gauge attached with the regulator at a visible position. If the air pressure in the cylinder becomes low, it may disturb the flow of the air from the cylinder to the regulator; this situation may create breathing problems for firefighters. Notably, an audible alarm is activated when the

(a)

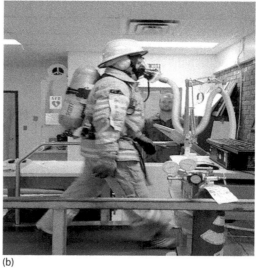

(b)

Figure 2.10

An SCBA (a) worn by a fighter (b).

2. Firefighters' Protective Clothing and Equipment

cylinder pressure decreases to approximately one-fourth of the maximum rated pressure in the cylinder.

A face-piece assembly provides some protection from facial and respiratory burns and holds the cool breathing air for firefighters. The face-piece assembly consists of the face-piece lens, a low-pressure hose to carry the air from the regulator to the face-piece mask if the regulator is separate, and an exhalation valve. The face-piece lens is made up of clear safety plastic and is connected to a flexible rubber mask. The face-piece is held snugly against the face by a head harness with adjustable straps, net, or some other arrangement. Some helmets have a face-piece lens bracket that connects directly to the helmet instead of using a head harness. The lens should be protected from scratches during use and storage and could have a speech diaphragm to make communication clearer for firefighters with their fire stations. The low-pressure corrugated hose attached with the face-piece assembly brings the air from the regulator to the face-piece mask; the corrugation is mainly to prevent collapse when a person is working in close quarters, breathing deeply, or leaning against a hard surface. The exhalation valve is a simple, one-way valve that releases an exhaled breath without admitting any of the contaminated outside atmospheres. Notably, dirt or foreign materials can cause the valve to become partially opened, which may permit excess air from the tank to escape the face-piece assembly. Therefore, it is important that the valve be kept clean and free of foreign material. It is also important that the exhalation valve be tested by the firefighter during face-piece-fit tests and before entering a hazardous atmosphere. As observed, an improperly sealed face-piece assembly or a fogged face-piece lens can cause problems for the firefighters. The different temperatures inside and outside the face-piece assembly can cause the face-piece lens to fog (the fogging inside the lens occurs when the lens is cool and condenses the highly humid exhaled breath of firefighters; external fogging occurs when condensation collects on the relatively cool lens during interior firefighting operations), which hampers the vision of working firefighters. The external fogging can be removed by wiping the lens or prevented by using some methods, for example, using nosecaps to deflect the firefighters' exhalations away from the lens, by applying some antifogging chemical on the face-piece lens, or by storing the lens in a packed case, bag, or coat pouch.

2.3.4 Gloves

Firefighters are recommended to use different types of high dexterity gloves to get protection from various hazards that they may encounter while fighting fires (Figure 2.11). For example, high-strength thermal protective gloves protect from physical and thermal hazards (e.g., flame, radiant heat, steam), chemical protective gloves protect from chemical hazards (e.g., liquids, air vapors), and surgical gloves protect from biological hazards (e.g., blood-borne pathogens) (Uniforms and Equipment, n.d.). The design and materials of firefighters' gloves are certified by ISO 11999-4:2015, NFPA 1971:2013, or EN 659:2003 standards. The designs and materials of the gloves depend on several factors: types of hazards handled (e.g., thermal, chemical), nature of the hazards (e.g., liquid based, vapor based), duration of the exposure to the hazards, area requiring protection (hand only, forearm, arm), and designs and size fit of the gloves (Personal Protective Equipment, 2004; Spahr, 2004).

According to ISO 11999-4:2015 and NFPA 1971:2013 standards, the design of gloves should cover the circumference and length of the hands and each

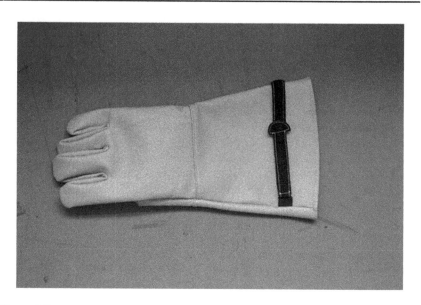

Figure 2.11

Firefighters' gloves.

finger of the firefighters' hand. Although ISO 11999-4:2015 recommends only six unique and distinct sizes of gloves, the gloves are available in seven different sizes, from XXS to XXL as per NFPA 1971:2013 standard. Thus, it is notable that inconsistencies in the glove sizes have been observed for different standards/manufacturers, and it could be a real challenge for firefighters to choose the right size of gloves while firefighting (Park et al., 2014). Also, the designs of the gloves are generally standardized based on the collected hand-size data from the military personnel 20–30 years ago. Although these hand-size data are suitable to design the contemporary gloves for firefighters, these designs may not be suitable to provide adequate protection to firefighters from various hazards. This is because the hand sizes of the world population have changed in the last 50 years, so the designs of the gloves based on the old hand-size data could not be properly fit for the present firefighters. Also, firefighters are generally taller and heavier; their hand sizes could be different from military personnel. Eventually, the present designs of the gloves may not be completely adequate for providing protection for firefighters. Furthermore, hand sizes are highly dependent on gender, race, ethnicity, and occupational status (volunteer firefighters or career firefighters) of firefighters; standard designs of the gloves (recommended by NFPA or CEN) may not be suitable for all firefighters. If the designs of the gloves are improper, the firefighters may opt for not wearing the gloves or they may wear improperly fit gloves; this situation can create injuries on firefighters' hand (Personal Protective Equipment, n.d.). Notably, a tightly fitted glove may provide better dexterity, but the compression of the glove layers may affect the protective performance under thermal hazards. Also, an oversized glove significantly limits the grip and dexterity and requires additional physical effort to complete a task. Overall, it is recommended to collect appropriate firefighters' hand-size data to improve the design of the present gloves.

Furthermore, it is recommended to use the appropriate materials depending on the types of the gloves (Components of Wildland Fire PPE, 1993; Structural

2. Firefighters' Protective Clothing and Equipment

Firefighting Gloves, 1993). For example, high-strength thermal protective gloves can be made by leather fabrics or aramid fiber–based fabrics or aluminized synthetic fiber–based fabrics; high dexterity chemical protective or surgical gloves can be made with different kinds of rubbers [natural (can protect against blood-borne pathogens, acids, alkalis, salts, and ketones), butyl (can protect against peroxide, rocket fuels, corrosive acids, string bases, alcohols, aldehydes, ketones, ester, and nitro-compounds), neoprene (can protect against hydraulic fluids, gasoline, alcohols, organic acids, alkalis), nitrile (can protect against oils, greases, acids, caustics, and alcohol), fluorocarbon], plastics (polyvinyl chloride, polyvinyl alcohol, polyethylene), or a blend of rubbers and plastics.

The above discussion indicates that there is no single material–based glove that can provide the protection from all types of hazards (Structural Firefighting Gloves, 1993). Considering this, NFPA 1971:2013 standard recommended the use of multilayered fabricer–based gloves for firefighters' protection from various hazards; nevertheless, proper training, experience, and supervision are needed to ensure the correct glove is being worn by firefighters while fighting fires. As per NFPA 1971:2013 standard, the gloves used for firefighting are generally configured as joined or continuous layers of fabrics: a shell fabric (made of a leather fabric or aramid fibers–based woven fabric to provide grip and integrity under various hazards), a moisture barrier (made of a polytetrafluoroethylene coated fire-retardant/resistant fibers–based woven fabrics to provide protection from chemical and biological hazards), and a thermal liner (made of a fire-retardant modacrylic fibers–based knitted fabric or an aramid fibers–based nonwoven fabric to provide protection from thermal hazard). This key standard also sets the performance requirements for the gloves, including flame and conductive heat resistance, liquid penetration, cut/puncture resistance, hand function, and grip. As a general rule, a thick glove can provide better protection from various physical, thermal, and chemical hazards; nevertheless, thick gloves may impair the dexterity and grip and that may have a negative impact on firefighters' safety (Browne, n.d.). Notably, the grip of the firefighters' gloves can be verified by following another standard method: ASTM F 2961: Standard Test Method for Characterizing Gripping Performance of Gloves Using a Torque Meter. A thick glove can also create the problem during its donning by firefighters. In this context, it is required that the donning of the gloves should be achievable within 10 and 30 s in the case of dry and wet hands of firefighters, respectively; the moisture barrier or thermal liner should not be detached from the gloves during the donning by firefighters. Last but not least, a thick glove may restrict the transmission of metabolic heat and sweat vapor from firefighters' hand to their surrounding environment, and this situation may create discomfort for firefighters. Although there is no standard available to evaluate the comfort performance of firefighters' gloves, sweating hand manikin has been installed by many organizations (Laboratories for Functional Textiles and Protective Clothing, Iowa State University, USA; Textile Protection and Comfort Laboratory, North Carolina State University, USA) to evaluate the comfort performance as per their requirements. Overall, in some cases, a thin glove could be more preferable by firefighters for better grip, wearability, and comfortability; however, this glove should not generate any first-degree or second-degree burns on firefighters' hand within 6 and 10 s, respectively.

The protective gloves for firefighters should be inspected before each use to ensure that they are not torn, punctured, or ineffective in any way. Every time, an overall visual inspection may help to detect cuts or tears in the gloves. A thorough inspection by filling the gloves with water and/or tightly rolling the cuff toward the fingers will help to reveal any pinhole leaks inside the gloves. A decision to reuse chemically exposed gloves should take into consideration the toxicity of the chemicals involved and factors such as duration of exposure, storage, and temperature. The reuse of chemical protective gloves should be evaluated carefully by taking into consideration the chemical absorptive qualities of the gloves. Discolored and stiff gloves (due to excessive use and/or chemical exposure) need to be discarded for the better protection of firefighters.

2.3.5 Boots

Firefighters' feet and legs can be frequently exposed to hot surfaces, falling debris, toxic liquids, and/or blood-borne pathogens. Keeping this in mind, firefighters' boots should be of high quality, should be well fitted, and can meet the recommended specification from ISO 11999-6:2015 or NFPA 1971:2013 (Figure 2.12). These boots should always be designed in such a way that they can provide the proper protection and comfort to firefighters from various hazards, for example, thermal, physical, chemical, and biological (Uniforms and Equipment, n.d.).

The toes and soles of firefighters' boot are generally impact-resistant and heat-resistant (Browne, n.d.; Components of Wildland Fire PPE, 1993). The metal-toed boots are frequently used by firefighters, and these boots can give better protection to them from sharp debris they might accidentally kick or step on. However, these metal-toed boots are not recommended for firefighters because the metal can easily transfer the heat from the hot surfaces to the firefighters' feet. Firefighters may also need to work on uneven terrain or slippery floors; non-slip soles with high grip are essential for firefighters' boots. Considering these, firefighters' boots should have a dual density, cushioned, fire-resistant rubber

Figure 2.12

Firefighters' boots.

2. Firefighters' Protective Clothing and Equipment

sole (like Vibram® sole) of 6 to 8 inches. These designs and materials aspects reduce the boot weight and provide orthotic superiority, greater ankle support, and reduced skeletal impact-related injuries. This allows maximum traction and prevents melting upon exposure to thermal hazards.

Additionally, the upper parts of the boots (e.g., toe cap, vamp) are manufactured by using high-strength fire-resistant and liquid-resistant leather materials to provide protection from various hazards (Personal Protective Equipment, 2004). Sometimes, firefighter activities involve the use of a chainsaw; in this case, the boots must be made by cut-resistant materials as per the guideline of Occupational Health and Safety Administration (OHSA) of the United States. To qualify as a cut-resistant boot, the cut-resistant material can be incorporated in the boot, a cut-resistant sock can be inserted in the boot, the boot could be covered by cut-resistant material, or any other means as approved to comply with being cut resistant. Sometimes, firefighters may also need to wear rubber boots while working in lowland or peat areas. In this case, firefighters need to be cautious when working around the hot areas. In a recent study, firefighters reported that rubber boots could be also stiff and oversized in use and that could lead to frequent fall-off of the boot during ascending and descending movements; this situation results in several injuries such as blister due to poor fitting, fall injuries, ankle sprains, stumbles, and slips.

Furthermore, the full interior of the boots are constructed by fire-retardant/resistant fibers–based liner to provide increased comfort to firefighters, by properly transmitting the metabolic heat and sweat vapor from their legs and feet to the surrounding environment. This liner can also provide increased protection from thermal hazards (Uniforms and Equipment, n.d.).

The laces of firefighters' boots should be made by leather or rawhides or aramid fibers. Similarly, stitches in the boot should be carried out by relatively thick waxed linen thread instead of nylon or other synthetic materials. Nylon or synthetic materials are generally smaller in diameter, usually shiny, and will melt or burn when heat is applied to a loose thread end.

Firefighters' boots should be kept in clean areas, and it is necessary to apply boot grease on a regular basis to keep the leather in soft condition (Personal Protective Equipment, n.d.). It is always necessary to inspect that boots are in good condition: this includes inspecting and replacing soles, ensuring all stitching is present, and checking the condition of laces.

2.4 Key Issues Related to Firefighters' Protective Clothing and Equipment

Based on the discussion in Sections 2.2 and 2.3, many innovative approaches have been applied to develop high-performance firefighters' protective clothing and equipment. Still, firefighters' injuries and fatalities occur worldwide. This may be due to the improper designs (non-ergonomically fit) and materials (heavyweight woven/nonwoven fabrics, impermeable moisture barrier, leather fabrics) used in firefighters' protective clothing and equipment.

The protective clothing (tunic and trouser) and equipment (helmet, flash hood, SCBA, gloves, boots) are generally designed based on the male anthropometric data collected by a group of researchers for a certain race, ethnicity, and/or nationality; even some of these data were collected decades ago. However, the structure (shape, size) of firefighters' body parts varies depending on race, ethnicity, gender, and nationality. In fact, the structure of the body parts also changes with time. Therefore, protective clothing and equipment designed for a particular race, ethnicity, gender, and nationality based on the old data may not be suitable for all firefighters now. Most importantly, the protective clothing and equipment of the present day are poorly fitted for female firefighters compared to their male colleagues. Recently, Park et al. (2014) reported that the gloves used by female firefighters in the United States are generally oversized especially at the fingers, and they may need to use some tools to improve the grip and dexterity of the gloves. In this context, Huck (1988), Akbar-Khanzadeh et al. (1995), Boorady et al. (2013), Sinden et al. (2013), and Park et al. (2014) mentioned that the protective clothing and equipment need to be designed by considering the proper human factors and the inputs from the firefighters about their firefighting-related activities. Park et al. (2014) suggested that by considering the proper human size, fit, and factors, it is possible to design protective equipment with improved mobility, protection, and comfort. Considering all these studies, it seems that there is a need to collect a new set of anthropometric data for firefighters who belong to different race, ethnicity, gender, and nationality. There is also a need to understand the human factors related to firefighters' activities. This will be helpful to redesign existing firefighters' protective clothing and equipment. While redesigning, the part of a newly designed protective clothing and equipment should be compatible (from design, protective/comfort performance aspects) to use in conjunction with the other parts of the existing/developed clothing or equipment as per the ISO 11999-2:2015 standard. For example, the interfacing area of helmet, flash hood, tunic, and SCBA should be compatible; the interfacing region between tunic/boots and trouser should be compatible; the interfacing area between gloves and jacket should be compatible. This approach could help to design more ergonomically fit protective clothing and equipment for firefighters. An ergonomically fit design could enhance the ability of a firefighter to visually inspect and recognize the risk factor in the fire field, where there may be a variety of hazards especially due to damaged building structures. Overall, the improvement in design aspects could provide the better protection to firefighters under various hazards such as physical, thermal chemical, or biological. This change in design could also lead to decreased size of some protective equipment such as the helmet and SCBA. It will ultimately help to reduce the weight of the equipment and give less physical burden or improved comfort to firefighters.

Furthermore, the materials used for firefighters' protective clothing and equipment are generally heavyweight, thick, and/or liquid-/air-impermeable to provide protection from various hazards. This kind of thick materials could store a lot of heat and chemicals when firefighters are exposed to different thermal and chemical hazards. Over time and/or with compression, the stored heat and chemicals could be transmitted toward firefighters' bodies and can generate injuries. The thick and heavyweight materials also impose physical burden on firefighters. For example, Coca et al. (2010) found that firefighters' motions are obstructed (e.g., limited mobility is observed in the head, neck, and arms; restricted the access to the tunic pocket) when they wear the heavyweight turnout gear along with helmet, SCBA, and boots. Helneman et al. (1989), Marshall (1980), and Park et al. (2010) corroborated that a heavy SCBA of nearly 9–13 kg weight can increase the risk of slip and fall injuries of firefighters, by changing their center of gravity of body mass and disturbing their body balance, and these injuries could be greater for lighter-weight and short-heighted firefighters. Neeves et al. (1989) and

Taylor et al. (2011) found that a heavy boot of nearly 5 kg weight can increase the physical burden of firefighters by lowering their leg movement and increasing their metabolic rate per unit mass. Additionally, thick and/or liquid-/air-impermeable materials may disturb the dissipation of the metabolic heat and sweat vapor generated from firefighters' bodies to their surrounding environment. Altogether, the transmission of stored heat/chemicals, physical burden, and disruption of metabolic heat and sweat vapor could generate a lot of discomfort or heat stress on firefighters and may result in more injuries and fatalities for them. Considering this, many researchers have put great efforts in developing new materials or implementing smart technologies for firefighters' protective clothing and equipment (Dadi, 2010; Donnelly et al., 2006; Hocke et al., 2000; Holme, 2004; Jin et al., 2011; Otsuka and Wayman, 1998; Song and Lu, 2013). These efforts have contributed in providing proper protection and comfort to firefighters. Based on their research, it has been found that the incorporation of some techniques (e.g., foamed silicon on vapor-permeable moisture barrier, macro-encapsulated nanoporous gels or aerogels or nano-clay reinforced resin coating on woven/nonwoven fabrics, nanofibers in nonwoven fabrics, nano finishes or phase change materials or shape memory alloy in woven/nonwoven fabrics) could enhance the protective and comfort performances of the materials used in firefighters' protective clothing and equipment. Additionally, it was evident that implementing smart technologies (electrical wires, sensors, gas detectors, safety alarm, cooling devices, two-way mobile, two-way portable communication devices) in the protective clothing and equipment could also passively control the injuries and fatalities of firefighters by monitoring their work situations (ambient temperatures, air pressure, heat stress level, toxic gas level). Altogether, it is evident that the development of new materials or implementation of new technologies has improved the protective and comfort performance of firefighters' protective clothing or equipment. In this context, it is necessary to remember that the performance requirement set by some standards (e.g., ISO 11999:2015) are mainly applicable when firefighters are wearing all their protective equipment along with the protective clothing. So, the consideration of the performance improvement of individual protective clothing or equipment could be misleading for firefighters. Furthermore, although new materials and smart technologies could provide the optimal protection and comfort to firefighters, these new materials and smart technologies are not yet cost effective and are confined to laboratory settings and/or are only applied in highly specialized circumstances such as aerospace, military, or defense. Development of these new materials and incorporation of smart technologies for cost-conscious mass markets are a real challenge at present. In the future, it is recommended to develop these new materials and technologies in a cost-effective way for firefighters' protective clothing and equipment so that they could be used in many countries by firefighters. The easy availability of newly developed high-performing protective clothing and equipment could provide better occupational health and safety to firefighters worldwide.

References

Abdel-Rehim, Z. S., Saad, M. M., Ei-Shakankery, M., & Hanafy, I. (2006). Textile fabrics as thermal insulators. *AUTEX Research Journal*, 6(3), 148–161.

Akbar-Khanzadeh, F., Bisesi, M. S., & Rivas, R. D. (1995). Comfort of personal protective equipment. *Applied Ergonomics*, 26(3), 195–198.

Au, K. F. (2011). *Advances in Knitting Technology*. England: Woodhead Publishing.

Bajaj, P. (1992). Fire-retardant materials. *Bulletin of Materials Science, 15*(1), 67–76.

Bajaj, P., & Sengupta, A. K. (1992). Protective clothing. *Textile Progress, 22*(2), 1–110.

Barker, R. L., & Heniford, R. C. (2011). Batting materials for use in firefighter turnout suit. *Journal of Engineered Fiber & Fabrics, 6*(1), 1–10.

Barker, R. L., & Lee, Y. M. (1987). Analyzing the transient thermophysical properties of heat-resistant fabrics in TPP exposures. *Textile Research Journal, 57*(6), 331–338.

Bennett, D. (2013). *Firefighters of Cambridge*. United Kingdom: Amberley Publishing.

Berglund, L. G., & Gonzalez, J. A. (2006). Clothing ventilation estimates from manikin measurements. In J. Fan (ed.), *Thermal Manikins and Modelling* (pp. 158–165). Hong Kong: 6th International Thermal Manikin and Modeling Meeting.

Boorady, L. M., Barker, J., Lee, Y.-A., Lin, S.-H., Cho, E., & Ashdon, S. P. (2013). Exploration of firefighter turnout gear; Part 1: Identifying male firefighter use needs. *Journal of Textile and Apparel, Technology and Management, 8*(1), 1–13.

Bourbigot, S. (2008). *Flame retardancy of textiles: New approaches*. In A. R. Horrocks & D. Price (eds), *Advances in Fire Retardant Materials* (pp. 9–40). United Kingdom: Woodhead Publishing Limited.

Bourbigot, S., & Flambard, X. (2002). Heat resistance and flammability of high performance fibres: A review. *Fire and Materials, 4*(4–5), 155–168.

Browne, C. (n.d.). What Kind of Gear Do Firefighters Wear? Retrieved 21 July 2017 from http://work.chron.com/kind-gear-firefighters-wear-9547.html

Celcar, D., Meinander, H., & Gersak, J. (2008). A study of the influence of different clothing materials on heat and moisture transmission through clothing materials, evaluated using a sweating cylinder. *International Journal of Clothing Science & Technology, 20*(2), 119–130.

Chitrphiromsri, P., & Kuznetsov, A. V. (2005). Modeling heat and moisture transport in firefighter protective clothing during flash fire exposure. *Heat and Mass Transfer, 41*(3), 206–215.

Coca, A., Williams, W. J., Roberge, R. J., & Powell, J. B. (2010). Effects of fire fighter protective ensembles on mobility and performance. *Applied Ergonomics, 41*(4), 636–641.

Components of Wildland Fire PPE. (1993). Retrieved 23 Aug. 2017 from https://www.fs.fed.us/eng/pubs/htmlpubs/htm93512851/com.htm

Crown, E. M., Ackerman, M. Y., Dale, D. J., & Tan, Y. (1998). Design and evaluation of thermal protective flightsuits. Part 2: Instrumented mannequin evaluation. *Clothing and Textile Research Journal, 16*(2), 79–87.

Dadi, H. H. (2010). *Literature Overview of Smart Textiles* (M.Sc. thesis). University of Borås, Borås.

Donnelly, M. K., Davis, W. D., Lawson, J. R., & Selepak, M. J. (2006). Thermal Environment for Electronic Equipment Used by First Responders, National Institute of Standards and Technology—Technical Note 1474, National Institute of Standards and Technology, USA, 1–36.

Fan, J. T., & Hunter, L. (2009). *Flammability of Fabrics and Garments*. United Kingdom: Woodhead Publishing.

Fan, J. T., Luo, Z., & Li, Y. (2000). Heat and moisture transfer with sorption and condensation in porous clothing assemblies and numerical simulation. *International Journal of Heat and Mass Transfer, 43*(16), 2989–3000.

Farnworth, B., & Dolhan, P. A. (1985). Heat and water transport through cotton and polypropylene underwear. *Textile Research Journal, 55*(10), 627–630.

Farnworth, B., Lotens, W. A., & Wittgen, P. (1990). Variation of water vapor resistance of microporous and hydrophilic films with relative humidity. *Textile Research Journal, 60*(1), 50–53.

Ferguson, L. H., & Janicak, C. A. (2015). *Fundamentals of Fire Protection for the Safety Professional.* United Kingdom: Rowman & Littlefield Publishing Group.

Frydrych, I., Dziworska, G., & Bilska, J. (2002). Comparative analysis of the thermal insulation properties of fabrics made of natural and man-made cellulose fibres. *Fibres and Textiles in Eastern Europe, 10*(4), 40–44.

Ghassemi, A., Mojtahedi M. R. M., & Rahbar, R. S. (2011). Investigation on the physical and structural properties of melt-spun multifilament yarns, drawn yarns and textured yarns produced from blend of PP and oxidized PP. *Fibers and Polymers, 12*(6), 789–794.

Gibson, P., & Lee, C. (2007). Application of nanotechnology to nonwoven thermal insulation. *Journal of Engineered Fiber & Fabrics, 2*(2), 1–8.

Glock, R. E., & Kunz, G. I. (1999). *Apparel Manufacturing: Sewn Product Analysis.* USA: Prentice Hall.

Hasenmeier, P. (2008). The History of Firefighter Personal Protective Equipment. Retrieved 23 Aug. 2017 from http://www.fireengineering.com/articles/2008/06/the-history-of-firefighter-personal-protective-equipment.html

Hatch, K. L. (1993). *Textile Science.* USA: West Publishing Company.

Hearle, J. W. S. (1991). Understanding and control of fibre structure. *Journal of Applied Polymer Science, 47*, 1–31.

Helneman, E. F., Shy, C. M., & Checkoway, H. (1989). Injuries on the fireground: Risk factors for traumatic injuries among professional fire fighters. *American Journal of Industrial Medicine, 15*(3), 267–282.

Hes, L. (2008). *Analysing the thermal properties of animal furs for the production of artificial furs.* In A. Abbott & M. Ellison (eds.), *Biologically Inspired Textiles* (pp. 150–167). United Kingdom: Woodhead Publishing Limited.

Ho, C. P., Fan, J., Newton, E., & Au, R. (2011). Improving thermal comfort in apparel. In G. Song (Ed.), *Improving Comfort in Clothing* (pp. 165–180). United Kingdom: Woodhead Publishing.

Hocke, M., Strauss, L., & Nocker, W. (2000). Firefighter garment with non textile insulation. In K. Kuklane & I. Holmer (Eds.), Proceedings of NOKOBETEF 6 and 1st European Conference on Protective Clothing, Stockholm, Sweden. European Society for Protective Clothing, Denmark, pp. 293–295.

Holme, I. (2004). Innovations in performance clothing and microporous film. *Technical Textiles International, 13*(4), 26–30.

Holmer, I., & Nilsson, H. (1995). Heated manikins as a tool for evaluating clothing. *Annals of Occupational Hygiene, 39*(6), 809–818.

Horrocks, A. R. (1986). Flame-retardant finishing of textiles. *Coloration Technology, 16*(1), 62–101.

Horrocks, A. R. (2003). Flame retardant finishes and finishing. In D. H. Heywood (ed.), *Textile Finishing* (pp. 214–220). United Kingdom: Society of Dyers and Colorists.

Huck, J. (1988). Protective clothing systems: A technique for evaluating restriction of wearer mobility. *Applied Ergonomics, 19*(3), 185–190.

Jin, L., Hong, K. A., Nam, H. D., & Yoon, K. J. (2011). Effect of the thermal barrier on the thermal protective performance of firefighter garment. In Y. Li, X. N. Luo, & Y. F. Liu (Eds.), Proceedings of TBIS2011, Beijing, China. Textile Bioengineering and Informatics Symposium Society, Hong Kong, pp. 1010–1014.

Kahn, S. A., Patel, J. H., Lentz, C. W., & Bell, D. E. (2012). Firefighter burn injuries: Predictable patterns influenced by turnout gear. *Journal of Burn Care & Research, 33*(1), 152–156.

Keiser, C., Becker, C., & Rossi, R. M. (2008). Moisture transport and absorption in multilayer protective clothing fabrics. *Textile Research Journal, 78*(7), 604–613.

Keiser, C., & Rossi, R. M. (2008). Temperature analysis for the prediction of steam formation and transfer in multilayer thermal protective clothing at low level thermal radiation. *Textile Research Journal, 78*(11), 1025–1035.

Kim, I. Y., Lee, C., Li, P., Corner, B. D., & Paquette, S. (2002). Investigation of air gaps entrapped in protective clothing systems. *Fire and Materials, 26*(3), 121–126.

Klein, W. (1987). *A Practical Guide to Ring Spinning*. England: Textile Institute.

Konarska, M., Soltynski, K., Sudol-Szopinska, I., Mlozniak, D., & Chojnacka, A. (2006). Aspects of standardisation in measuring thermal clothing insulation on a thermal manikin. *Fibres and Textiles Eastern Europe, 14*(4), 58–63.

Lawson, L. K., Crown, E. M., Ackerman, M. Y., Dale, D. J. (2004). Moisture effects in heat transfer through clothing systems for wildland firefighters. *International Journal of Occupational Safety and Ergonomics, 10*(3), 227–238.

Lee, Y. M., & Barker, R. L. (1986). Effect of moisture on the thermal protective performance of heat-resistant fabrics. *Journal of Fire Sciences, 4*(5), 315–330.

Lu, Y., Song, G., & Li, J. (2013). Analyzing performance of protective clothing upon hot liquid exposure using instrumented spray manikin. *Annals of Occupational Hygiene, 57*(6), 793–804.

Mah, T., & Song, G. (2010). Investigation of the contribution of garment design to thermal protection. Part 1: Characterizing air gaps using three-dimensional body scanning for women's protective clothing. *Textile Research Journal, 80*(13), 1317–1329.

Mao, N., & Russel, S. J. (2007). The thermal insulation properties of spacer fabrics with a mechanically integrated wool fiber surface. *Textile Research Journal, 77*(12), 914–922.

Mandal, S., & Abraham, N. (2010). Fabric properties, sewing thread properties and sewing parameters influencing the seam quality. *Asian Textile Journal, 19*(9), 70–75.

Mandal, S., Lu, Y., Wang, F., & Song, G. (2014). Characterization of thermal protective clothing under hot water and pressurized steam exposure. *AATCC Journal of Research, 1*(5), 7–16.

Mandal, S., & Song, G. (2014). An empirical analysis of thermal protective performance of fabrics used in protective clothing. *The Annals of Occupational Hygiene, 58*(8), 1065–1077.

Mandal, S., Song, G., Ackerman, M., Paskaluk, S., & Gholamreza, F. (2013). Characterization of textile fabrics under various thermal exposures. *Textile Research Journal*, *83*(10), 1005–1019.

Marks, R., & Robinson, A. T. C. (1976). *Principles of Weaving*. England: Textile Institute.

Marom, G., & Weinberg, A. (1975). Effect of fiber critical length on thermal expansion of composite materials. *Journal of Material Science*, *10*(6), 100–1010.

Marshall, S. L. A. (1980). *The Soldier's Load and the Mobility of a Nation*. Quantico, Virginia: Marine Corps Association.

Matusiak, M., & Sikorski, K. (2011). Influence of the structure of woven fabrics on their thermal insulation properties. *Fibres and Textiles in Eastern Europe*, *19*(5), 46–53.

Morozumi, Y., Akaki, K., & Tanabe, N. (2012). Heat and moisture transfer in gaps between sweating imitation skin and nonwoven cloth: Effect of gap space and alignment of skin and clothing on the moisture transfer. *Heat and Mass Transfer*, *48*(7), 1235–1245.

Nayak, R., & Padhye, R. (2015). *Garment Manufacturing Technology*. England: Woodhead Publishing.

Neeves, R., Barlow, D. A., Richards, J. G., Provost-Craig, M., & Castagno, P. (1989). Physiological and Biomechanical Changes in Fire Fighters Due to Boot Design Modifications. USA: International Association of Fire Fighters and the Federal Emergency Management Agency.

Otsuka, K., & Wayman, C. M. (1998). *Shape Memory Materials*. Cambridge University Press, Cambridge.

Park, H., Park, J., Lin, S., & Boorady, L. (2014). Assessment of firefighters' need for personal protective equipment. *Fashion & Textiles*, *1*(8), 1–13.

Personal Protective Equipment. (n.d.). Retrieved 21 July 2017 from files.dnr.state.mn.us/forestry/wildfire/rxfire/protective_equip.pdf

Personal Protective Equipment. (2004). USA: Occupational Health and Safety Administration. Retrieved 21 July 2017 from https://www.osha.gov/Publications/osha3151.pdf

Psikuta, A., Frackiewicz-Kaczmarek, J., Frydrych, I., & Rossi, R. (2012). Quantitative evaluation of air gap thickness and contact area between body and garment. *Textile Research Journal*, *82*(14), 1405–1413.

Qashou, I., Tafreshi, H. V., & Pourdeyhimi, B. (2009). An investigation of the radiative heat transfer through nonwoven fibrous materials. *Journal of Engineered Fiber & Fabrics*, *4*(1) 9–15.

Ramachandran, T., Manonmani, G., & Vigneswaran, C. (2010). Thermal behaviour of ring-and compact-spun yarn single jersey, rib and interlock knitted fabrics. *Indian Journal of Fibre & Textile Research*, *35*(3), 250.

Rao, D. R., & Gupta, V. B. (1992). Thermal characteristic of wool fiber. *Journal of Macromolecular Science - Part B*, *31*(2), 149–162.

Reischl, U., & Stransky, A. (1980). Assessment of ventilation characteristics of standard and prototype firefighter protective clothing. *Textile Research Journal*, *50*(3), 193–201.

Rengasamy, R. S., & Kawabata, S. (2002). Computation of thermal conductivity of fibre from thermal conductivity of twisted yarn. *Indian Journal of Fibre & Textile Research*, *27*, 342–345.

Rezazadeh, M., & Torvi, D. A. (2011). Assessment of factors affecting the continuing performance of firefighters' protective clothing: A literature review. *Fire Technology*, *47*(3), 565–599.

Rossi, R. (2003). Firefighting and its influence on the body. *Ergonomics*, *46*(10), 1017–1033.

Samanta, A. K., Baghchi, A., & Biswas, S. K. (2011). Fire retardant finishing of jute fabric and its thermal behaviour using phosphorous and nitrogen based compound. *Journal of Polymer Materials*, *28*(2), 149–169.

Shabaridharan, S., & Das, A. (2012). Study on heat and moisture vapour transmission characteristics through multilayered fabric ensembles. *Fibers and Polymers*, *13*(4), 522–528.

Shalev, I., & Barker, R. L. (1983). Analysis of heat transfer characteristics of fabrics in an open flame exposure. *Textile Research Journal*, *53*(8), 475–482.

Shalev, I., & Barker, R. L. (1984). Protective fabrics: A comparison of laboratory methods for evaluating thermal protective performance in convective/radiant exposures. *Textile Research Journal*, *54*(10), 648–654.

Shekar, R. I., Yadav, A., Kasturiya, N., & Raj, H. (1999). Studies on combined flame retardant and water-repellent treatment on cotton drill fabric. *Indian Journal of Fibre and Textile Research*, *24*(3), 197–207.

Sinden, K., MacDermid, J., Buckman, S., Davis, B., Matthews, T., & Viola, C. (2013). A qualitative study on the experiences of female firefighters. *Work: A Journal of Prevention, Assessment and Rehabilitation*, *45*(1), 97–105.

Song, G. (2007). Clothing air gap layers and thermal protective performance in single layer garment. *Journal of Industrial Textiles*, *36*(3), 193–205.

Song, G. (2009). Thermal insulation properties of textiles and clothing. In J. T. Williams (ed.), *Textiles for Cold Weather Apparel* (pp. 19–30). United Kingdom: Woodhead Publishing.

Song, G., Chitrphiromsri, P., & Ding, D. (2008). Numerical simulations of heat and moisture transport in thermal protective clothing under flash fire conditions. *International Journal of Occupational Safety and Ergonomics*, *14*(1), 89–106.

Song, G., & Lu, Y. (2013). Structural and proximity firefighting protective clothing: Textiles and issues. In F. S. Kilinc (Ed.), *Handbook of Fire Resistant Textiles* (pp. 520–548). United Kingdom: Woodhead Publishing.

Song, G., & Mandal, S. (2016). Testing and evaluating thermal comfort of clothing ensembles. In L. Wang (Ed.), *Performance Testing of Textiles: Methods, Technology, and Applications* (pp. 39–64). United Kingdom: Woodhead Publishing.

Song, G., Mandal, S., & Rossi, R. (2016). *Thermal protective clothing: A critical review*. United Kingdom: Woodhead Publishing.

Song, G., Paskaluk, S., Sati, R., Crown, E. M., Dale, J. D., & Ackerman, M. (2011). Thermal protective performance of protective clothing used for low radiant heat protection. *Textile Research Journal*, *81*(3), 311–323.

Spahr, J. S. (2004). Glove fit for firefighters—An accommodation comparison between U.S. NFPA 1971 and European EN 659 glove size schemes with a contemporary (CAESAR) anthropometric hand size database. The 7th World Conference on Injury Prevention and Safety Promotion, Vienna, Austria, 6–9th June, 2004.

Spencer, D. J. (2001). *Knitting Technology*. England: Pergamon Press.

Structural Firefighting Gloves. (1993). Retrieved 21 July 2017 from https://www
.dhs.gov/sites/default/files/publications/Fire-Gloves-TN_1113-508.pdf

Sun, G., Yoo, H. S., Zhang, X. S., & Pan, N. (2000). Radiant protective and transport properties of fabrics used by wildland firefighters. *Textile Research
Journal*, *70*(7), 567–573.

Tasukada, M., Khan, M. M. R., Tanaka, T., & Morikawa, H. (2011). Thermal characteristics and physical properties of silk fabrics grafted with phosphorous
flame retardant agents. *Textile Research Journal*, *81*(15), 1541.

Taylor, N. A. S., Lewis, M. C., Notley, S. R., & Peoples, G. E. (2011). The Oxygen
Cost of Wearing Firefighters' Personal Protective Equipment: Ralph Was
Right! In S Kounalakis & M Koskolou (Eds.), ICEE 2011 XIV International
Conference on Environmental Ergonomics: Book of Abstracts (pp. 236–
239). Greece: National and Kapodestrian University of Athens.

Torvi, D. A. (1997). Heat transfer in thin fibrous materials under high heat flux
conditions. PhD Thesis. University of Alberta, Edmonton, Canada.

Types of Breathing Apparatus. (n.d.). http://bronksman.tripod.com/typescba
.htm

Ukponmwan, J. O. (1993). The thermal-insulation properties of fabrics. *Textile
Progress*, *24*(4), 1–54.

Uniforms and Equipment. (n.d.). Retrieved 26 Sept. 2017 from https://www.fire
.nsw.gov.au/page.php?id=164

U.S. Fire Administration Emergency Incident Rehabilitation. (2008). USA:
International Association of Firefighters. Retrieved 21 July 2017 from
https://www.usfa.fema.gov/downloads/pdf/publications/fa_314.pdf

Varshney, R. K., Kothari, V. K., & Dhamija, S. (2011). A study on thermophysiological comfort properties of fabrics in relation to constituent fibre fineness
and cross-sectional shapes. *The Journal of Textile Institute*, *101*(6), 495–505.

Wang, C., & Chen, C. (1997). Some physical properties of various amine-
pretreated nomex aramid yarns. *Journal of Applied Polymer Science*, *96*(1),
70–76.

Ward, M. (2014). *Fire Officer Principles and Practice*. USA: Jones and Barttett
Learning.

Yoda, N. (1962). Studies of the structure and properties of aromatic polyamide.
I. physical properties of poly(m-xylyleneadipamide). *Bulletin of Chemical
Society of Japan*, *35*(8), 1349–1353.

3

Human Thermoregulation System and Comfort

Xiao Liao and Chuansi Gao

3.1 Body Temperature

The human body can be considered conceptually in two parts, a core and a shell (skin). Normally, core temperature is regulated and maintained in a narrow range around 37°C. Skin temperature, or peripheral temperature, varies depending on different internal and external conditions and clothing. Classic studies on body temperature defined core temperature as the hypothalamic temperature (Djongyang, Tchinda, & Njomo, 2010; Hammel & Pierce, 1968; Parsons, 2014c; Taylor, Tipton, & Kenny, 2014). In the following content, the term "core temperature" is used for describing the deep-body temperature, while the term "skin temperature" is used for describing the temperature of skin (i.e., real skin temperature). It should be noted that core temperature and skin temperature used in general discussion are usually both conceptual convenience concepts. The term "mean body temperature" is also used to describe the mean temperature of the whole body mass calculated from weighted average of core and skin temperatures. The mean body temperature is usually calculated by the following equation (Parsons, 2014c):

$$T_b = 0.8T_{core} + 0.2T_{skin} \ (in \ the \ heat) \tag{3.1}$$

$$T_b = 0.65T_{core} + 0.35T_{skin} \ (in \ the \ cold) \tag{3.2}$$

3.1.1 Core Temperature

Measurement sites of core temperature began from rectal temperatures in the 1800s (Pembrey & Nicol, 1898). There are now at least 12 measurement sites reported, including axilla, sublingual, auditory canal, liver, stomach, esophagus, tympanic membrane, zero-gradient auditory canal, urine, rectum, gastrointestinal tract, and bladder (Taylor et al., 2014). Rectum was first chosen because of its relative thermal stability. It was also believed to be the hottest tissue in the human body, while the brain is believed to be the hottest site in recent years (Taylor et al., 2014). Cross-comparisons between the temperatures recorded from different sites show that temperature variations exist throughout the body. Therefore, it is important to notice the core temperature measurement site and depth when conducting relevant studies.

Thermometer pills are now commonly used in thermal physiological studies. The ingestible telemetric pill, swallowed a few hours before temperature recording, goes through the human digestive tract and wirelessly transmits the measured absolute temperature for monitoring and recording. Because of this nature, it is difficult to determine the actual site of core temperature measurement as it depends on the user's normal rate of motility. Rectal temperature is still commonly measured as core temperature in laboratory and clinical settings. Non-intrusive methods such as estimation of core temperature using skin temperature, skin heat flow, and heart rate to predict the core temperature were also reported (Buller et al., 2013; Niedermann et al., 2013). But the accuracy needs to be improved because a number of factors can affect the prediction, such as physical fitness, age, body composition, clothing, physical activity intensity, and thermal environment conditions.

Core temperature is often considered as the set point, a regulated variable, in human thermoregulatory studies. Most publications describe it as a temperature near 37°C (Jendritzky, de Dear, & Havenith, 2011; Parsons, 2014c). Recently, it has been reported whether the conventional discussion of a regulated variable should be revised (Romanovsky, 2006). Classic studies describe thermoregulation as a process including integrating local temperatures into an overall body temperature, comparing the overall body temperature to a set point, forming a thermal sensation, and finally sending orders to thermoeffectors. Arguments were raised regarding whether the "set point" actually exists and how exactly the comparison is conducted in human thermoregulation system (Kobayashi, Okazawa, Hori, Matsumura, & Hosokawa, 2006; Okazawa et al., 2002).

3.1.2 Skin Temperature

Heat exchanges between the human body and the environment always occur at skin, except respiration heat exchange. Therefore, skin temperature plays an important role in thermoregulation. Its value, again, depends on the measurement sites and methods.

Commonly skin temperature is measured via attaching sensors onto the skin surface. Sensors that measure absolute temperature are used. Local skin temperature depends on factors including core temperature, the internal heat production and exchange (e.g., blood flow, sweating), environmental condition and external heat exchange (e.g., convective, radiant, evaporative, and conductive heat exchange, and heat transfer coefficient; see Section 3.1.4), and clothing thermal properties of specified body regions. Mean skin temperature is usually used to provide an overall thermal status of the whole body skin. There are many methods to evaluate a mean skin temperature with a different number of measurement sites and weighting coefficients (ISO, 2004; Liu, Wang, Di, Liu, & Zhou, 2013). The number of locations that should be measured is determined by environmental conditions (ISO, 2004).

3.1.3 Neurophysiological Mechanism of Thermoregulation

The thermoregulation system functions as any typical neural system: receiving external stimulation, generating neural signals, and reacting to autonomic and behavioral responses.

3.1.3.1 Thermoreceptor

There are generally two types of thermoreceptors (or thermosensors): central and peripheral thermoreceptors. Central thermosensors are located in several brain stem neuronal groups, including the spinal cord and the preoptic anterior hypothalamus (POA). Those located in the POA are mostly responsible for autonomic responses, being further divided into warm- and cold-sensitive responses. Classic understandings find they function reciprocally (Bligh, 2005), while the latest studies show warm-sensitive neurons respond to both heat and cold stimulations. Warm-sensitive neurons activity would increase as heat-defense response and would decrease as cold-defense response (Chen, Hosono, Yoda, Fukuda, & Kanosue, 1998; Zhang, Yanase-Fujiwara, Hosono, & Kanosue, 1995).

Peripheral thermosensors, located in the skin and in the oral and urogenital mucosa detect the "skin" temperature (see Section 3.1.2). The superficial sensors are mainly cold-sensitive neurons. Their neurons activity would be active when the temperature is changing, and quickly adapt to a stable temperature. Currently, their thermosensitivities are believed to be due to changes in the resting membrane potential; to be more specific, thermal transient receptor potential (TRP) ion of the ThermoTRP channels. It should also be noticed that there are other groups of peripheral thermoeffectors, located in the esophagus, stomach, large intra-abdominal veins, etc., that respond to core body temperature.

3.1.3.2 Stimulation and Transmission

For stimulations detected by central thermosensors, signals are transmitted from the sensitive neurons to the hypothalamus via the periventricular stratum and medial forebrain bundle. Thermosensitivity of warm-sensitive neurons are due to currents that determine the rate of spontaneous depolarization between successive action potentials. Theses sensors are regarded as pacemaker. However, the sensitivity of hypothalamic thermosensitivity is still disputed (Romanovsky, 2006).

With regard to peripheral thermosensors, neural signals of cold-sensitive neurons are conveyed by thin myelinated Aδ fiber while those of warm-sensitive neurons are travelled via unmyelinated C fiber. The mammalian TRP superfamily consists of 30 channels divided in six subfamilies known as the TRPC (canonical), TRPV (vanilloid), TRPM (melastatin), TRPML (mucolipin), TRPP (polycystin), and TRPA (ankyrin). Of these, the heat-activated TRPV1-V4, M2, M4, and M5 and the cold-activated TRPM8 and A1 are often referred as the thermoTRP channels. They cover a sensible temperature range in an overlapping fashion, and their activities have different sensitivities to temperature. Activation of thermoTRP channels will generate inward nonselective cationic current and consequently increase the resting membrane potential.

3.1.3.3 Responses

Stimulation on the neural system would lead to physiological responses and behavior responses. These include sweating, shivering, and change of blood distribution. Change of blood distribution could change the skin temperature and consequently affect the heat exchange between the human body and external environments. Sweating is an effective thermoregulatory response to increase evaporative heat loss from the human body to the environment in the heat, while shivering is a common physiological response to increase heat production for maintaining body core temperature in the cold.

Sweating from eccrine glands, distributed all over the body, is one of the key and effective thermoregulatory responses to heat stress in a hot environment. Sweat glands would be activated integrating neural responses from the human thermoregulation system (Shibasaki & Crandall, 2010). Sweat rates are not evenly distributed in body regions (Havenith et al., 2008; Smith et al., 2012). The body mapping of such sweat patterns is used for clothing development to facilitate evaporative cooling (Gao, Lin, Halder, Kuklane, & Jou, 2017b; Jiao et al., 2017; Lin, Gao, Halder, Kuklane, & Jou, 2017; Wang, Del Ferraro, Molinaro, Morrissey, & Rossi, 2014).

In response to different stimulations, the threshold and sensitivity of sweating may change. Observed factors that can influence sweating include dehydration, heat acclimatization, exposure to altitude, physical training, age, exercise intensity, sleep loss, sex hormones, and circadian rhythm (see summary in Cheuvront et al., 2009).

The aforementioned thermal sensation and thermoregulatory responses are modulated by clothing. The following section introduces heat balance and heat exchange in the human–clothing–environment system.

3.1.4 Heat Balance and Heat Exchange in the Human–Clothing–Environment System

Humans are always exposed to thermal stimuli that could prompt thermoregulation. Heat could be gained or lost through mass and heat exchange with the external environment. The human body can generate energy itself from food through oxidation. When the energy is in balance, the energy consumed, proportional to oxygen uptake, is converted to thermal energy and external work.

Metabolic rate (M, in W) is the rate of metabolic energy expenditure of the human body to maintain basic functions and perform activities by consuming energy sources inside the human body. Thermogenesis is the term used to describe internal heat production in this process. There are two types of thermogenesis: shivering thermogenesis and nonshivering thermogenesis. Metabolic rate is always positive. At rest, there is a basal metabolic rate. During work and exercise, a large amount of energy expenditure is mainly caused by the activities of skeletal muscles. A large fraction of the energy consumed is converted to heat. A small fraction goes to external work (W, in W), depending on the type of activities. The excess heat should be dissipated from the body to the environment. Otherwise, heat will be stored (S, in W) in the human body leading to increased core and skin temperatures. There are four avenues of heat exchange between the human body and external environments: conduction (K, in W), convection

(C, in W), radiation (R, in W), and evaporation (E, in W). Depending on environmental conditions, the human body can lose or gain heat to/from the external environment. In hot environments, for example, with strong radiant heat from sun and fire, the body may gain heat. Respiration (Res) is another way of heat exchange. Therefore, the heat balance equation is expressed as

$$S = (M - W) - (K + C + R + E + Res)$$ (3.3)

The importance of each of the heat exchange pathways differs in hot, moderate, and cold environments. For example, in hot environments, evaporative heat loss is more important, while in cold environments, convection and radiation are more significant.

3.1.4.1 Heat Production

Body heat production is determined by metabolic rate. Whether all energy consumed is released as heat from the body or part of energy consumed is used to perform external mechanical work is determined by the type of work.

Heat can be stored inside the human body if heat dissipation is restricted, for example, by protective clothing, and if heat is gained from the external environment. The human body temperature would rise at about 1°C per hour for a resting person if there is no heat exchange between the human and the environment (Parsons, 2014c). The amount of energy stored and the rate of storage are the key aspects of human heat balance studies, as they reflect the challenge and consequence of human body of the combined effect of all factors (see Section 3.1.4.2).

3.1.4.2 Heat and Mass Transfer between the Human Body and the Environment

Conduction heat transfer (K) is driven by the temperature difference between two contacting objects. This is often assumed to be negligible in the human–clothing–environment system. Meanwhile, it should be noticed that from the inside of the human body to the skin surface, heat is exchanged through conduction. There are also conduction heat exchanges between the human body and the clothes contacting skin. In circumstances such as when human hands or feet are in contact with hot or cold solid surfaces and when phase change material cooling (Gao, Kuklane, & Holmér, 2010a) is applied, conduction heat transfer plays an essential role. The fundamental formula (Parsons, 2014c) to calculate conduction heat exchange in steady state is

$$K = \frac{k(T_2 - T_1)}{d}$$ (3.4)

where K is the conduction heat transfer in W, k is the thermal conductivity of the medium in W/(m*K), T_2 and T_1 are the temperature of two contact objects, respectively, in K, and d is the distance of heat transferred in m.

Non-steady-state is used to describe a condition where temperature dynamically changes with time (t). The rate of heat transfer $\left(\dfrac{dK}{dt}\right)$ and rate of temperature change $\left(\dfrac{dT}{dt}\right)$ with time are commonly addressed (Kerslake, 1972) using the formula:

$$\frac{dK}{dt} = k\frac{d^2T}{dx^2} \tag{3.5}$$

$$\frac{dT}{dx} = \frac{k}{pc}\frac{d^2T}{dx^2} \tag{3.6}$$

where $\dfrac{dK}{dt}$ is the rate of heat transfer in W/(m²*s), T is the temperature in K, k is the thermal conductivity in W/(K*m), is the heat transferred distance in m, p is the object density in kg/m³, and c is the heat capacity per unit mass in J/(kg*K).

Convection is the heat transfer pathway through movement of a fluid or air. Convection is a major avenue for heat transfer between the human body and external environment. In principle, the heat exchange (C) could be calculated by the formula (Parsons, 2014c)

$$C = h_c\left(T_2 - T_1\right) \tag{3.7}$$

where h_c is the convective heat exchange coefficient in W/(m²*K), and T_2 and T_1 are the temperature of object (e.g., skin) and fluid (e.g., ambient air), respectively, in K.

In practice, T_2 could be relatively easily obtained through measuring skin temperature and T_1 could be obtained through measuring air temperature, or water temperature if in the case where water is the medium surrounding the human body. Determining h_c, however, is worthy of investigating. Factors affecting the convection coefficient include the shape of the human body, velocity of fluid flow, and physical properties of the fluid. Convection condition can be classified into forced convection when external air velocity is dominant (e.g., due to wind) and natural convection when the surrounding air is relatively still.

Regarding forced convection, Reynolds number (Re) is a dimensionless number used to describe the flow pattern of fluid around the object. It is defined as proportional to the ratio of inertial forces to viscous forces. Nusselt number (Nu) is used to describe heat exchange. It is defined as proportional to the ratio of the heat transfer by convection to that by conduction in the same fluid at rest. Prandtl number (Pr) is used to describe fluid's thermal properties. It is defined as the ratio of the kinematic viscosity to the thermal diffusivity. These three dimensionless numbers can be calculated by (Parsons, 2014b)

$$\mathrm{Re} = \frac{VL}{v}; v = \frac{\mu}{p} \tag{3.8}$$

$$\mathrm{Nu} = \frac{h_c L}{K} \tag{3.9}$$

3. Human Thermoregulation System and Comfort

$$Pr = \frac{\mu/p}{K/pc} = \frac{\mu c}{K} \qquad (3.10)$$

where V is the fluid velocity in m/s, v is the kinematic velocity of the fluid in m²/s, μ is the fluid dynamic viscosity in kg/(m*s), and p is the fluid density in kg/m³. K is the conduction heat transfer in the same fluid at rest in W/m², and c is the specific heat capacity in J/(kg*K). L is the feature characteristic to describe the size of the objects in the same shape in m.

Regarding natural convection, Grashof number (Gr) is defined considering different factors related. It is calculated by (Parsons, 2014b)

$$Gr = \frac{agL^3(T_s - T_a)}{v^2} \qquad (3.11)$$

where a is the coefficient of expansion of air in K⁻¹, g is the acceleration due to gravity in m/s², T_s is the surface temperature and T_f is the temperature of the fluid in K, L is the object size dimension in m, and v is the kinematic velocity of the fluid in m²/s.

The fundamental knowledge of radiation is that wavelengths from visible light to radio waves provide heat; all bodies above absolute zero emit and absorb radiation. In the human–clothing–environment system, there are two types of radiation heat exchange: solar radiation gained and surrounding radiation emitted and gained. One strong surrounding radiation heat source is fire, which gives intense radiant heat to firefighters. The basic calculation formula of radiation heat exchange can be expressed with the following formula (Parsons, 2014b). The heat exchange coefficient, h_r, is affected by the distance between two objects. Generally speaking, we will receive more radiative heat transfer when closer to the heat source, that is, fire.

$$R = h_r(T_2^4 - T_1^4) \qquad (3.12)$$

where h_r is radiative heat transfer coefficient in W/(m²*K), and T_2 and T_1 are temperature of two objects respectively in K.

The above heat exchange equations describe the calculations in unit area. Heat exchange between the human body and external environment mostly occurs through the human skin. The body area affects the heat exchange. Dubois formula is commonly used for calculating the skin area (in m²) of humans.

$$A_D = 0.202(m)^{0.425}(H)^{0.725} \qquad (3.13)$$

where m is the body weight in kg and H is the body height in m.

"Sensible heat loss," defined by ASHRAE (1997), includes the above two types of heat exchange, convection and radiation. The outer surface area of a clothed person, A_{cl}, is greater than the surface area of a nude body, A_D. The ratio of these is the clothing area factor, f_{cl} (ISO 9920, 2007; Parsons, 2014). It could be calculated by

$$f_{cl} = \frac{A_{cl}}{A_D} \tag{3.14}$$

ASHRAE (1997) provides derivations of Equations 3.7 and 3.12 as

$$C = f_{cl} h_c \left(t_{cl} - t_{air} \right) \tag{3.15}$$

$$R = f_{cl} h_r \left(t_{cl} - t_r \right) \tag{3.16}$$

$$C + R = f_{cl} h \left(t_{cl} - t_o \right) \tag{3.17}$$

where

$$h = h_c + h_r \tag{3.18}$$

$$t_0 = \frac{\left(h_c t_{air} + h_r t_r \right)}{h_c + h_r} \tag{3.19}$$

Regarding the heat exchange between the skin and clothing, an overall clothing thermal resistance (R_{cl}) was introduced. The heat exchange can be calculated by

$$C + R = \frac{\left(t_{sk} - t_{cl} \right)}{R_{cl}} \tag{3.20}$$

The sensible heat loss can therefore be calculated by

$$C + R = \frac{\left(t_{sk} - t_o \right)}{\left(R_{cl} + \dfrac{1}{f_{cl} h} \right)} \tag{3.21}$$

where C is the conductive heat loss in W, R is the radiation heat loss in W, h_c is the convective heat transfer coefficient in W/(m²*K), h_r is the linear radiative heat transfer coefficient in W/(m²*K), h is the combined heat transfer coefficient in W/(m²*K), t_o is the operative temperature in °C, t_r is the mean radiant temperature in °C, t_{air} is the air temperature in °C, t_{cl} is the mean temperature over the clothed body in °C, t_{sk} is mean skin temperature in °C, and R_{cl} is the thermal resistance of clothing in (K*m²)/W. This is also termed basic or intrinsic clothing thermal insulation in international and European standards (EN, 2004; ISO, 2005, 2007a, 2007b, 2013) and in the literature (e.g., Parsons, 2014c).

Evaporation heat loss is the heat transfer due to the change of state from liquid sweat on human skin to vapor. Evaporation heat loss includes two steps as sweating and evaporation. Dripped sweat had no contribution to body heat loss. Sweating and subsequent evaporation is an effective way to dissipate heat from the human body to the external environment particularly in hot environments, while moisture diffusion serves as a pathway of evaporative heat exchange when not sensible sweating. ASHRAE (1997) provides the following equation for calculating the evaporation heat loss:

$$E = \frac{w\left(P_{sk,s} - P_a\right)}{\left[R_{e,cl} + \dfrac{1}{f_{cl}h_e}\right]} \tag{3.22}$$

where w is the skin wetness (the fraction of wet skin); $P_{sk,s}$ is the water vapor pressure at the skin in kPa, which commonly presented the same as the saturated water vapor pressure at skin temperature; P_a is the water vapor pressure in the ambient air in kPa; $R_{e,cl}$ is the basic (intrinsic) evaporative resistance of clothing in (m²*kPa)/W; and h_e is the evaporative heat transfer coefficient in W/(m²*kPa).

Evaporative heat transfer coefficient can be calculated using Lewis relation as (ASHRAE, 1997)

$$h_e = LRh_c \tag{3.23}$$

where LR is the Lewis relation (approximate value at sea-level atmospheric pressure: 16.5), h_e is the evaporative heat transfer coefficient, and h_c is the convective transfer coefficient.

Respiration is one type of "special" heat exchange as it occurs through the human respiration system. This also includes "dry" convective heat transfer (C_{res}) and evaporation heat transfer (E_{res}). Air inhaled inside the human body would be heated to core temperature (t_{cr}) and consequently exhaled to the environment. The heat exchanged in this process is through convection heat transfer (C_{res}). Meanwhile, the air inhaled would be moistened by the respiratory tract and lungs before being exhaled to the environment, forming the evaporation heat transfer (E_{res}). A formula was developed by ASHRAE (1997) to quantify the heat exchange amount:

$$C_{res} + E_{res} = \left[0.0014M\left(34 - t_a\right) + 0.0173M\left(5.87 - P_a\right)\right] \tag{3.24}$$

where M is the metabolic rate in W/m², t_a is the ambient air temperature in °C, and P_a is the water vapor pressure in the ambient air in kPa.

In summary, Figure 3.1 illustrates the body heat exchange and heat transfer avenues.

The human body thermoregulatory system is sometimes described as two interacting subsystems (Jendritzky et al., 2011). The controlling active system contains the physiological responses due to thermoregulation, that is, shivering

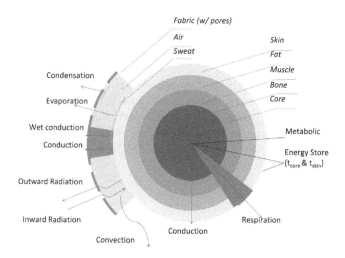

Figure 3.1

Body heat exchange avenues.

thermogenesis, sweating, and peripheral blood flow regulation. The passive system includes the physical human body and heat transfer (e.g., convection, radiation) via the human body surface.

These two subsystems function in such a manner that the passive system would continuously work; the active system would work at a higher activity level along with increasing discomfort and at the lowest activity level in a comfort condition. The next section discusses issues regarding thermal comfort.

3.2 Thermal Comfort

Depending on the integrated impacts from the above factors and heat exchange avenues, the body temperature (either core temperature and/or skin temperature) can be changed, consequently affecting the physiological responses and psychological comfort perception. Here, the responses are divided into survival, physiological, and psychological levels. It should be noticed that there are no absolute borderlines between these three levels. Due to the dynamic nature of heat balance between the human body and external environment, human thermal responses may change with time as well.

3.2.1 Thermophysiological Responses and Injuries in Extreme Environments

This section introduces conditions that can be harmful to human health and can induce damage to the human body.

3.2.1.1 Hypothermia and Hyperthermia

Hypothermia (HT) is defined as when body core temperature drops below 35°C. For the purpose of rescue, five stages of hypothermia are defined according to the core temperature, as shown in Table 3.1 (Brown, Brugger, Boyd, & Paal, 2012). Atrial fibrillation is usually observed when core temperature drops below 32°C (Danzl, 2012). It was also reported that patients engaged in paradoxical undressing when core temperature dropped below 28°C (Brändström, Eriksson,

Table 3.1 Five Stages of Hypothermia

	Typical Body Core Temperature (°C)
HT I: Clear consciousness with shivering	35–32
HT II: Impaired consciousness without shivering	<32–28
HT III: Unconsciousness	<28–24
HT IV: No vital signs	<24

Sources: Modified from Brown, D. J. A. et al., *New England Journal of Medicine*, 367(20), 1930–1938, doi:10.1056/NEJMra1114208, 2012; Durrer, B. et al., *High Altitude Medicine & Biology*, 4(1), 99–103, doi:10.1089/152702903321489031, 2003.

Giesbrecht, Ängquist, & Haney, 2012). Risk of cardiac arrest would also increase when core temperature drops below 32°C (Danzl & Pozos, 1994).

Hyperthermia is defined as uncontrolled body temperature increase that exceeds the heat loss ability of the human thermoregulatory system, which will lead to body temperature increases above maximum normal temperature, for example, oral temperature 37.2°C at 6 am and 37.7°C at 4 pm (Dinarello & Gelfand, 2005). Hyperpyrexia describes the condition when body temperature increases above 40°C. Hyperpyrexia could be life-threatening. It should be noticed that hyperthermia does not change the set point, differing from fever (Axelrod & Diringer, 2008; Laupland, 2009).

3.2.1.2 Cold-Pain Sensation

Cold-pain sensation would be perceived when skin temperature drops below 15°C (Filingeri, 2016). However, it seems that the boundaries between nocuous cold and innocuous cold have not been clearly defined yet (Foulkes & Wood, 2014).

3.2.1.3 Heat-Pain Sensation and Heat Injuries

Threshold of heat pain inducing temperature was experimentally found to be above 40°C in the literature (44.7°C, Hashmi & Davis, 2010; 42.6°C, Tousignant-Laflamme, Pagé, Goffaux, & Marchand, 2008; 40.3°C, Greffrath, Baumgärtner, & Treede, 2007; 40.7°C, Weidner et al., 1999). These thresholds were determined using contacted thermode as the stimulator. The thermode could gradually increase its temperature at a predefined rate. Temperature higher than 40°C could eventually lead to damage to protein and therefore injury.

Continuous exposure to the high temperature would lead to burn injury. The severity of burn injury depends also on heat transfer properties of the contact media, size of the contact area, contact duration, contact pressure, etc. Burn wound depths are internationally classified into superficial (I), superficial partial (II), deep partial (IIb), and deep (III). Table 3.2 presents conditions of exposure duration and temperature under which full thickness burn would occur.

Table 3.2 Full Thickness Burn Due to Temperature and Duration of Exposure

Temperature in °C	Duration of Exposure in s
45.0	3600
54.4	30
60.0	10
69.0	1

Source: Evers, L. H. et al., *Experimental Dermatology*, 19(9), 777–783, doi:10.1111/j.1600-0625.2010.01105.x, 2010.

3.2.1.4 Heat Stress and Heat Strain

Heat stress is used to describe conditions under which the human body would face overheating challenges. As described in Section 3.1.4.2, many factors could impact the human thermoregulatory systems and the interactions between these factors are relatively complex. Heat stress indices were therefore developed to provide integrated evaluation of contributing factors to heat load on the human body in different conditions and working scenarios. A comprehensive heat stress index calculation needs inputs from six factors, including four thermal climate factors (air temperature, humidity, air velocity, and heat radiation) and two human-clothing factors (human activity intensity and clothing). A single parameter such as air temperature usually cannot capture the whole picture of heat stress, for example, in firefighting conditions. The test of firefighters' turnout gear in a hot and humid environment (55°C and 30% relative humidity) by Holmér, Kuklane, and Gao (2006) showed that heat stress level and thermophysiological responses are directly associated with the thermal environment conditions (compared with room temperature environment), insulation of the firefighting ensembles, and body metabolic heat production. The study showed that light to moderate physical work in the hot and humid environment imposed extremely high levels of heat stress on the firefighters when they wore a full set of protective clothing and compressed air respirator. The weight of the protective ensemble and equipment including helmet, full-face mask, compressed air cylinder, pulse belt, and watch ranged from 19.8 to 21.2 kg.

However, there is no universal heat stress index so far that can be applied to all conditions, to cover various activity intensities, durations, protective clothing, local thermal environment, etc. A recent review on occupational heat stress assessment and protective strategies in the context of climate change summarized the advantages and shortcomings when making use of available meteorological data (Gao, Kuklane, Östergren, & Kjellstrom, 2017a). The historical thermal indices and recent developments of thermophysiological modeling for heat stress in relation to different applications were reviewed by Havenith and Fiala (2015). All indices have their own advantages and limitations. It is suggested the choice of heat stress index should be made according to actual applications. Two established ISO standards, Wet Bulb Globe Temperature (WBGT) (ISO 7243) and Predictive Heat Strain (PHS) (ISO 7933), could be used as a screening tool and analytical method of heat stress, respectively (Gao et al., 2017a; Parsons, 2014a). In the situations of firefighting, PHS is limited by integrating highly insulated firefighting clothing (Gao et al., 2017a; Wang, Kuklane, Gao, & Holmér, 2011).

For evaluating heat stress in connection to climate change, Gao et al. (2017a) suggested five dimensions that should be clarified:

1. Target group for heat stress evaluation, that is, working people (e.g., firefighters), general population, professional athletes, etc.
2. Individual- or population-based assessment, which determines the needs to estimate individual or population characteristics (e.g., age, gender, body weight and height, etc.), clothing, and activity intensity.
3. Underlying evaluation principles: rational indices based on heat transfer and heat balance, empirical indices based on physiological and psychological responses quantification, or direct indices based on direct measurements of environmental variables.

4. Index used for describing current situation or to predict future challenges. The latter would need estimation of future environment conditions, e.g., future climate scenarios.
5. Calculated environmental index value or limit exposure time; it should be clear which value is used.

It is common that the terms "heat strain" and "heat stress" are mixed up. Heat strain is the internal human thermophysiological response to external environment conditions (e.g., external heat stress encountered by firefighters). Real-time monitoring of heat strain parameters of firefighters including core and skin temperatures, sweat rate and dehydration, and heart rate (ISO 9886) can help to prevent risk and support work performance. Besides, oxygen uptake (VO_2) could be measured for assessing the metabolic rate and physical workload. Direct evaluation of the psychological responses such as thermal sensation and thermal comfort, as well as perceived physical exertion (Borg, 1982), could also be added for better understanding of subjective responses. Measurements of heat strain of individuals would be useful for directly assessing their individual responses due to individual variations, rather than estimating average thermophysiological response, accordingly being able to provide timely and necessary personalized coping strategies such as personal cooling measures and breaks (Gao, Kuklane, Wang, & Holmér, 2012; Zhao et al., 2013), for example, for firefighting in extremely hot environments (Gao et al., 2010a; Gao, Kuklane, & Holmér, 2010b).

Total weight of clothing system is another key factor that contributes to heat strain, especially in situations that heavy protective clothing system is required, for example, for firefighters. In studies comparing different ensembles of protective clothing of firefighters, results show that an increase in clothing system weight can lead to higher metabolic rate for the same environment condition and activity intensity (Wang, Gao, Kuklane, & Holmér, 2013; Wang et al., 2011). Experimental results show that metabolic cost of walking can be expressed as a linear function of the total weight, including body weight and external loads (Dorman & Havenith, 2008). In these studies, the total weight of firefighter clothing system could be as high as 7.0 kg (Dorman & Havenith, 2008).

3.2.2 Thermal Sensation and Thermal Comfort

3.2.2.1 Individual Responses

Studies on human thermal sensation usually are population or group based: predicted average perception (e.g., predicted mean vote, PMV) is provided, regardless of individual characteristics. It is not surprising since investigation on individual differences requires relatively larger resources and more comprehensive experiments. Havenith (2001) introduced individual differences indicated by body mass, body fat layer thickness, acclimation state, maximal physical fitness (Maximum O_2 uptake capability), and body surface area to a modified classic two-node model. This "individualized" modification was found to have significant improvements to the individual heat strain prediction. However, it is also noticed that there are still factors that impact on individual thermal sensations remaining unknown.

The effects of ethnic and culture differences on thermal sensation (e.g., Westerners versus Asians) were also found to be significant (Liao et al., 2017; Nakano, Tanabe, & Kimura, 2002; Zhou, Lian, & Lan, 2013; Zhou, Zhang, Lian, & Zhang, 2014). Skin color is also considered as one factor as it may influence the radiation heat exchange (Zhang, Huizenga, Arens, & Yu, 2001).

There is another approach to model individual responses (Takada, Kobayashi, & Matsushita, 2009) by considering individual characteristics including different set point temperatures of core and skin, and individual coefficients in the dynamic model of sweating and skin blood flow rate, which are preset as constants as for an average population in a two-node model (Gagge, Stolwijk, & Nishi, 1971). These individual characteristics were calculated based on experiment data for each subject. The thermal sensation model of each subject was then modified using their "own constants." Results show the improved prediction of the two-node model. However, the study only involved four subjects and conducted validation experiment in only one transient thermal condition. Nevertheless, this approach, though application to mass population may be difficult, is feasible for studying the individual difference of thermal sensation.

3.2.2.2 Comfort Indices

There are standards established to describe the thermal sensation and thermal comfort, including widely used ones like ASHRAE standard 55, ISO standard 7730, and CEN 15251. These standards present methods for assessing and predicting general thermal sensation and discomfort. In ASHRAE 55-2010, thermal comfort is defined as "the condition of mind that expresses satisfaction with the thermal environment." Thermal comfort is the result of a comprehensive cognitive process of integrating physical, physiological, and psychological factors.

The most classic thermal comfort prediction model may be Fanger's PMV model introduced in 1970 (Fanger, 1970), though its application in different conditions was questioned (van Hoof, 2008). A large number of indices have been developed in the past for predicting the thermal comfort (or sensation) of humans in specified environment and clothing condition. A comprehensive summary of these indices was reported by de Freitas and Grigorieva (2014). There are mainly two approaches to evaluate thermal comfort: rational (heat-balance approach) and adaptive approach.

Rational comfort indices showed that cold discomfort is related to skin temperature and hot discomfort is related to skin wetness and skin temperature (Djongyang et al., 2010); in contrast, comfort condition should be the condition under which skin temperature is above the cold-discomfort threshold and there is neither sweating on skin nor skin temperature above the hot-discomfort threshold. As a typical example, the PMV value can be calculated by (Fanger, 1970):

$$PMV = \left[0.303 exp(-0.036M) + 0.028 \right] L \qquad (3.25)$$

where M is the metabolic heat production in W/m^2 and L (in W/m^2) is the heat load. Heat load is defined as the differences calculated by the heat balance equation (i.e., heat stored) with predetermined "comfort" skin temperature and respiration evaporation heat loss (i.e., breath). It was proposed to be calculated by the following equation (Fanger, 1970):

$$L = \left[\left(M - W \right) - 3.05 \times 10^{-3} \left\{ 5733 - 6.99 \left(M - W \right) - P_a \right\} \right.$$
$$- 0.42 \left\{ \left(M - W \right) - 58.15 \right\} - 1.7 \times 10^{-5} M \left(58667 - P_a \right)$$
$$\left. -0.0014 M \left(34 - t_a \right) - 3.96 \times 10^{-8} f_{cl} \left[\left(t_{cl} + 273 \right)^4 - \left(t_r + 273 \right)^4 \right] - f_{cl} h_c \left(t_{cl} - t_a \right) \right]$$

$$(3.26)$$

3. Human Thermoregulation System and Comfort

and

$$t_{cl} = 35.7 - 0.028(M - W)$$
$$-0.155I_{cl}\left[3.96\times10^{-8}f_{cl}\left\{(t_{cl}+273)^4 - (t_r+273)^4\right\} + f_{cl}h_c(t_{cl}-t_a)\right] \quad (3.27)$$

where W is the external work in W/m^2, P_a is the water vapor pressure in the ambient air in kPa, t_a is the ambient air temperature in °C, f_{cl} is the clothing area factor, t_{cl} is the mean temperature over the clothed body in °C, t_r is the mean radiant temperature in °C, h_c is the convection heat exchange coefficient in W/(m^2*K), and I_{cl} is the clothing insulation values (clo).

Furthermore, predicted percentage of dissatisfied (PPD) is defined as the percentage of the people who felt more than slightly warm or slightly cool, calculated by the formula (Fanger, 1970)

$$PPD = 100 - 95\exp\left[-0.03353PMV^4 + 0.2179PMV^2\right] \quad (3.28)$$

Adaptive comfort indices are originated from field study in different thermal comfort environments, mostly in free-running building environments. Occupants of the buildings can adapt to prevailing outdoor thermal conditions to adjust their comfort temperatures, for example, through adjustment of clothing, acclimatization, opening/closing windows, or personal ventilation. The adaptive comfort temperatures allow a wider range of indoor temperatures than a fixed set point, such as 22°C. The wider range of adaptive comfort temperature in the built environments provides the possibility to reduce energy use for heating in cold season and cooling in hot season (Nicol & Humphreys, 2002). The adaptive comfort indices established are based on specific buildings in certain climate zones, making it difficult to cross-compare and evaluate the effectiveness of their applications.

3.3 Sensorial Comfort

Touch sensation is usually used to describe sensorial comfort. Studies were conducted using descriptors to identify aspects of sensorial comfort. The descriptor list is quite long and varied from studies. It is evident that these descriptors are all correlated to each other to some extent. Their similarities and differences, meanwhile, should also be clarified. In Table 3.3, a summary is provided on the previous conclusions on the dimensions of the descriptors.

3.3.1 Tactile Comfort

3.3.1.1 Softness

Force is always applied onto a fabric when we touch it. The reactions would subsequently be exerted onto human skin. Such reaction forces could stimulate skin mechanoreceptors and consequently generate a touch sensation. It is common to differentiate reaction forces into compression and bending forces in textiles research. Fabric bending properties were concluded to be highly correlated to "softness" or "stiffness."

Table 3.3 Summary of Sensorial Comfort Dimensions

Study	Dimension 1	Dimension 2	Dimension 3	Dimension 4
(Howorth, 1964)	Smoothness	Stiffness	Thermal	Bulkiness
(Hollins, Faldowski, Rao, & Young, 1993)	Rough/Smooth	Hard/Soft		
(Li, 1998)		Tactile	Thermal-Wet	Pressure
(Hollins, Bensmaia, Karlof, & Young, 2000)*	Rough/Smooth	Hard/Soft		Sticky
(Picard, Dacremont, Valentin, & Giboreau, 2003)	Rough	Hard/Soft		Relief
(Ballesteros, Reales, de Leon, & Garcia, 2005)	Rough/Smooth	Hard/Soft		Sticky
(Shirado & Maeno, 2005)	Rough/Smooth	Hard/Soft	Cold/Warm	Moist/Dry
(Tiest & Kappers, 2006)	Smooth/Rough	Hard/Soft		
(Ju & Ryu, 2006)*	Roughness	Softness		Bulkiness
(Tanaka, Tanaka, & Chonan, 2006)	Rough/Smooth	Hard/Soft	Cold/Warm	Moist/Dry
(Yoshioka, Bensmaia, Craig, & Hsiao, 2007)	Rough/Smooth	Hard/Soft		Sticky
(Guest et al., 2012)	Rough/Smooth	Hard/Soft		Moist/Dry
(Asaga, Takemura, Maeno, Ban, & Toriumi, 2013)	Roughness	Softness		

Source: Modified from Liao, X., *Neuropshyological mechanisms of fabric touch sensations*, (Ph.D.), Hong Kong, 2014.
* There are more dimensions that are not shown in the table.

3.3.1.2 Smoothness

Objects' form and texture information could be perceived as cutaneous stimuli and generate sensations (Johnson & Yoshioka, 2002; Liao, Hu, Li, Li, & Wu, 2011). Form information of fabrics is limited due to its flat and flexible nature. Therefore, texture information should be mainly addressed. It is commonly believed that friction force and surface roughness are two major stimuli of "smoothness" perception, despite various terms that may otherwise be used. Friction force is commonly reported to be empirically correlated to smoothness sensation. However, a study by Morley, Goodwin, and Dariansmith (1983) found that relative movement is not crucial for texture perception. This suggests that fabric smoothness could be sensed without friction force, or relative movement between fabric and skin.

3.3.1.3 Measurement of Tactile Comfort

Well-known fabric sensorial comfort measurement methods include Fabric Assurance by Simple Testing (FAST) developed by the Commonwealth Scientific and Industrial Research Organisation (CSIRO) (Minazio, 1995) and the Kawabata Evaluation System (KES) devised by Kawabata research group (Kawabata, 1980, 1982, 1984). Recently, Fabric Touch Tester (FTT) was introduced as an alternate method to the conventional measures (Liao, Li, Hu, Wu, & Li, 2014).

For compression properties measurement, vertical forces are usually applied onto the test sample. KES used a metal circle with area of 2 cm^2, with normal force changing from 0 to 100 gf at a change rate of 0.02 gf/s. FAST used a square pressure device with an area of 10 cm^2, with a normal pressure at two levels as 2 gf/cm^2 and 100 gf/cm^2. FTT used a square metal with a relative large area of 121 cm^2, with normal pressure changing from 0 to 70 gf/cm^2.

KES measures the bending properties of fabrics with two vertically installed bending bars (one fixed and one movable), which hold the samples vertically.

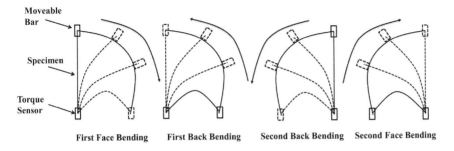

Figure 3.2

Movement of KES Bending Measurement. (Adapted from Liao, X., *Neuropsychological mechanisms of fabric touch sensations*, (Ph.D.), Hong Kong, 2014.)

The fabric would be bent along with the moving bar. Figure 3.2 shows the movement of two bars during KES measurement. Fabric sample is actually bent twice (one toward each side). The bending properties were calculated as average results from these two bending processes with indexes including bending rigidity and moment of hysteresis. FAST examines fabric bending properties through the cantilever method, which is similar to the British Standard Method (BSI, 1990). Fabric samples freely bend due to gravity, passing over a horizontal platform. FTT uses a unique method to measure the fabric bending properties together when measuring fabric compression properties (Liao et al., 2014).

Regarding fabric surface properties, KES includes measurements on both surface friction and surface roughness. Fabric sample is made to move across the friction force sensor and roughness sensor. The normal force placed on the sample for measuring friction force is 50 gf and that for measuring roughness is 10 gf. Identified parameters include coefficient of friction, standard deviation of friction coefficient, and mean deviation of surface roughness. FTT applied a similar measurement principle. There is no measurement on fabric surface properties in FAST. The connection between the above-mentioned measurement method and subjective touch sensation is covered in Section 3.3.3.

3.3.2 Wetness Properties

Wetness is one of the important dimensions of sensorial comfort, as shown in Table 3.3. A number of psychophysical studies on fabric wetness perception have been conducted. Some actively added liquid water onto fabric samples as input (Tang, Kan, & Fan, 2014), and some measured (or estimated) water content of clothes after exercise. The former one may provide a better investigation of the sensitivity of wetness perception, while the latter is more closely aligned to reality. Though the study found that there is no significant difference in wetness perception among body sites (Ackerley, Olausson, Wessberg, & McGlone, 2012), the distribution of sweat glands and sweating rates do differ in body sites and consequently varied wetness comfort perception (Havenith, Fogarty, Bartlett, Smith, & Ventenat, 2007; Smith & Havenith, 2010).

To quantify the possible wetness of clothing, physical measurement methods were developed. AATCC 195 (i.e., using Moisture Management Tester, MMT) is one of the widely used methods to measure the liquid moisture (i.e., sweat when wearing clothing) distribution on both the inner side and outer side of fabric, with liquid input initiated from next-to-skin side (Hu, Li, Yeung, Wong, & Xu, 2005). Comparison across different measurement methods also showed its advantage in characterizing in-plane and transplanar water transport properties. New instruments were recently developed to overcome limitations of the MMT method (Tang, Chau, Kan, & Fan, 2015; Tang, Wu, Chau, Kan, & Fan, 2015).

3.3.3 Prediction of Sensorial Comfort

Statistical regression prediction methods of fabric sensorial sensation are the most commonly used method, for example, used in KES (Kawabata, 1982) and FTT systems (Liao, Li, Hu, Li, & Wu, 2016). However, there are limitations in using linear regression. Regression analysis methods build prediction equations based on a group of samples. Subject population and sample selection could largely affect the expanded application of the built models. In addition, most regression models neglect the underlying mechanism of neuroreceptors, which is important for revealing why and how we perceive sensations.

Psychophysical laws were proposed to correlate sensory-related stimuli and perceived sensation. The basic assumption of psychophysical laws is that such correlations could be expanded to any type of sensation. Fechner et al. (1966) defined sensation perception as "outer psychophysics" and "inner psychophysics" and raised a question: Is there a universal law that could explain how we subjectively sense the world? They believed that general laws exist for "inner psychophysics" (Johnson, Hsiao, & Yoshioka, 2002). Weber–Fechner law incorporates two findings from the work of Weber and Fechner. Weber's finding indicated that the just-noticeable difference (JND), that is, sensitivity of a stimulus, would be proportional to the magnitude of the stimuli. Fechner stated that there is a relationship between stimulus intensity and sensation perceived by the subject, with JND as the connecting factor. Based on the above, Weber–Fechner law describes this relationship as a logarithmic function, while later Stevens's power law was proposed to describe the relationship as a power function (Stevens, 1957, 1961). In comparison with the mentioned psychophysical laws with regression models, it is found that Stevens's power law has the highest predictive accuracy (Hu, Chen, & Newton, 1993; Li, 2005; Mazzuchetti, Demichelis, Songia, & Rombaldoni, 2008; Rombaldoni, Demichelis, & Mazzuchetti, 2010).

The assumption in the psychophysical laws that there are equal JNDs of all the senses has also been questioned recently (Heidelberger, 2004; Masin, Zudini, & Antonelli, 2009). Some studies even proved that there is no consistent relationship (Johnson, Turner, Zwislocki, & Margolis, 1993; Schroder, Viemeister, & Nelson, 1994).

Meanwhile, advanced statistical techniques have been adapted for predicting the sensorial comfort, including artificial neural network (ANN) and fuzzy logic. Compared to the prediction capacity of regression models, an advantage of ANN over regression predictions was found (Gao, 2003; Meng, 2010; Park, Hwang, Kang, & Yeo, 2000; Zhang, 1993). It should also be noticed that the accuracy of ANN prediction models is largely affected by the number of layers and training cases.

With the recent development of identifying neuroreceptors for different sensorial stimuli and the signal detection and transmission mechanisms of these receptors, the study of sensorial comfort should move toward implying underlying neurophysiological and physiological mechanisms, just as we did for thermal sensation. Cross-disciplinary research has recently been reported combining sensorial comfort and neurophysiological studies (Zhang, Yue, Jia, & Wang, 2016). It is believed that this would be the main direction in future research and the outcomes would bring new insights into the field of sensorial and clothing comfort.

3.4 Movement Comfort

The above discussion regarding sensorial comfort is mainly on the basis of fabric level. Comfort perception of worn garment would be different from that of fabric. On one hand, there would be dynamic interactions between the garment fabrics and human skin. Such interactions would also be affected by human activities. This makes it quite difficult to simulate the process and evaluate the wear comfort perception. On the other hand, garments do not strictly stick to the human body when worn, which creates air gaps between the garment and the human body. The amount of ease allowance is commonly used to describe such gaps. Evaluation of the movement comfort and preferences of garment ease allowance is mostly conducted through a subjective questionnaire survey (Yu, 2004).

3.4.1 Standard Ease Allowance

Standard ease allowance is defined as the differences of maximum and minimum perimeters of the human body. This is generally obtained from standard body shape in still posture. A commonly applied garment construction method, 2D pattern making, is based on predefined standard ease allowance. The techniques of determining ease allowance on the pattern according to different garment styles and fabric types are commonly based on experiences and know-hows of pattern makers. Radial ease allowance (REA) is defined as the distance between the human body and clothes. Girth ease allowance (GEA) is defined as the additional girth measurement in a garment compared to the body measurement.

Studies have been conducted to evaluate the relationship between the amount of ease allowance from garment patterns and wear comfort, while impacts were found in qualitative term (Ashdown & DeLong, 1995; Hirokawa & Miyoshi, 1997). Body scanning technology was used to quantify the radial ease allowance amount and the distribution on the body (Miyoshi & Hirokawa, 2001; Petrova & Ashdown, 2008).

3.4.2 Fabric Mechanical Properties

Fabrics in nature are generally flexible and more or less extensible. The mechanical properties of fabric would surely affect the garment ease allowance. Hunter and Fan (2004) grouped fabric physical properties that will affect the garment making process, or "tailoring performance". These properties are mass per unit area, relaxation, hygral expansion, extensibility, bending properties, and shearing properties.

Relaxation and *hygral expansion* are terms used to describe the fabric dimension changes in moisture. Relaxation (shrinkage) releases the tension of fabrics during the manufacturing process and let it return to its original dimension. Hygral expansion represents the dimension changes when fabric is exposed to changing moisture. These two properties together could lead to garment fit problems including sizing, waviness, and pucker in garment.

Extensibility means fabric tensile properties, which describe the fabric dimension changes under external forces. Fabric tensile properties should be identified before making the garment so that appropriate garment manufacturing methods could be applied to avoid seam stitching problems and multi-pieces attaching issues. Shearing and bending properties are also related to fabric dimension changes under external forces, through applying external forces in different directions.

The aforementioned KES system can measure fabric tensile and shear properties as well. Indices including tensile linearity, tensile energy, and tensile resistance are identified for tensile properties. Shearing rigidity, presenting the slope of force-strain curve, and shearing hysteresis, presenting the differences between the deforming curve and recovery curve, are identified for shearing properties. FAST measures fabric dimensions changing properties including extensibility (in warp, weft, and 45° directions) and shearing rigidity, as well as relaxation shrinkage and hygral expansion.

3.4.3 Dynamic Ease Allowance

Dynamic ease allowance expanded from standard ease allowance in two ways: (1) it considers different body shapes, and (2) it deals with movements during wearing. There are studies conducted to evaluate the dynamic ease allowance for several specified postures while comprehensive investigation is limited. For modeling the dynamic ease allowance, fuzzy logic technique has been adopted and provided positive outcomes (Chen et al., 2006, 2009). It has been proposed that the air gap between the human body and clothing is important to the garment's functions (e.g., thermal insulation, evaporative resistance, protection from skin burn) of firefighting protective clothing (Mert, Psikuta, Bueno, & Rossi, 2015). Research in dynamic ease allowance would help to estimate the amount of air gap (ease allowance) and in turn help protective clothing development.

3.4.4 Equipment and Garment Styles

The whole wearing system should be considered when evaluating the movement comfort. For intensive work like that engaged in by firefighters, the movement allowance would also be affected by the equipment assembles (including helmet, gloves, boots, and self-contained breathing apparatus) and the garment styles (e.g., long or short pants). An interview with 54 firefighters in the United States indicated their concerns about limited mobility of head and arms due to wearing the helmet and self-contained breathing apparatus (SCBA) and limited mobility and dexterity due to poor fit of gloves (Park, Park, Lin, & Boorady, 2014). A comparison study between firefighters wearing rubber and leather boots implies preference of leather ones, as rubber boots resulted in balance decrement (Garner, Wade, Garten, Chander, & Acevedo, 2013). A study comparing firefighter uniforms of long-sleeved shirt and pants versus T-shirt and shorts showed less restricted movement when the latter was worn, as reported by the participants (Malley et al., 1999).

3.5 Conclusions

The human body detects external environment, regulates internal systems, and responds to the external environment. The clothing system serves as a barrier between the human body and the external environment. It can adjust the microenvironment between the skin and outer environment and protect the body from strong radiant heat and skin burn for a short period in terms of firefighting. It can also prevent the body from cold stress. Clothing itself, because of direct contact with the skin, is one kind of detectable external stimuli. Therefore, it is essential to understand the mechanism of how human body perceives stimuli from the external environments and responds to them.

Specific for firefighters, one key environment feature in firefighting scene is thermal threats. The human thermoregulation system interacts between the human body and external thermal and related stimuli. Studies in this topic have built solid foundations and principles regarding body temperature, physiological mechanisms, and heat exchange between human body and external environment, though our understandings are still frequently updated by new findings. Recent development in neurophysiological mechanism of thermoregulation revealed more detailed processes inside the human body. Normally, the output of the human thermoregulation leads to body heat balance and perception of thermal comfort. In the worst case where external thermal stress exceeds the capacity of thermoregulation, thermal load can cause heat strain, health risks, reduced thermal comfort and work performance, skin burn, or threatened life. In those severe conditions that we have to face, such as firefighting, it is important to accurately estimate potential heat stress and take proper actions. A large number of indices to characterize and/or predict the thermal sensation have been developed. There is no clear preference in one single index. Generally speaking, each of them seems to have advantages and weaknesses in specific situations.

Apart from thermal comfort, sensorial comfort and movement comfort can also affect the work effectiveness and performance of firefighters. Tactile sensation and skin wetness sensation have been found to be correlated with thermal comfort and overall comfort perception. Similar to thermal comfort, predictive models have been developed for sensorial comfort while recent progress in cross-disciplinary study with neural science is promising. We found most movement comfort research focused on garment production techniques, including pattern making and fabric properties evaluation. Nevertheless, the effect of worn equipment, weight, and garment styles on performance of firefighters should not be neglected. Cross-disciplinary research on the related topics is likely to make significant contributions to further understanding of human thermoregulation and comfort.

References

Ackerley, R., Olausson, H., Wessberg, J., & McGlone, F. (2012). Wetness perception across body sites. *Neuroscience Letters, 522*(1), 73–77. doi:10.1016/j.neulet.2012.06.020

Asaga, E., Takemura, K., Maeno, T., Ban, A., & Toriumi, M. (2013). Tactile evaluation based on human tactile perception mechanism. *Sensors and Actuators a-Physical, 203*, 69–75. doi:10.1016/j.sna.2013.08.013

Ashdown, S. P., & DeLong, M. (1995). Perception testing of apparel ease variation. *Applied Ergonomics, 26*(1), 47–54. doi:10.1016/0003-6870(95)95750-t

ASHRAE. (1997). Thermal comfort. In *ASHRAE Handbook of Fundamentals*. Atlanta: ASHRAE.

Axelrod, Y. K., & Diringer, M. N. (2008). Temperature management in acute neurologic disorders. *Neurologic Clinics, 26*(2), 585–603. doi:10.1016/j.ncl.2008.02.005

Ballesteros, S., Reales, J. M., de Leon, L. P., & Garcia, B. (2005). The perception of ecological textures by touch: Does the perceptual space change under bimodal visual and haptic exploration? *World Haptics Conference: First Joint Eurohaptics Conference and Symposium on Haptic Interfaces for Virutual Environment and Teleoperator Systems, Proceedings*, 635–638.

Bligh, J. (2005). A theoretical consideration of the means whereby the mammalian core temperature is defended at a null zone. *Journal of Applied Physiology, 100*(4), 1332–1337. doi:10.1152/japplphysiol.01068.2005

Borg, G. (1982). Psychophysical bases of perceived exertion. *Medicine & Science in Sports & Exercise, 14*(5), 377–381.

Brändström, H., Eriksson, A., Giesbrecht, G., Ängquist, K.-A., & Haney, M. (2012). Fatal hypothermia: An analysis from a sub-arctic region. *International Journal of Circumpolar Health, 71*(1). doi:10.3402/ijch.v71i0.18502

Brown, D. J. A., Brugger, H., Boyd, J., & Paal, P. (2012). Accidental hypothermia. *New England Journal of Medicine, 367*(20), 1930–1938. doi:10.1056/NEJMra1114208

BSI. (1990). Method for determination of bending length and flexural rigidity of fabrics. In (Vol. BS 3356).

Chen, X.-M., Hosono, T., Yoda, T., Fukuda, Y., & Kanosue, K. (1998). Efferent projection from the preoptic area for the control of non-shivering thermogenesis in rats. *The Journal of Physiology, 512*(3), 883–892. doi:10.1111/j.1469-7793.1998.883bd.x

Chen, Y., Zeng, X., Happiette, M., Bruniaux, P., Ng, R., & Yu, W. (2006). Estimation of ease allowance of a garment using fuzzy logic. In *Fuzzy Applications in Industrial Engineering* (pp. 367–379).

Chen, Y., Zeng, X., Happiette, M., Bruniaux, P., Ng, R., & Yu, W. (2009). Optimisation of garment design using fuzzy logic and sensory evaluation techniques. *Engineering Applications of Artificial Intelligence, 22*(2), 272–282. doi:10.1016/j.engappai.2008.05.007

Cheuvront, S. N., Bearden, S. E., Kenefick, R. W., Ely, B. R., DeGroot, D. W., Sawka, M. N., & Montain, S. J. (2009). A simple and valid method to determine thermoregulatory sweating threshold and sensitivity. *Journal of Applied Physiology, 107*(1), 69–75. doi:10.1152/japplphysiol.00250.2009

Danzl, D. (2012). Accidental hypothermia. In P. Auerbach (Ed.), *Wilderness medicine*. Philadelphia: Mosby. (Reprinted from 6th).

Danzl, D., & Pozos, R. (1994). Accidental hypothermia. *New England Journal of Medicine, 331*(26), 1756–1760.

de Freitas, C. R., & Grigorieva, E. A. (2014). A comprehensive catalogue and classification of human thermal climate indices. *International Journal of Biometeorology, 59*(1), 109–120. doi:10.1007/s00484-014-0819-3

Dinarello, C. A., & Gelfand, J. A. (2005). Fever and hyperthermia. *HARRISONS PRINCIPLES OF INTERNAL MEDICINE, 16*(1), 104.

Djongyang, N., Tchinda, R., & Njomo, D. (2010). Thermal comfort: A review paper. *Renewable and Sustainable Energy Reviews, 14*(9), 2626–2640. doi:10.1016/j.rser.2010.07.040

Dorman, L. E., & Havenith, G. (2008). The effects of protective clothing on energy consumption during different activities. *European Journal of Applied Physiology, 105*(3), 463–470. doi:10.1007/s00421-008-0924-2

Durrer, B., Brugger, H., & Syme, D. (2003). The medical on-site treatment of hypothermia: ICAR-MEDCOM recommendation. *High Altitude Medicine & Biology, 4*(1), 99–103. doi:10.1089/152702903321489031

EN. (2004). Protective clothing—Ensembles and garments for protection against cold. In (Vol. CSN EN 342).

3. Human Thermoregulation System and Comfort

Evers, L. H., Bhavsar, D., & Mailänder, P. (2010). The biology of burn injury. *Experimental Dermatology, 19*(9), 777–783. doi:10.1111/j.1600-0625.2010 .01105.x

Fanger, P. (1970). *Thermal comfort.* New York: McGraw-Hill Inc.

Fechner, G. T., Boring, E. G., & Howes, D. H. (1966). *Elements of psychophysics* (H. E. Alder, Trans. D. H. Howes & E. G. Boring Eds.). New York: Holt, Rinehart and Winston.

Filingeri, D. (2016). Neurophysiology of skin thermal sensations. In *Comprehensive Physiology* (pp. 1429–1491).

Foulkes, T., & Wood, J. (2014). Mechanisms of cold pain. *Channels, 1*(3), 154–160. doi:10.4161/chan.4692

Gagge, A., Stolwijk, J., & Nishi, Y. (1971). An effective temperature scale based on a simple model of human physiological regulatory response. *ASHRAE Transactions, 77*, 247–262.

Gao, C., Kuklane, K., & Holmér, I. (2010a). Cooling vests with phase change material packs: The effects of temperature gradient, mass and covering area. *Ergonomics, 53*(5), 716–723. doi:10.1080/00140130903581649

Gao, C., Kuklane, K., & Holmér, I. (2010b). Cooling vests with phase change materials: The effects of melting temperature on heat strain alleviation in an extremely hot environment. *European Journal of Applied Physiology, 111*(6), 1207–1216. doi:10.1007/s00421-010-1748-4

Gao, C., Kuklane, K., Östergren, P.-O., & Kjellstrom, T. (2017a). Occupational heat stress assessment and protective strategies in the context of climate change. *International Journal of Biometeorology.* doi:10.1007/s00484-017-1352-y

Gao, C., Kuklane, K., Wang, F., & Holmér, I. (2012). Personal cooling with phase change materials to improve thermal comfort from a heat wave perspective. *Indoor Air, 22*(6), 523–530. doi:10.1111/j.1600-0668.2012.00778.x

Gao, C., Lin, L.-Y., Halder, A., Kuklane, K., & Jou, G.-T. (2017b). *Thermophysiological responses of exercising in body mapping T-shirts in a warm and humid environment: Subject test and predicted heat strain.* Paper presented at the 17th International Conference on Environmental Ergonomics Kobe, Japan.

Gao, J. (2003). *The principle of the artifical neural network and simulation.* Beijing: Mechanical Engineering Press.

Garner, J. C., Wade, C., Garten, R., Chander, H., & Acevedo, E. (2013). The influence of firefighter boot type on balance. *International Journal of Industrial Ergonomics, 43*(1), 77–81. doi:10.1016/j.ergon.2012.11.002

Greffrath, W., Baumgärtner, U., & Treede, R.-D. (2007). Peripheral and central components of habituation of heat pain perception and evoked potentials in humans. *Pain, 132*(3), 301–311. doi:10.1016/j.pain.2007.04.026

Guest, S., Mehrabyan, A., Essick, G., Phillips, N., Hopkinson, A., & Mcglone, F. (2012). Physics and tactile perception of fluid-covered surfaces. *Journal of Texture Studies, 43*(1), 77–93. doi:10.1111/j.1745-4603.2011.00318.x

Hammel, H. T., & Pierce, J. (1968). Regulation of internal body temperature. *Annual Review of Physiology, 30*(1), 641–710.

Hashmi, J. A., & Davis, K. D. (2010). Effects of temperature on heat pain adaptation and habituation in men and women. *Pain, 151*(3), 737–743. doi:10.1016/j .pain.2010.08.046

Havenith, G. (2001). Individualized model of human thermoregulation for the simulation of heat stress response. *Journal of Applied Physiology, 90*(5), 1943–1954.

Havenith, G., & Fiala, D. (2015). Thermal indices and thermophysiological modeling for heat stress. In *Comprehensive Physiology* (pp. 255–302).

Havenith, G., Fogarty, A., Bartlett, R., Smith, C. J., & Ventenat, V. (2007). Male and female upper body sweat distribution during running measured with technical absorbents. *European Journal of Applied Physiology, 104*(2), 245–255. doi:10.1007/s00421-007-0636-z

Heidelberger, M. (2004). *Nature from within: Gustav Theodor Fechner and his psychophysical worldview.* Pittsburgh, PA: University of Pittsburgh Pre.

Hirokawa, T., & Miyoshi, M. (1997). Relations between drafting factors of jacket pattern and wearing feeling of clothing (Part 2): Ease quantity and wearing feeling of clothing. *Journal-Japan Research Association for Textile End Uses, 38*, 42–51.

Hollins, M., Bensmaia, S., Karlof, K., & Young, F. (2000). Individual differences in perceptual space for tactile textures: Evidence from multidimensional scaling. *Perception & Psychophysics, 62*(8), 1534–1544. doi:10.3758/Bf03212154

Hollins, M., Faldowski, R., Rao, S., & Young, F. (1993). Perceptual dimensions of tactile surface texture: A multidimensional-scaling analysis. *Perception & Psychophysics, 54*(6), 697–705. doi:10.3758/Bf03211795

Holmér, I., Kuklane, K., & Gao, C. (2006). Test of firefighter's turnout gear in hot and humid air exposure. *International Journal of Occupational Safety and Ergonomics, 12*(3), 297–305. doi:10.1080/10803548.2006.11076689

Howorth, W. S. (1964). The handle of suiting, lingere and dress fabric. *Journal of the Textile Institute, 55*, 251–260.

Hu, J., Li, Y., Yeung, K.-W., Wong, A. S. W., & Xu, W. (2005). Moisture management tester: A method to characterize fabric liquid moisture management properties. *Textile Research Journal, 75*(1), 57–62. doi:10.1177/004051750507500111

Hu, J. L., Chen, W. X., & Newton, A. (1993). A psychophysical model for objective hand evaluation: An application of Steven's law. *Journal of the Textile Institute, 84*, 354–363.

Hunter, L., & Fan, J. (2004). Fabric properties related to clothing appearance and fit. In J. Fan, W. Yu, & L. Hunter (Eds.), *Clothing Appearance and Fit: Science and Technology.* Cambridge: Woodhead Publishing.

ISO. (2004). Ergonomics—Evaluation of thermal strain by physiological measurements. In (Vol. 9886). Geneva: ISO.

ISO. (2005). Ergonomics of the thermal environment—Analytical determination and interpretation of thermal comfort using calculation of the PMV and PPD indices and local thermal comfort criteria. In (Vol. 7730). Geneva: ISO.

ISO. (2007a). Ergonomics of the thermal environment—Determination and interpretation of cold stress when using required clothing insulation (IREQ) and local cooling effects. In (Vol. 11079). Geneva: ISO.

ISO. (2007b). Ergonomics of the thermal environment—Estimation of thermal insulation and water vapour resistance of a clothing ensemble. In (Vol. 9920). Geneva: ISO.

ISO. (2013). Clothing—Physiological effects—Measurement of thermal insulation by means of a thermal manikin. In (Vol. 15831). Geneva: ISO.

Jendritzky, G., de Dear, R., & Havenith, G. (2011). UTCI—Why another thermal index? *International Journal of Biometeorology, 56*(3), 421–428. doi:10.1007/s00484-011-0513-7

Jiao, J., Li, Y., Yao, L., Chen, Y., Guo, Y., Wong, S. H. S.,... Hu, J. (2017). Effects of body-mapping-designed clothing on heat stress and running performance in a hot environment. *Ergonomics, 60*(10), 1435–1444. doi:10.1080/0014013 9.2017.1306630

Johnson, J. H., Turner, C. W., Zwislocki, J. J., & Margolis, R. H. (1993). Just noticeable differences for intensity and their relation to loudness. *The Journal of the Acoustical Society of America, 93*(2), 983–991. doi:10.1121/1.405404

Johnson, K. O., Hsiao, S. S., & Yoshioka, T. (2002). Neural coding and the basic law of psychophysics. *Neuroscientist, 8*(2), 111–121. doi:10.1177/10738584 0200800207

Johnson, K. O., & Yoshioka, T. (2002). Neural mechanisms of tactile form and texture perception. In R. J. Nelson (Ed.), *The somatosensory system: Deciphering the brain's own body image* (pp. 73–101). Boca Raton, FL: CRC Press.

Ju, J., & Ryu, H. (2006). A study on subjective assessment of knit fabric by ANFIS. *Fibers and Polymers, 7*(2), 203–212.

Kawabata, S. (1980). *The standardization and analysis of hand evaluation.* Paper presented at the The Textile Machinery Society of Japan, Osaka, Japan.

Kawabata, S. (1982). *The development of the objective measurement of fabric handle.* Paper presented at the First Japan-Australia Symposium on Objective Specification of Fabric Quality, Mechanical Properties and Performance, Osaka, Japan.

Kawabata, S. (1984). Development of a device for measuring heat moisture transfer properties of apparel fabrics. *Journal of Textile Machinery Society of Japan, 37*(8), 130–141.

Kerslake, D. M. (1972). *The stress of hot environment.* Cambridge: Cambridge University Press.

Kobayashi, S., Okazawa, M., Hori, A., Matsumura, K., & Hosokawa, H. (2006). Paradigm shift in sensory system—Animals do not have sensors. *Journal of Thermal Biology, 31*(1–2), 19–23. doi:10.1016/j.jtherbio.2005.11.011

Laupland, K. B. (2009). Fever in the critically ill medical patient. *Critical Care Medicine, 37*(Supplement), S273–S278. doi:10.1097/CCM.0b013e3181 aa6117

Li, Y. (1998). Sensory comfort: Consumer responses in three countries. *Journal of China Textile University.*

Li, Y. (2005). Perceptions of temperature, moisture and comfort in clothing during environmental transients. *Ergonomics, 48*(3), 234–248. doi:10.1080 /0014013042000327715

Liao, X. (2014). *Neuropsychological mechanisms of fabric touch sensations.* (Ph.D.), Hong Kong.

Liao, X., Hu, J. Y., Li, Y., Li, Q. H., & Wu, X. X. (2011). A review on fabric smoothness-roughness sensation studies. In Y. Li, Y. F. Liu, X. N. Luo, & J. S. Li (Eds.), *Textile Bioengineering and Informatics Symposium Proceedings, Vols 1–3* (pp. 1150–1156). Hong Kong Sar: Textile Bioengineering & Informatics Society Ltd.

Liao, X., Li, Y., Hu, J., Ding, X., Zhang, X., Ying, B.,... Wu, X. (2017). Effects of contact method and acclimation on temperature and humidity in touch perception. *Textile Research Journal.* doi:10.1177/0040517517705628

Liao, X., Li, Y., Hu, J. Y., Li, Q. H., & Wu, X. X. (2016). Psychophysical relations between interacted fabric thermal-tactile properties and psychological touch perceptions. *Journal of Sensory Studies, 31*(3), 181–192. doi:10.1111/joss.12189

Liao, X., Li, Y., Hu, J. Y., Wu, X. X., & Li, Q. H. (2014). A simultaneous measurement method to characterize touch properties of textile materials. *Fibers and Polymers, 15*(7), 1548–1559. doi:10.1007/s12221-014-1548-2

Lin, L.-Y., Gao, C., Halder, A., Kuklane, K., & Jou, G.-T. (2017). *Validation of body-mapping sports shirts on thermal physiological responses and comfort in warm and humid environment.* Paper presented at the 17th International Conference on Environmental Ergonomics, Kobe, Japan.

Liu, Y. F., Wang, L. J., Di, Y. H., Liu, J. P., & Zhou, H. (2013). The effects of clothing thermal resistance and operative temperature on human skin temperature. *Journal of Thermal Biology, 38*(5), 233–239. doi:10.1016/j.jtherbio.2013.03.001

Malley, K., Goldstein, A., Aldrich, T., Kelly, K., Weiden, M., Coplan, N.,... Prezant, D. J. (1999). Effects of fire fighting uniform (modern, modified modern, and traditional) design changes on exercise duration in New York City Firefighters. *Journal of Occupational and Environmental Medicine, 41*(12), 1104–1115.

Masin, S. C., Zudini, V., & Antonelli, M. (2009). Early alternative derivations of fechner's law. *Journal of the History of the Behavioral Sciences, 45*(1), 56–65. doi:10.1002/jhbs.20349

Mazzuchetti, G., Demichelis, R., Songia, M. B., & Rombaldoni, F. (2008). Objective measurement of tactile sensitivity related to a feeling of softness and warmth. *Fibres and Textiles in Eastern Europe, 16*(4), 67–71.

Meng, X. L. (2010). *The objective evaluation model on wearing touch and pressure sensation based on GRNN.* Paper presented at the Information and Computing (ICIC), Third International Conference, Wuxi, China.

Mert, E., Psikuta, A., Bueno, M.-A., & Rossi, R. M. (2015). Effect of heterogenous and homogenous air gaps on dry heat loss through the garment. *International Journal of Biometeorology, 59*(11), 1701–1710. doi:10.1007/s00484-015-0978-x

Minazio, P. G. (1995). FAST–Fabric Assurance by Simple Testing. *International Journal of Clothing Science and Technology, 7*(2/3), 43–48. doi:10.1108/09556229510087146

Miyoshi, M., & Hirokawa, T. (2001). Study on the method of measuring a vacant space distance in a worn jacket for clothing pattern design: Using the three-dimensional measuring system. *Journal-Japan Research Association for Textile End Uses, 42*(4), 37–46.

Morley, J. W., Goodwin, A. W., & Dariansmith, I. (1983). Tactile discrimination of gratings. *Experimental Brain Research, 49*(2), 291–299.

Nakano, J., Tanabe, S.-i., & Kimura, K.-i. (2002). Differences in perception of indoor environment between Japanese and non-Japanese workers. *Energy and Buildings, 34*(6), 615–621. doi:10.1016/s0378-7788(02)00012-9

Nicol, J. F., & Humphreys, M. A. (2002). Adaptive thermal comfort and sustainable thermal standards for buildings. *Energy and Buildings, 34*(6), 563–572. doi:10.1016/s0378-7788(02)00006-3

Niedermann, R., Wyss, E., Annaheim, S., Psikuta, A., Davey, S., & Rossi, R. M. (2013). Prediction of human core body temperature using non-invasive measurement methods. *International Journal of Biometeorology, 58*(1), 7–15. doi:10.1007/s00484-013-0687-2

Okazawa, M., Takao, K., Hori, A., Shiraki, T., Matsumura, K., & Kobayashi, S. (2002). Ionic basis of cold receptors acting as thermostats. *Journal of Neuroscience, 22*(10), 3994–4001.

Park, H., Park, J., Lin, S.-H., & Boorady, L. M. (2014). Assessment of firefighters' needs for personal protective equipment. *Fashion and Textiles, 1*(1), 8. doi:10.1186/s40691-014-0008-3

Park, S. W., Hwang, Y. G., Kang, B. C., & Yeo, S. W. (2000). Applying fuzzy logic and neural networks to total hand evaluation of knitted fabrics. *Textile Research Journal, 70*(8), 675–681. doi:10.1177/004051750007000804

Parsons, K. (2014a). Heat stress. In *Human Thermal Environments* (3rd ed.). London, UK: Taylor & Francis.

Parsons, K. (2014b). Human thermal environments. In *Human Thermal Environments*. London, UK: Taylor & Francis.

Parsons, K. (2014c). Human thermal physiology and thermoregulation. In *Human Thermal Environments* (3rd ed.). London, UK: Taylor & Francis.

Pembrey, M., & Nicol, B. (1898). Observations upon the deep and surface temperature of the human body. *The Journal of Physiology, 23*(5), 386–406.

Petrova, A., & Ashdown, S. P. (2008). Three-dimensional body scan data analysis. *Clothing and Textiles Research Journal, 26*(3), 227–252. doi:10.1177/0887302x07309479

Picard, D., Dacremont, C., Valentin, D., & Giboreau, A. (2003). Perceptual dimensions of tactile textures. *Acta Psychologica, 114*(2), 165–184. doi:10.1016/j.actpsy.2003.08.001

Romanovsky, A. A. (2006). Thermoregulation: Some concepts have changed. Functional architecture of the thermoregulatory system. *AJP: Regulatory, Integrative and Comparative Physiology, 292*(1), R37–R46. doi:10.1152/ajpregu.00668.2006

Rombaldoni, F., Demichelis, R., & Mazzuchetti, G. (2010). Prediction of human psychophysical perception of fabric crispness and coolness hand from rapidly measurable low-stress mechanical and thermal parameters. *Journal of Sensory Studies, 25*(6), 899–916. doi:10.1111/j.1745-459X.2010.00312.x

Schroder, A. C., Viemeister, N. F., & Nelson, D. A. (1994). Intensity discrimination in normal-hearing and hearing-impaired listeners. *The Journal of the Acoustical Society of America, 96*(5), 2683. doi:10.1121/1.411276

Shibasaki, M., & Crandall, C. G. (2010). Mechanisms and controllers of eccrine sweating in humans. *Frontiers in Bioscience (Scholar edition), 2*, 685–696.

Shirado, H., & Maeno, T. (2005). Modeling of human texture perception for tactile displays and sensors. *World Haptics Conference: First Joint Eurohaptics Conference and Symposium on Haptic Interfaces for Virutual Environment and Teleoperator Systems, Proceedings,* 629–630.

Smith, C. J., & Havenith, G. (2010). Body mapping of sweating patterns in male athletes in mild exercise-induced hyperthermia. *European Journal of Applied Physiology, 111*(7), 1391–1404. doi:10.1007/s00421-010-1744-8

Stevens, S. S. (1957). On the psychophysical law. *Psychological Review, 64*(3), 153–181. doi:10.1037/h0046162

Stevens, S. S. (1961). To honor Fechner and repeal his law: A power function, not a log function, describes the operating characteristic of a sensory system. *Science, 133*(3446), 80–86. doi:10.1126/science.133.3446.80

Takada, S., Kobayashi, H., & Matsushita, T. (2009). Thermal model of human body fitted with individual characteristics of body temperature

regulation. *Building and Environment, 44*(3), 463–470. doi:10.1016/j
.buildenv.2008.04.007

Tanaka, Y., Tanaka, M., & Chonan, S. (2006). Development of a sensor system for measuring tactile sensation. *2006 Ieee Sensors, Vols 1–3*, 554–557.

Tang, K.-p. M., Kan, C.-w., & Fan, J.-t. (2014). Assessing and predicting the subjective wetness sensation of textiles: Subjective and objective evaluation. *Textile Research Journal, 85*(8), 838–849. doi:10.1177/0040517514555799

Tang, K. P. M., Chau, K. H., Kan, C. W., & Fan, J. T. (2015). Characterizing the transplanar and in-plane water transport properties of fabrics under different sweat rate: Forced Flow Water Transport Tester. *Scientific Reports, 5*(1). doi:10.1038/srep17012

Tang, K. P. M., Wu, Y. S., Chau, K. H., Kan, C. W., & Fan, J. T. (2015). Characterizing the transplanar and in-plane water transport of textiles with gravimetric and image analysis technique: Spontaneous Uptake Water Transport Tester. *Scientific Reports, 5*(1). doi:10.1038/srep09689

Taylor, N. A. S., Tipton, M. J., & Kenny, G. P. (2014). Considerations for the measurement of core, skin and mean body temperatures. *Journal of Thermal Biology, 46*, 72–101. doi:10.1016/j.jtherbio.2014.10.006

Tiest, W. M. B., & Kappers, A. M. L. (2006). Analysis of haptic perception of materials by multidimensional scaling and physical measurements of roughness and compressibility. *Acta Psychologica, 121*(1), 1–20. doi:10.1016/j
.actpsy.2005.04.005

Tousignant-Laflamme, Y., Pagé, S., Goffaux, P., & Marchand, S. (2008). An experimental model to measure excitatory and inhibitory pain mechanisms in humans. *Brain Research, 1230*, 73–79. doi:10.1016/j.brainres.2008.06.120

van Hoof, J. (2008). Forty years of Fanger's model of thermal comfort: Comfort for all? *Indoor Air, 18*(3), 182–201. doi:10.1111/j.1600-0668.2007.00516.x

Wang, F., Del Ferraro, S., Molinaro, V., Morrissey, M., & Rossi, R. (2014). Assessment of body mapping sportswear using a manikin operated in constant temperature mode and thermoregulatory model control mode. *International Journal of Biometeorology, 58*(7), 1673–1682. doi:10.1007
/s00484-013-0774-4

Wang, F., Gao, C., Kuklane, K., & Holmér, I. (2013). Effects of various protective clothing and thermal environments on heat strain of unacclimated men: The PHS (predicted heat strain) model revisited. *Industrial Health, 51*(3), 266é274. doi:10.2486/indhealth.2012-0073

Wang, F., Kuklane, K., Gao, C., & Holmér, I. (2011). Can the PHS model (ISO7933) predict reasonable thermophysiological responses while wearing protective clothing in hot environments? *Physiological Measurement, 32*(2), 239–249. doi:10.1088/0967-3334/32/2/007

Weidner, C., Schmelz, M., Schmidt, R., Hansson, B., Handwerker, H., & Torebjork, H. (1999). Functional attributes discriminating mechano-insensitive and mechano-responsive C nociceptors in human skin. *Journal of Neuroscience, 19*, 10184–10190.

Yoshioka, T., Bensmaia, S. J., Craig, J. C., & Hsiao, S. S. (2007). Texture perception through direct and indirect touch: An analysis of perceptual space for tactile textures in two modes of exploration. *Somatosensory and Motor Research, 24*(1–2), 53–70. doi:10.1080/08990220701318163

Yu, W. (2004). Subjective assessment of clothing fit. In J. Fan, W. Yu, & L. Hunter (Eds.), *Clothing Appearance and Fit: Science and Technology*. Cambridge: Woodhead Publishing.

Zhang, H., Huizenga, C., Arens, E., & Yu, T. (2001). Considering individual physiological differences in a human thermal model. *Journal of Thermal Biology, 26*(4-5), 401–408. doi:10.1016/s0306-4565(01)00051-1

Zhang, L. M. (1993). *The model and application of the Artifical Neural Network*. Shanghai: Fudan University Press.

Zhang, X., Yue, J., Jia, J., & Wang, G. (2016). An electroencephalogram study on softness cognition of silk fabric hand. *The Journal of The Textile Institute, 107*(12), 1601–1606. doi:10.1080/00405000.2015.1130958

Zhang, Y. H., Yanase-Fujiwara, M., Hosono, T., & Kanosue, K. (1995). Warm and cold signals from the preoptic area: Which contribute more to the control of shivering in rats? *The Journal of physiology, 485*(1), 195–202. doi:10.1113/jphysiol.1995.sp020723

Zhao, M., Gao, C., Wang, F., Kuklane, K., Holmér, I., & Li, J. (2013). A study on local cooling of garments with ventilation fans and openings placed at different torso sites. *International Journal of Industrial Ergonomics, 43*(3), 232–237. doi:10.1016/j.ergon.2013.01.001

Zhou, X., Lian, Z., & Lan, L. (2013). An individualized human thermoregulation model for Chinese adults. *Building and Environment, 70*, 257–265. doi:10.1016/j.buildenv.2013.08.031

Zhou, X., Zhang, H., Lian, Z., & Zhang, Y. (2014). A model for predicting thermal sensation of Chinese people. *Building and Environment, 82*, 237–246. doi:10.1016/j.buildenv.2014.08.006

4

Protective Performance of Firefighters' Clothing

*Sumit Mandal, Martin Camenzind,
Simon Annaheim, and René M. Rossi*

4.1 Introduction

National Fire Protection Association (NFPA) statistics revealed that more than one million fire incidences happened in the United States in 2016 (Haynes, 2017). These fire incidences resulted in 62,085 firefighter injuries (Haynes & Molis, 2017). In France, firefighters are part of military personnel; therefore, firefighter injuries are systematically reported there among all the European countries. This report states that nearly 133 firefighter injuries occur every year in France (Brushlinsky et al., 2017). As clothing is the only barrier between firefighters and their work hazards, it has been found that the inadequate protective performance of firefighters' clothing plays a major role in causing these injuries (Kahn et al., 2012). Along with the high protective performance, it is also required that the clothing should not exert burden on firefighters and should effectively transmit the metabolic heat and sweat vapor from their bodies toward the ambient environment. Thus, firefighters' clothing should possess both high protective (by providing hazards protection) as well as comfort (by providing less burden and heat stress) performances. As the title suggests, this chapter addresses the protective performance only, and the comfort performance of firefighters' clothing is presented in other chapters of this book.

It has been stated that the types of hazard encountered by firefighters in a fire incidence strongly influence the protective performance of their clothing (Lawson, 1996). Traditionally, it was believed that firefighters get exposed to thermal hazards in a fire incidence. These thermal hazards could be flame, radiant heat, flash fires, hot surface contact, molten metal contact, hot liquids, and/or steam (Benisek & Phillips, 1981; Lu et al., 2014; Rossi, Indelicato, & Bolli, 2004; Rossi & Zimmerli, 1994; Shalev & Barker, 1984). However, after the tragic terror attack of 9/11 in the United States, it was clear that firefighters' exposure is not limited to the thermal hazards (Barker et al., 2010). Today, terror incidents involving improvised explosive devices (designed to distribute biological or chemical weapons of mass destruction) could expose firefighters to various nonthermal hazards, namely physical (e.g., sharp objects, falling debris), biological (e.g., blood-borne pathogens), chemical (e.g., toxic liquids and air vapors), and radiological (e.g., aerosols). In both thermal and nonthermal hazards, a significant amount of heat (e.g., dry heat) and/or mass (e.g., liquids, chemicals, bloods) transfer occurs from the source of the hazards to the firefighters' clothing or multilayered fabrics (a combination of shell fabric, moisture barrier, and/or thermal liner) used in their clothing. And, this heat and/or mass could be transmitted through the fabrics/clothing and can generate significant injuries on firefighters' bodies. Eventually, firefighters' clothing should possess high protective performance under thermal and nonthermal hazards by lowering the heat and mass transfer through the fabrics used in the clothing. This lowering of heat and mass transfer could help to reduce the injuries for firefighters. Therefore, firefighters' clothing should be such that it slows down the heat and mass transfer through the fabrics, thus causing reduced burn injuries.

Research and development has been initiated in the field of firefighters' protective clothing by considering the modes of heat and mass transfer (e.g., convection, conduction, diffusion) under various thermal and nonthermal hazards (Ackerman et al., 2012; Jalbani et al., 2012; Kemp et al., 2016; Mandal et al., 2017a; Seed et al., 2008; Shalev & Barker, 1984; Song, 2004; Wen, 2014). This research has mainly focused on developing the test methods to evaluate the protective performance of existing state-of-the-art firefighters' clothing and/or setting the standard requirements for the protective performance. As an outcome of this research, several methods have been developed by standard organizations— for example, International Organization for Standardization (ISO), European Committee for Standardization (EN), American Society for Testing Materials (ASTM), and Canadian General Standards Board (CGSB)—for evaluating the protective performance of fabrics and/or clothing under thermal and nonthermal hazards. Based on these evaluation methods, some standard organizations (e.g., ISO, EN, NFPA) have also set the minimum requirements for the protective performance of commercially available fabrics and/or clothing. Some research institutes—for example, Swiss Federal Laboratories for Materials Science & Technology (Empa), Switzerland; Protective Clothing and Equipment Research Facility (PCERF), University of Alberta, Canada; and French Navy clothing, the steam laboratory of the Institut de Médecine Navale du Service de Santé des

Armées (IMNSSA)—have also developed their customized test methods for evaluating the protective performance of fabrics or clothing.

By using the standardized and/or customized methods, many researchers have further assessed the protective performance of the fabrics and/or clothing (Benisek & Phillips, 1981; Lu et al., 2014; Mandal et al., 2013; Mandal et al., 2017a; Rossi et al., 2004; Rossi & Zimmerli, 1994; Shalev & Barker, 1984). Their studies infer that many factors such as fabric construction (e.g., number of layers, type of layers) and properties (e.g., weight, thickness), clothing designs (e.g., size, fit), and/or features of the test methods (e.g., intensity, duration, and/or angle of impingement of the hazards) could significantly affect the protective performance.

This chapter discusses the evaluation and assessment of the protective performance of firefighters' protective fabrics and/or clothing under various thermal and nonthermal hazards. The evaluation and assessment sections mainly highlight the modes of heat and mass transfer (e.g., convection, conduction, diffusion) toward firefighters' protective fabrics/clothing under the hazards, standardized/customized test methods for evaluating the protective performance, and various factors (e.g., fabric construction and properties, clothing designs) influencing the protective performance. These sections could be helpful to understand the engineering involved in developing and testing the high protective performance-based firefighters' clothing. Finally, the key issues related to the evaluation and assessment of the protective performance are highlighted. In the future, by resolving these key issues, the textile and/or materials engineers can modify the existing test methods or develop new test methods for evaluating the protective performance of firefighters' protective fabrics and/or clothing. These modified or new test methods will help to more realistically assess the protective performance under various hazards so as to reduce the firefighters' injuries. This attempt may also uncover other factors affecting the protective performance of firefighters' clothing. In the long run, this new understanding may contribute to develop new firefighters' protective fabrics and/or clothing and could contribute to providing better protection to firefighters worldwide.

4.2 Evaluation and Assessment of the Protective Performance of Fabrics and Clothing under Thermal Hazards

In a fire incidence, firefighters could be exposed to various thermal hazards such as flame, radiant heat, flash fire (a combination of flame, radiant heat, fire balls, and hot gases), hot surface, molten metal, hot liquids, and/or steam; the modes of heat and mass transfer toward firefighters' protective fabrics/clothing are different under these hazards. Considering these modes of heat and mass transfer, many standard organizations have developed test methods for evaluating the protective performance of fabrics or clothing under these thermal hazards (e.g., ISO 9151:2016 for flame, ISO 6942:2015 for radiant heat). Some research institutes (e.g., Empa, Switzerland; PCERF, Canada; IMNSSA, France) have also developed customized test methods for evaluating the protective performance of fabrics or clothing (Ackerman et al., 2012; Jalbani et al., 2012; Mandal et al., 2017a). It is notable that some standardized or customized test methods have also been developed for evaluating the protective performance of fabrics under a combined set of thermal hazards. For example, ISO 17492:2003 (or ASTM F 2703:2013 or ASTM F 2700:2000) standard test method has been developed to evaluate the protective performance of fabrics under 50% flame and 50% radiant heat hazard of 80 kW/m² intensity (ASTM F 2703, 2013; ASTM F 2700, 2000; ISO 17492, 2003). Nevertheless, ISO 11999-3:2015 and ISO 11613:2012 (ISO 11999-3:2015; ISO 11613, 2012) standards recommend that ISO 17492:2003 test could be used as an alternative of ISO 9151:2016 and ISO 6942:2015 test standards for evaluating the protective performance of fabrics under flame and radiant heat. Table 4.1 presents the available standardized or customized test methods for evaluating the protective performance of fabrics and/or clothing under thermal hazards.

In most of the test methods (except molten metal contact test of ISO 9185:2007; the ASTM F 955:2015 counterpart of this test uses copper sensor instead of PVC film) in Table 4.1, one of the common and prime requirements is a small, lightweight, inexpensive, and highly sensitive heat flux sensor that can properly simulate human skin (Barker et al., 1999; Dale et al., 1992). To date, five different types of heat flux sensors (epoxy resin sensor, copper sensor, PyroCal copper sensor, aluminum sensor, skin simulant sensor) have been commonly used in these standard or customized test methods (Figure 4.1) (Mandal & Song, 2015; Song, Mandal, & Rossi, 2016). During the testing, a sensor or a group of sensors is placed on the back side of the fabric or clothing (that will be aligned with the wearer's skin), and the face side of the fabric or clothing is exposed to a particular thermal hazard such as flame, radiant heat, or hot surface. During and after the exposure to thermal hazard, temperature increase of the sensor(s) with respect to time is measured by the thermocouple. This temperature rise can further be used to calculate the thermal energy (in terms of heat flux) transmitted to the

Table 4.1 Standardized or Customized Test Methods for Evaluating the Protective Performance under Thermal Hazards

Thermal Hazards		Test Methods	Means of Protective Performance Data Collection	Applicable for
Flame	Gas Fuel Based	ISO 9151:2016	Copper Sensor	Fabrics
	Liquid Fuel Based	Empa, Switzerland	Epoxy Resin Sensor	Fabrics
Radiant Heat		ISO 6942:2015	Copper Sensor	Fabrics
Flash Fire		ISO 13506-1:2017 & ISO 13506-2:2017	Epoxy Resin, PyroCal, or Skin Simulant Sensor	Clothing
		Empa, Switzerland	Epoxy Resin Sensor	Fabrics
Hot Surface Contact		ISO 12127-1:2015	Aluminum Sensor	Fabrics
Molten Metal Contact		ISO 9185:2007	Polyvinyl Chloride (PVC) Film	Fabrics
Hot Liquids	Splash	ASTM F 2701:2008	Copper Sensor	Fabrics
		PCERF, Canada	Copper or Skin Simulant Sensor	Fabrics
	Immersion with Compression	PCERF, Canada	Skin Simulant Sensor	Fabrics
Steam	Spray	PCERF, Canada	Skin Simulant Sensor	Clothing
		PCERF, Canada	Skin Simulant Sensor	Fabrics
		IMNSSA, France	Thermocouple	Clothing

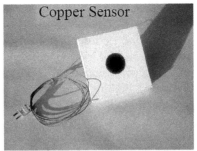
Epoxy Resin Sensor

Copper Sensor

Aluminum Sensor

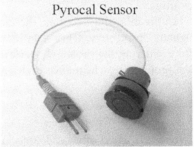
Pyrocal Sensor

Skin Simulant Sensor

Figure 4.1

Heat flux sensors for evaluating the protective performance of fabrics under thermal hazards. (Courtesy of Mr. Pat Kavanaugh, Precision Products, USA, for epoxy, copper, and PyroCal sensors; Ms. Pauline Weisser, DuPont, Switzerland, for aluminum sensor; Mr. Mark Ackerman, MYAC-Eng., Canada, for skin simulant sensor.)

sensor (a firefighter's skin) through clothing using different equations; a summary of these equations is provided by Mandal and Song (2015). Based on the heat flux data, the time to generate different degree of burns (first degree, second degree, or third degree) on firefighters' skin can be calculated according to the Stoll Curve (this curve is only applicable to predict the second-degree burn) or Henriques Burn Integral (HBI) equation (this equation could help to predict different degrees of burns) (Mandal & Song, 2015; Song et al., 2016). The temperature rise, heat flux, and/or time provide the basis to assess the protective performance of fabrics under various thermal hazards.

The above discussion confirms that a great deal of standardized and customized test methods have been developed for evaluating the protective performance of fabrics and clothing under various thermal hazards. In the last few decades, many researchers have extensively used these test methods for assessing the protective performance of fabrics and clothing under various thermal hazards (Benisek & Phillips, 1981; Benisek & Edmondson, 1981; Lu et al., 2014; Mandal et al., 2013; Mandal et al., 2017a; Rossi et al., 2004; Rossi & Zimmerli, 1994; Shalev & Barker, 1984). Their assessments concluded that different factors—namely, fabric construction and properties, clothing designs, and/or features of the test methods—could influence the thermal protective performance of fabrics and clothing.

In the following sections, the modes of heat and mass transfer toward firefighters' protective fabrics/clothing, standardized/customized test methods for evaluating the protective performance, and various factors associated with the protective performance under a particular thermal hazard are thoroughly discussed.

4.2.1 Flame Hazard

Firefighters could often encounter a flame hazard while working in a fire incidence. Under the flame hazard, hot gaseous molecules move toward the exposed fabric and form a boundary layer on the fabric surface (Song et al., 2016). At a particular point on the fabric surface, the thickness of the boundary layer (Δx) depends on several factors, namely, v = velocity of the gaseous molecules (m/s), ρ = density of the gaseous molecules (g/cm^3), and μ = dynamic viscosity of the gaseous molecules (m^2/s) (Equation 4.1). As the temperature of the boundary layer is higher than the temperature of the fabric surface, convective heat energy transfer (Q) occurs from the boundary layer to the fabric surface. Here, convective heat transfer coefficient (h_c) is dependent on several factors: v, ρ, μ, k = thermal conductivity of the gaseous molecules (K·m·W^{-1}), c_p = specific heat capacity of the gaseous molecules (J/K), s_g = surface geometry of the fabric, and f_c = flow conditions of the gaseous molecules (Equation 4.2) (Arpaci & Larsen, 1984; Burnmeister, 1993). In this context, it is notable that flame comprises carbon particles of different diameters. During the flame hazard, these carbon particles also radiate some amount of heat onto the fabric surface. Therefore, both convective and radiative heat impose on the fabric surface under the flame hazard (Shalev & Barker, 1983).

$$\Delta x = \int (v, \rho, \mu) \qquad (4.1)$$

$$h_c = \int (v, \rho, \mu, k, c_p, f_c, s_g, f_c) \qquad (4.2)$$

For evaluating the protective performance of a fabric under a flame hazard, ISO 9151:2016 or EN 367:1992 standard is used (ISO 9151, 2016). According to ISO 9151:2016 standard, a horizontally placed fabric specimen is exposed to the 80 kW/m^2 flame hazard (Figure 4.2). Then, thermal energy (dry heat) passed through the fabric specimen is measured by a copper sensor in terms of time (in seconds) required to increase the 24±0.2°C temperature of the sensor; this time [i.e., heat transfer index at 24°C (HTI$_{24}$)] is interpreted as the flame protective

Figure 4.2

Flame hazard tester.

performance of the fabric. Notably, the minimum flame protective performance required for two different category fabrics (Fabrics with Performance Level A1 and Fabrics with Performance Level A2) that are commercially used in firefighters' clothing are also set by ISO 11999-3:2015 and ISO 11613:2012 (ISO 11999-3, 2015; ISO 11613, 2012). Their values are $HTI_{24} \geq 13$ s for A1 fabrics and $HTI_{24} \geq 17$ s for A2 fabrics.

Similar to ISO 9151:2016, ASTM D 4108:1987 standard was also available for determining the protective performance of fabrics under flame hazard (ASTM D 4108, 1987). However, this ASTM D 4108:1987 standard document has been presently withdrawn by the ASTM committee. Although ASTM standard does not support the flame hazard test of fabrics anymore, ASTM F 2703:2013 (considers the stored energy inside the specimen after the flame hazard has ceased) and ASTM F 2700:2013 (does not consider the stored energy inside the specimen after the flame hazard has ceased) standards are presently available for determining the protective performance of fabrics under the combined flame and radiant heat hazards (ASTM F 2703, 2013; ASTM F 2000, 2013).

Recently, ISO 9151:2016 tester was also modified by replacing the copper sensor with a skin simulant sensor (Mandal, 2016). By using the skin simulant sensor, the time required to generate the second-degree burns on firefighters' bodies was calculated using HBI equation (Mandal & Song, 2015); this burn time was interpreted as the flame protective performance of fabrics (Mandal, 2016). Furthermore, ISO 9151:2016 standard mainly uses the flame generated from the propane gas fuel. As no standard method is available for determining the protective performance of a fabric under liquid fuel–based flame hazard, Empa, Switzerland has recently developed a customized method for evaluating the protective performance under liquid fuel–based flame using epoxy resin sensor

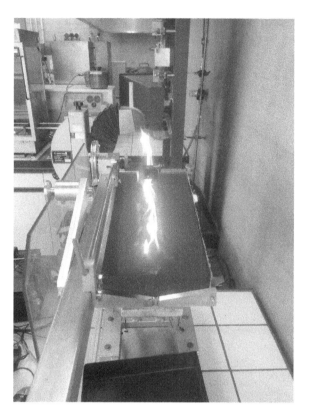

Figure 4.3

Liquid fuel-based flame hazard tester.

(Figure 4.3) (Kemp et al., 2016). These modified testers of Mandal (2016) and Kemp et al. (2016) can be effectively used for determining the flame protective performance of different single- and multilayered fabrics.

Many researchers (Barker & Lee, 1987; Behnke, 1984; Benisek, Edmondson, & Philips, 1979; Ghazy, 2014, 2017; Ghazy & Bergstrom, 2010, 2011, Lee & Barker, 1986; Shalev & Barker, 1983, 1984; Torvi & Dale, 1998) have studied the flame protective performance of fabrics. They corroborated that mainly convective and radiative modes of heat transfer occur from the flame to the fabric surface. Nevertheless, the heat transfer characteristics through the fabrics under these two modes are quite different. In the convective mode, heat imposes on a fabric's surface and blows through it; in the radiant mode, heat directly penetrates through the fabric. Contextually, Shalev and Barker (1984) explained that heat transfer through fabric under flame hazard can be a complex combination of absorption, re-radiation, conduction, and perhaps forced convection.

Furthermore, a group of researchers found fabric properties affect the protective performance. In the early 1980s, Benisek and Phillips (1979) analyzed the flame protective performance of single- and double-layered fabrics in high-intensity flame hazards. They found that the thickness and weight of fabrics affected their protective performance, and the protection of double-layered fabrics was much higher than that of single-layered fabrics (Shalev & Barker, 1984). Barker and Lee (1987) explained that a fabric with low density possesses higher performance than a fabric with high density. Here, the low-density fabric possesses more air within its structure, thus leading to less thermal energy transfer through the fabric. However, there is no simple linear relationship between the density and thermal protective performance. If the density of a fabric is gradually increased, the protective performance gradually decreases; however, after a certain density (~60 kg/m³), the thermal protective performance decreases rapidly. This is because the trapped dead air inside the fabric starts the conductive heat transfer toward wearers and that lowers the thermal protective performance of the fabrics. Moreover, Torvi and Dale (1998) found that a fabric with high thermal conductivity and low specific heat could quickly transfer the thermal energy through it and lowers the flame protective performance. They also mentioned that the fabric could decompose in a flame hazard, and this decomposition can generate a lot of exothermic thermal energy depending on the intensity and duration of the hazard. This thermal energy generated could lower the flame protective performance of the fabric. Furthermore, Shalev and Barker (1984) and Torvi and Dale (1998) stated that fabric properties such as air permeability, fabric surface frictional coefficient, surface optical properties, fiber to air ratio, and air void distribution mainly affect thermal energy transfer through fabrics under flame hazards, which ultimately affect the flame protective performance of fabrics.

In this context, Ghazy and Bergstrom (2010, 2011) mentioned that the air gap between the fabrics and firefighters' skin could play a crucial role on the flame protective performance of fabrics. In this air gap, the conductive and radiative heat transfer depend on various factors, especially the emissivity of the fabrics aligned with wearers' bodies and the shrinkage property of the fabrics. If the emissivity of the fabric is high, it will increase the radiative heat transfer toward wearers' bodies, which ultimately lowers the protective performance. Also, if the fabric shrinks, this will ultimately reduce the air gap and contribute toward more heat transfer toward wearers' bodies (Ghazy, 2014). Recently, Ghazy (2017) found that the reduction of the air gap size could lower the radiative heat transfer, but the chances of conductive heat transfer become high. This situation ultimately lowers the flame protective performance of fabrics.

Although fabric properties affect the flame protective performance, many researchers explained that features of the flame hazard test methods also have a significant impact on the protective performance. Shalev, Barker, and Lee (1983, 1987) demonstrated that the protective performance of single-layered fabrics is affected by the changes in intensity (heat flux) and/or duration of the flame hazard. It was also observed that the thermal energy transmissions through fabrics are different at high and low intensity. This is because thermal energy does not directly contact the fabric at low intensity, which results in a lower convective heat transfer coefficient through fabric. In contrast, the thermal energy directly contacts the tested fabric at high intensity; in this situation, turbulent hot air movement on the fabric enhances the thermal energy absorptivity of the fabric. Contextually, Shalev and Barker (1984) found that the configuration of testers such as fabrics' surface exposure area, angle of flame impingement, flame turbulence, and distance of the exposed fabrics from burner top have a negligible effect on the performance of thermal protective fabrics.

4.2.2 Radiant Heat Hazard

Firefighters are most likely exposed to radiant heat in a fire incidence. In this situation, electromagnetic waves are mainly responsible for transferring the heat energy from the radiant heat source to the fabric (Song et al., 2016). Here, the transfer of heat energy from the source to the fabric surface can be denoted by Equation 4.3, where Q = net heat energy transfer (kW/m²), σ = Stefan Boltzmann constant (5.670373×10⁻⁸ W/m²/K⁴), A = surface area of the fabric exposed (m²), ε_s = the effective emissivity of the radiant heat source (dimensionless), ε_r = emissivity of the radiant heat source at temperature T_r (°C); that is, the ratio of source's emissive power at T_r (°C) to that of black body at T_r (°C); α_{rfs} = absorptivity of fabric at T_r (°C) for incident waves from a black body at temperature T_s (°C); that is, the ratio of the absorptive power at T_r (°C) to that of a black body at T_r (°C) for incident waves from a black body at T_s (°C) (Hsu, 1963; Siegel & Howell, 2002).

$$Q = \sigma A \varepsilon_s \left(\varepsilon_r T_r^4 - \alpha_{rfs} T_s^4 \right) \qquad (4.3)$$

For evaluating the protective performance of a fabric under the radiant heat hazard, ISO 6942:2015 or ASTM F 1939:2015 standard is used (ISO 6942, 2015; ASTM F 1939, 2015). As per ISO 6942:2015 standard, a nonmoving vertically oriented fabric specimen is exposed to the radiant heat at different intensity levels (low intensity level: 5 and 10 kW/m²; medium intensity level: 20 and 40 kW/m²; and high intensity level: 80 kW/m²) (Figure 4.4). Then, thermal energy (dry heat) passed through the fabric specimen is measured by a copper sensor in terms of time (in seconds) required to increase the 24 ± 0.2°C temperature of the sensor; this time [radiant heat transfer index (RHTI$_{24}$)] is interpreted as the radiant heat protective performance of the fabric. Notably, ISO 11999-3:2015 and ISO 11613:2012 standards have also set the minimum radiant heat protective performance (at 40 kW/m²) required for two different category fabrics (Fabrics with Performance Level A1 and Fabrics with Performance Level A2) that are commercially used in firefighters' clothing (ISO 11999-3, 2015; ISO 11613, 2012). Fabrics with RHTI$_{24}$ ≥ 18 s are categorized as A1 fabrics and in case of RHTI$_{24}$ ≥ 26 s as A2 fabrics. Although ISO 6942:2015 standard is widely used for evaluating

Figure 4.4

Radiant heat hazard tester.

4. Protective Performance of Firefighters' Clothing

the radiant heat protective performance of fabrics, Rossi and Bolli (2000) developed an instrumented manikin-based test method in Empa, Switzerland for evaluating the radiant heat protective performance of whole clothing. As per this method, clothing was donned on a polyvinyl chloride (PVC) based adult-sized manikin (instrumented with eight temperature sensors) and exposed to 5–10 kW/m^2 radiant heat. This heat was generated from a radiant heat panel and that was placed parallel to the manikin. The temperature rise of the sensor was predicted as the radiant heat protective performance of the clothing. Notably, the manikin used in this study was not made up of flame-resistant material such as ceramic or glass-reinforced vinyl ester resin. Eventually, this manikin was not able to resist the high heat flux that could be generated from the heating rods. Considering this, a new test method in combination with the flame-resistant material-based manikin and radiant heat panel has been developed and being internally used by Empa, Switzerland (Figure 4.5).

Related to the radiant heat intensity of the test methods, Sun et al. (2000), Perkins (1979), and Song et al. (2011) reported that both low (<20 kW/m^2) and high (>20 kW/m^2) intensities of radiant heat hazards have a significant effect on the radiant heat protective performance of the fabrics. In both low- and high-intensities radiant heat hazards, a significant amount of thermal energy transfers through the fabrics, which can lower the radiant heat protective performance of the fabrics by generating burns on wearers' bodies. Although both low and high intensities have a significant effect on the radiant heat protective performance of the fabrics, this effect depends on various properties related to fabrics—thickness, weight, density, and moisture content. Contextually, by using the instrumented manikin test, Rossi and Bolli (2000) found that the radiant heat flux of 5–10 kW/m^2 could significantly damage the moisture barrier present in firefighters' clothing, and this damaging situation could change the properties of the fabrics and lower the radiant heat protective performance.

Shalev and Barker (1984) observed that the thermal energy transfer rate was lower in thick fabrics than in thin fabrics, and air permeability did not

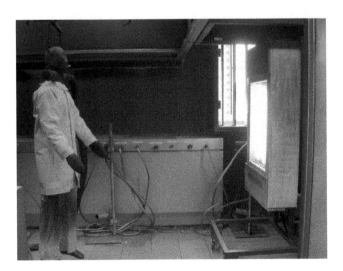

Figure 4.5

Radiant heat protective performance test for clothing using instrumented manikin.

significantly affect the transfer of thermal energy. They concluded that air permeability has little or no impact on radiant heat protective performance. Perkins (1979) concluded that fabric weight and thickness are the main properties to consider while analyzing fabric performance in low-intensity radiant heat hazards. Through statistical analysis, he confirmed that fabric weight and thickness are positively associated with radiant heat protective performance. Fabrics with high weight and thickness entrap more dead air than thinner fabrics, and this air could help to insulate the wearer (Song, Cao, & Gholamreza, 2011; Sun et al., 2000; Torvi & Dale, 1999; Zhu, Zhang, & Chen, 2007). Furthermore, Mandal et al. (2013) and Song et al. (2011) established that fabrics with low density and high thermal resistance show higher protective performance than fabrics with high density and low thermal resistance; this is because low density and high thermal resistance fabrics trap more still or dead air inside their structures than high density and low thermal resistance fabrics (Sun et al., 2000).

Although fabrics with high weight, high thickness, and low density show high protective performance, Song et al. (2011) observed that weighty and thick fabrics store more energy than thin fabrics in low-intensity radiant heat hazard, and this stored energy releases naturally or due to the compression during and after the hazard. This energy released causes quick burn injury on the wearer's skin and consequently lowers the radiant heat protective performance of the firefighters' clothing (Eni, 2005; Barker et al., 2006). Recently, He et al. (2017) indicated that the amount of thermal energy released naturally is not proportionately related to the amount of energy stored inside the fabrics. However, this is quite proportional to the amount of energy released due to the compression. Additionally, this natural and compressed release of thermal energy is dependent on several factors including the number (e.g., single-, double-layered), type (e.g., thermal liner, moisture barrier), and position (e.g., inner layer, outer layer) of the fabric layers in a multilayer fabric system; a thermal liner plays the most critical role in this thermal energy release as it stores most of the energy. Furthermore, the air gap between the clothing and firefighters' bodies could act as an insulator and enhance the radiant heat protective performance during the natural release of thermal energy. However, this air gap diminishes during the compression, and this ultimately lowers the radiant heat protective performance of the fabric.

Perkins (1979) also found that a fabric with open constructions and/or in touch with wearers' bodies allows more thermal energy to pass through the fabric through convection and/or conduction. If the surface of the fabric is opaque, the absorption of thermal energy into the fabric could be high and it may ignite the fabric. This situation of more transfer of thermal energy through the fabrics and/or ignition of the fabrics could lower the radiant heat protective performance. Contextually, it is notable that the transmission of this thermal energy is not much dependent on the surface color of the fabric. Therefore, fabric color has a negligible effect on the radiant heat protective performance of the fabrics.

Additionally, Barker et al. (2006) stated that textile fabrics absorb moisture due to perspiration from a sweating firefighter; this moisture absorbed increases the thermal conductivity of textile fibers, lowering the protective performance of the fabric at a low-intensity radiant heat hazard (Lee & Barker, 1986; Lu et al., 2013b). However, this phenomenon is dependent on the amount of moisture present, especially in the case of single-layered fabrics. It has been concluded that the protective performance could be high if the amount of moisture is more than

15%. This is because the heat capacity of the moisture is high, and it could store the thermal energy inside the fabrics. This situation results in less transmission of the thermal energy toward firefighters' bodies and thus increases the radiant heat protective performance of the fabrics. At high intensity, the moisture could evaporate from the fabrics and may transmit toward the firefighters' bodies and lower the protective performance. Furthermore, as the firefighters' protective fabrics are multilayered, the effect of moisture on the radiant heat protective performance is dependent on the location (inner layer and/or outer layer) of the moisture present in the fabrics and/or intensity of the radiant heat hazard. According to Lawson et al. (2004), the moisture present in the outer and inner layers could enhance and lower the protective performance at a high-intensity radiant heat hazard, respectively. Although this phenomenon is inconclusive for the outer layer at low-intensity radiant heat, it has been observed that the moisture present in the inner layer could enhance the protective performance of the fabrics. Recently, Su, Li, and Song (2017) concluded that the impact of moisture on the protective performance of a fabric could be a combined effect of the amount of moisture in the fabric and duration of the radiant heat hazard at a particular intensity. This study shows that the protective performance decreases with the increase in the duration of the hazards at a particular amount of moisture; the protective performance also decreases with the increase in the moisture content, when the duration of exposure to radiant heat hazard is kept fixed. Regarding the impact of moisture on the protective performance, an interesting phenomenon is also observed by Zhang et al. (2017). They found that the moisture present in the outer layer of a fabric could increase the protective performance under the combined flame and radiant heat hazards. They concluded that the moisture present in the fabric could store the thermal energy, which ultimately helps to increase the protective performance of the fabrics.

4.2.3 Flash Fire Hazard

It has been observed that the flash fire is generally a combination of flame, radiant heat, fire balls, and hot gases (ASTM F 1930:2015; ISO 13506, 2008; Song, 2004). As the flash fire comprises radiant heat and flame, the radiative and convective heat transfer mechanisms from the fire source to the fabric surface could similarly be described as in the above sections. Additionally, fire balls could significantly damage or degrade the surface of the fabrics, which will ultimately affect the radiative and convective heat transfer through the fabrics. Notably, hot gases could blow through the pores of the fabric and move toward the wearers' body. This situation will reduce the thermal insulation capacity of the fabric.

For evaluating the protective performance of firefighters' clothing under flash fire hazard, ISO 13506-1:2017 and ISO 13506-2:2017 or ASTM F 1930:2017 standard method is used (ISO 13506-1, 2017; ISO 13506-2, 2017: ASTM F 1930, 2017). According to ISO 13506-1:2017 standard, an adult (average height: 1830 mm, average chest: 1025 mm, average waist: 850 mm, average hip: 1015 mm), upright, thermally stable (made of flame-resistant materials such as ceramics or glass-reinforced vinyl ester resin), and sensor instrumented [with at least 100 same type heat flux sensors (e.g., epoxy resin sensor, PyroCal, or skin simulant) having a capacity to measure the heat flux intensity over a range of 0–200 kW/m²] clothed manikin housed in a fire chamber is exposed to flash fire (84 kW/m²) for certain duration (Figure 4.6). During and after the flash fire hazard (a total data acquisition time of 60 s for the single-layered and 120 s for the

Figure 4.6

Flash fire hazard manikin tester.

multilayered clothing), the heat flux of each sensor is calculated as per ISO 13506-2:2017 standard (ISO 13506-2:2017). These heat flux values are further used in the HBI equation for calculating the time required to generate second- and third-degree burns on different parts of the manikin. The percentages of burn injury on different parts of the manikin are also calculated. These times or percentages of burn injuries are interpreted as the flash fire protective performance of the clothing.

ISO 13506-1:2017 and ISO 13506-2:2017 or ASTM F 1930:2017 standards are extensively used to evaluate the flash fire protective performance of firefighters' clothing under a high-intensity flash fire hazard (Crown et al., 1998; Camenzind, Dale, & Rossi, 2007; Rossi, Bruggmann, & Stämpfli, 2005; Rossi, Schmid, & Camenzind, 2014; Song, 2007). However, these standard methods involve the testing of whole clothing; therefore, these methods are expensive, cumbersome, and difficult to carry out on a routine basis (Song et al., 2016). The repeatability of these standard test methods could also be substantially low due to the variance in the draping condition of the clothing (Dale, Paskaluk, & Crown, 2016; Saner, 2013). To overcome these problems, Mandal et al. (2017a, 2017b, 2017c) recently developed a method to evaluate the flash fire protective performance of the

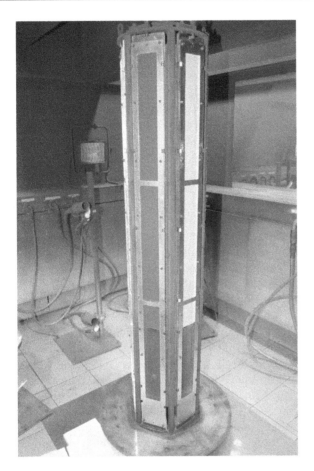

Figure 4.7

Flash fire hazard hexagon tester.

fabrics used in firefighters' clothing (Figure 4.7). This method consists of a hexagonal, fire-resistant, and epoxy resin sensor instrumented panel, which replaced the recommended manikin of the ISO 13506-1:2017 or ASTM F 1930:2017 standard (ASTM F 1930, 2017; ISO 13506-1, 2017). This hexagonal panel can hold the fabrics during and after the flash fire hazard for evaluating the protective performance as per the methods described in ISO 13506-2:2017 or ASTM F 1930:2017 standard (ISO 13506-2, 2017: ASTM F 1930, 2017).

In the 1990s, Behnke et al. (1992), Pawar (1995), and Crown et al. (1998) assessed the flash fire protective performance of clothing, and they identified that the design of the clothing has a significant effect on the protective performance. They found that clothing design such as size, fit, style, as well as amount and number of closures (e.g., zippers, cuff, collars) has a significant impact on the protective performance. In this context, Pawar (1995) mentioned that clothing size is a significant factor depending on the types of fabrics used in the clothing. Researchers concluded that clothing size did not significantly affect the protective performance of para-aramid/polybenzimidazole fabric–based clothing; however, this effect was significant when firefighters were wearing meta-aramid fabric–based clothing.

Song et al. (2004), Song (2007), and Mah and Song (2010a, 2010b) investigated the impact of clothing fit on the flash fire protective performance. They mentioned that an air gap is created between the clothing and firefighters' bodies, depending on the fit of the clothing. As a key factor, this air gap acts as a thermal insulator by storing the thermal energy (dry heat), and that slows down heat transfer to the skin. Notably, the size of the air gap also significantly varies depending on the firefighters' body structure and gender and fabrics used in the clothing. A small air gap in between a firefighter's body and clothing can lead to absorbing little thermal energy within the air gap. In this case, most of the thermal energy transmits toward the firefighter's body and generates burn injuries. Contextually, the heat capacity/storage and heat conductivity of the small air gap is also low and high, respectively; this situation can help to generate quick burns on firefighters' bodies. Furthermore, Song et al. (2004) and Song (2007) concluded that if the size of the air gap becomes large, it cannot properly trap the dead air. This situation causes natural convection in the air gap between a firefighter's body and clothing, which ultimately reduces the protective performance. Altogether, an optimum fitting of the clothing is important to create a right-sized air gap in between the clothing and firefighters' bodies, and this can provide a better protection to firefighters. In this context, Wang, Li, and Li (2015) recently indicated that air gap size could highly depend on the shrinkage of the fabric during the exposure to flash fire. If the fabric shrinks, it will ultimately reduce the size of the air gap and thus could lower the protective performance of the clothing. Contextually, Wang and Li (2016) mentioned that para-aramid/polybenzimidazole fabric-based clothing may not shrink even after repeated exposure to flash fire; however, the meta-aramid fabric-based clothing shrinks after repeated flash fire exposure. This continuous shrinkage contributes to lower the flash fire protective performance of clothing with aging. Recently, Mert et al. (2017) found that the posture of the wearer and fabric properties have a significant contribution toward the size of the air gap. For example, the air gap size diminishes in the sitting posture and that can ultimately increase the conductive heat transfer from the fabric to the wearers' bodies and can lower the protective performance.

Although the design of the clothing has a significant impact on the flash fire protective performance, it is notable that underwear worn below the clothing also plays a significant role on the protective performance. Crown et al. (1993) analyzed the flash fire protective performance of different clothing in combination with different types of underwear to evaluate the protective performance. They identified that the underwear made from inherently fire-resistant aramid fabrics can provide better protection than the underwear made from fire-retardant (FR) finished cotton fabrics. Contextually, Schmid et al. (2016) recommended that underwear worn by firefighters should be made of inherently fire-resistant or FR finished fabrics to avoid any injury during the exposure to flash fire hazard.

Dale et al. (1992) examined the protective performance of clothing related to the features of the flash fire hazard test. They concluded that the percentage of body burns increases with increased intensity of the flash fire. This is because the transmission of thermal energy from the environment toward the wearers' body increases. Recently, Wang et al. (2015) studied the thermal energy transfer from the flash fire source to the naked manikin by applying a three-dimensional transient computational fluid dynamics (CFD) modeling technique. By means of the CFD simulation model, the temperature and velocity fields on the manikin surface and the fire chamber are further used to calculate the average heat flux distribution during the flash fire hazard. It has been found that the average heat flux obtained from the 135 sensors on the manikin and the CFD simulation model is quite similar. Nevertheless, by implementing the 3D transient CFD simulation technique, the heat and mass transfer from a flash fire source (at 84 kW/m²) to the clothed manikin was also studied by Tian and Li (2016). Tian and Li (2016) concluded that the average heat fluxes obtained from the actual clothed sensors on the manikin and the CFD simulation model are quite different, depending on the fabric materials used in the clothing. To effectively calculate the thermal energy transmission from the flash fire to manikin, a new procedure has also been proposed by Rossi et al. (2014). This proposal is currently in consideration for the ISO standardization (ISO DIS 13506.1).

Recently, Mandal et al. (2017a, 2017b) studied and compared the protective performance of fabrics under flash fire, flame, and radiant heat hazards. They found that flash fire protective performance of a fabric is generally higher in comparison to the flame or radiant heat protective performance. It has been observed that the flash fire is mainly a combination of flame, radiant heat, hot gases, and fire balls. As a result, thermal energy did not directly penetrate through the fabric; rather, the thermal energy mainly deflected away from the surface of the fabric. This deflection of thermal energy enhances the flash fire protective performance of the fabric. Notably, the configurations of the source of the flash fire (angle, distance from the fabric) have significant effect on the flash fire protective performance of the fabric. It has also been found that stored energy inside the fabrics is high under the flash fire hazard, and this stored energy significantly affects the protective performance of the fabrics.

4.2.4 Hot Surface Contact Hazard

While crawling on the ground, firefighters' body parts, especially knees and shoulders, may come in contact with many hot solid surfaces (e.g., iron or steel furniture/windows/doors). In this situation, firefighters' clothing gets compressed with the hot surfaces and may generate burns. During the exposure to the hot surface contact, conductive heat energy [$q'(t)$] transfer occurs from the hot surface to the outer side of a fabric surface (surface that is exposed to hot surface) as per Equation 4.4, where Δ_{PF} = temperature difference between the hot surface plate and fabric surface (°C), ΔX_p = thickness of the hot surface plate (cm), k_p = thermal conductivity of the hot surface plate (W/m·K), A = area of contact between hot surface plate and fabric surface (m²), h_{PF} = thermal conductance coefficient between the hot surface plate and fabric surface (W/m/K), ΔX_F = thickness of the fabric (cm), V_A = fabric's air volume (cm³), V_F = fabric's total volume (cm³), k_γ = thermal conductivity of fabric's solid phase (W/m·K), and k_α = thermal conductivity of fabric's gaseous air phase (W/m·K) (Mandal & Song, 2016).

$$q'(t) = \frac{\Delta_{PF1}}{\Delta X_P / (k_P A) + 1 / (h_{PF1} A) + \Delta X_F / \left[\left(\left(1 - \frac{V_A}{V_F}\right) k_\gamma + \frac{V_A}{V_F} k_\alpha \right) A \right]} \qquad (4.4)$$

For evaluating the protective performance of a fabric in contact with hot surface, ISO 12127-1:2015 or ASTM F 1060:2016 standard is used (ASTM F 1060, 2016; ISO 12127-1, 2015). As per ISO 12127-1:2015, a fabric specimen is

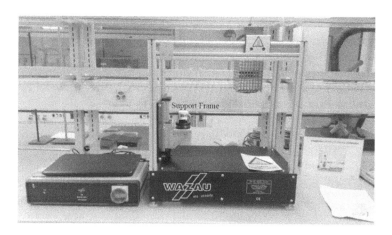

Figure 4.8

Hot surface contact hazard tester. (Courtesy of Ms. Pauline Weisser, DuPont, Switzerland.)

compressed (at 55 and 14 kPa, simulating the compression in the knee and shoulder regions of a firefighter's body, respectively) between a heating cylinder (100–500°C) and an aluminum sensor (sensor is fixed on a polyamide 66 circular plate) (Figure 4.8). This test starts recording the temperature rise of the sensor when the distance between the cylinder and sensor is 10 mm; the temperature recording continues until the temperature of the sensor reaches 24°C above its starting value. The time between the start of timing and the moment when the temperature of the sensor is 24°C above its starting value is called as a Conductive Compressive Heat Resistance (CCHR) rating. This CCHR rating can be used as hot surface contact protective performance of fabrics used in firefighters' clothing. Furthermore, the minimum hot surface contact protective performance required for two different categories of fabrics (Fabrics with Performance Level A1 and Level A2) that are commercially used in firefighters' clothing are also set by ISO 11999-3:2015 and ISO 11613:2012 (ISO 11999-3, 2015; ISO 11613, 2012). These A1 and A2 fabrics should have a CCHR rating of ≥13.5 when tested with a cylinder temperature of 180±5°C and 260±5°C, respectively.

According to Mandal and Song (2016b), conductive heat transfer mainly occurs from the hot surface to the fabric when they are in contact with each other. This conductive heat transfer substantially depends on the surface roughness of the fabrics. If the surface roughness of the fabric is high, it can trap a lot of dead air in the contact area between the hot surface and fabrics. This dead air acts as a thermal insulator; eventually, thermal contact resistance between the hot surface and fabric becomes high. This situation lowers the heat transfer from the hot surface to the fabrics and can increase the hot surface protective performance of the fabrics.

Although the conductive modes of heat transfer play a crucial role on the hot surface protective performance of fabrics, the construction of the fabrics also significantly affects the protective performance. Mandal and Song (2016b) mentioned that a multilayered fabric used in firefighters' clothing should possess a high thickness-based thermal liner to enhance the protective performance of the clothing under hot surface contact. Additionally, they indicated that the moisture barriers present in the fabric are prone to ignite during the contact with a hot surface. This can ultimately lower the hot surface protective performance of the fabric.

In relation to the fabric properties, Mandal et al. (2013) mentioned that fabrics become compressed in between hot surfaces and firefighters' bodies during the hot surface contact hazard. Due to this compression, the gaseous air phases inside the fabrics reduce and solid fiber phases predominate. As the thermal conductivity of a fiber is greater than air, most of the imposed thermal energy (dry heat) on fabrics is absorbed inside the fabrics and/or transmitted toward the firefighters' bodies. This transmitted thermal energy generates burns on firefighters' bodies. In this context, Mandal and Song (2011) stated that the ratio of gaseous air phase to solid fiber phase inside the fabrics varies depending on the compression characteristics of the fabrics. They observed that the ratio of gaseous air phase to solid fiber phase of compressible fabrics is lower than noncompressible fabrics. Consequently, the amount of thermal energy transfer toward wearers is greater in compressible fabrics. As a result, hot surface protective performance is reduced in the case of compressible fabrics.

Rossi and Zimmerli (1994) investigated the impact of moisture on the protective performance of multilayered fabrics under hot surface contact. It was evident that the presence of water in the outer layer (exposed to the hot surface) of the fabric can enhance the thermal conductivity of the fabric; as a result, the hot surface protective performance of the fabric can drop by 50–60%. In this context, it has been found that the multilayered fabrics comprising separate moisture barrier in the inner layer have better hot surface protective performance than the multilayered fabrics comprising laminated moisture barrier on the outer shell fabric. However, these two fabrics may have the same performance when their inner layer is wet. If the inner layer of the fabric is wet, the hot surface protective performance can drop by 10–25% for all types of fabrics. Here, the decrease in the protective performance was prominent at lower temperature; this is because water could store in the fabric layers without significant evaporation, and this stored water could enhance the thermal conductivity or lower the protective performance of the fabrics.

4.2.5 Molten Metal Contact Hazard

In a structural fire incidence, the temperature of the noncombustible solid substances (e.g., steel/iron furniture, door, or window) could increase when exposed to flame, radiant heat, flash fire, and/or hot surface. This situation could increase the thermal expansion and decrease the elasticity of the substances, which helps to make the substances viscous in nature. Eventually, the substances start melting, and they may drip and fix on firefighters' clothing. Thereafter, the conductive thermal energy (namely heat) transfer occurs from the fixed molten metal drop to the fabrics used in firefighters' clothing, depending on the temperature difference and contact resistance between the drops and fabrics (Makinen, Laiho, & Pajunen, 1997; Sharkey, 1997).

For evaluating the protective performance of a fabric in contact with molten metal substances, ISO 9185:2007 or ASTM F 955:2015 standard is used (ASTM F 955, 2015; ISO 9185, 2007) (Figure 4.9). As per ISO 9185:2007 standard, the protective performance of the fabric can be measured in a laboratory by using various molten metals such as aluminum, copper, iron, steel, and cryolite. For this test, a PVC film (300±30 gm/m²) sheet is mounted against an inclined rectangular pin frame of (248±2) mm × (160±2) mm; then a conditioned (at 20±2°C temperature

Figure 4.9

Molten metal contact hazard tester. (Courtesy of Ms. Britta Gortan, Hohenstein Laboratories, Germany.)

4. Protective Performance of Firefighters' Clothing

and 65±5% relative humidity) fabric specimen is mounted over the PVC film. Here, it needs to be ensured that there should not be any crease/wrinkles on the PVC film and fabric specimen, and the length [(260±2) mm × (100±2) mm] of both the PVC film and fabric specimen should be longer than the length of the rectangular pin frame. After mounting the PVC film and fabric specimen, the molten metal at a particular temperature (this temperature can vary between 780°C and 1550°C depending on the used metal) is poured from a pouring ring (50 gm metal is melted in a crucible, and this crucible transferred into the pouring ring) that is placed above the specimen at a certain height (height is 225–300 mm depending on the used metal). For pouring the metal, the pouring ring is rotated in such a way that the crucible turns through at least 130° from the horizontal at a constant rate of 36±2.5°/s (this rate should be 18±2.5°/s in the case of cryolite metal). After 30 s of the completion of the pouring, it is recorded by the test operator whether any molten metal has solidified and adhered to the surface of the test specimen.

Next, the tested fabric specimen is further removed from the pin frame and the damage that occurred on the PVC film is examined. If no damage is observed on the film, each time the quantity of metal in the crucible is increased by 50 gm, and the test is repeated with new fabric specimen and PVC film until the damage is observed. While increasing the quantity of the metal in the crucible, if it reaches the capacity of the crucible, it is concluded that the test is not sufficiently severe to obtain the damage on the PVC film. If the damage on the PVC film occurs within the capacity of the crucible, molten metal test is repeated for the new fabric specimen and PVC film with 10 gm less metal in the crucible than that used in the previous test; this process continues until no damage is visible on the PVC film. Then, the test is repeated with the metal quantity at which no damage occurred; this repetition is continued until four consecutive tests show no damage to the PVC film. Among these four tests, the minimum mass of the metal poured (i.e., actual metal mass in the crucible before the test–residual metal mass in the crucible after the test) to cause the damage on the PVC film is called the "Molten Metal Splash Index." This index can be used as the molten metal contact protective performance of the fabrics.

Benisek and Edmondson (1981) evaluated the protective performance of fabrics (FR wool, FR cotton, novoloid, aramid, glass, and asbestos) used in firefighters' clothing against various molten metal substances such as cast iron, steel, copper, aluminum, zinc, lead, and tin. Results showed that FR wool fabric had the best performance against any molten metal substance, and that fabric properties significantly affected the molten metal protective performance (Mehta & Willerton, 1997). It has been found that a fabric with high thermoplasticity, high thermal conductivity, high air permeability, low softening temperature, and high flammability has a low protective performance. On the other hand, a fabric with high weight and thickness can demonstrate a high molten metal protective performance (Barker & Yener, 1981). In this context, it is evident that a fabric possessing char formation features on its surface can provide better protection than a fabric with no char formation features.

Additionally, a highly smooth fabric cannot trap molten metal on its surface and possess high protective performance. As a consequence, Barker and Yener (1981) concluded that aluminized smooth fabrics, including fabrics made with fire-retardant cotton, rayon, glass, or carbon as base fibers, perform particularly well in deflecting the molten metal, resulting in a higher protective performance.

It was also found that heavier fabrics made from inorganic materials such as ceramic and silica fibers may have equivalent performance levels to aluminized smooth fabrics. Recently, Bruck Textiles have developed a blended fabric of merino wool and Lenzing FR®, which is called PR97®. This fabric has superior protective performance against a variety of molten metals including aluminum, cryolite, iron, steel, copper, magnesium, and nickel.

4.2.6 Hot Liquid Hazard

It was found that water used by firefighters becomes hot, and this hot water may cause burn injuries. It has also been observed that pipelines of hot water may burst in a structural fire hazard (especially those in the chemical, oil, and gas industries), and the hot water released can pose a great threat to firefighters (Mandal et al., 2014: Song et al., 2016). Along with the water, firefighters could be exposed to various other hot liquids such as drilling muds and cooking oils. When firefighters come in contact with the hot liquids, an interface is created between the surfaces of the liquids and fabrics used in the clothing. At this interface, a fixed contact angle (i.e., the tangent of the liquid surface makes with the fabrics) is formed at a particular temperature and pressure, depending on the surface tension of the hot liquids and the surface frictional coefficient of the fabric. If the contact angle is low, especially when the surface tension of the hot liquids is low and the surface frictional coefficient of the fabric is high, the hot liquids absorption level of the fabric becomes high. As the fabric is generally a porous medium, the absorbed hot liquids may also diffuse or penetrate through the fabrics (mass transfer) and come in contact with the firefighters' bodies. If firefighters get exposed to the high pressurized hot liquids, the liquids may heat and compress the fabrics with firefighters' skin. In this situation, conductive heat transfer occurs from the hot fabrics to the firefighters' skin especially in the case of nonporous fabric.

For evaluating the hot liquids protective performance of fabrics, ASTM F 2701:2008 standard test method is mainly used (ASTM F 2701, 2008). As per this standard, a fabric specimen (placed on a sensor board) is exposed to 1 L of hot liquids from a funnel. The amount of thermal energy (dry heat and hot liquid mass) transmitted through the fabric specimen during and after the hot liquid hazard is measured using the two instrumented copper slug sensors. The amount of transmitted thermal energy to cause a second-degree skin burn injury is measured using the Stoll second-degree burn criterion. This thermal energy for the second-degree burn is interpreted as the hot liquid splash protective performance of the fabric.

Contextually, it is notable that the hot liquid pouring procedure in this ASTM F 2701:2008 standard test is quite awkward, and that could affect the hot liquid pouring rate or repeatability of the test. This pouring procedure may also occasionally cause injuries to the test operator. Considering this, a group of researchers from the PCERF have modified the hot liquid pouring procedure of this test method. According to this modified procedure, hot liquid is poured over the fabric specimen with a low-pressure hot liquid jet (Jalbani et al., 2012). The funnel cone in the ASTM F 2701:2008 standard is replaced with a small pipe directly fed by a temperature-controlled circulating hot liquid bath via a small pump and through a hose and valve system. This modified device can allow a consistent application of a given quantity of liquid at a controlled temperature and flow rate. In this modified device, the liquid temperature, flow rate, and pressure can be controlled as per the requirements of the experimenters. Later, Mandal et al. (2013) modified the sensor board of this device and replaced the two copper slug sensors by two skin simulant sensors in an attempt to more accurately simulate human skin for predicting burn injuries using an HBI equation-based software package. Finally, in the ASTM F 2701:2008 test method, the fabric specimen can be exposed to hot liquid at one configuration (45°) only; however, there is a need to conduct tests at different fabric positions (horizontal, vertical, etc.) to accurately simulate the exposure of fabrics to hot liquid. Considering this, Mandal, Song, and Gholamreza (2016) recently modified this device by horizontally positioning the sensor board with skin simulant sensors.

Furthermore, ASTM F 2701:2008 is focused on the hot liquid splash only. However, on-duty firefighters have to kneel and crawl on the floor while extinguishing fires and rescuing victims. While performing these activities, their clothing is compressed with the floor specifically at the knees, elbows, and lower legs. Their clothing may also immerse in hot liquid. This hot liquid immersion with compression can cause skin burns at firefighters' arms, hands, legs, and feet (Barker, 2005; Lawson, 1996; Lawson et al., 2000). Burn injury statistics indicated that nearly 38% of burn injuries occurred on firefighters' arms/hands and legs/feet during the period 2007–2011 in the United States (Karter, 2013). Considering this, the PCERF research team has developed a test method for evaluating the protective performance of fabrics under exposure to hot water immersion with compression (Mandal et al., 2016; Mandal & Song, 2016a). Practically, this hot water immersion and compression test is carried out in a hot water bath. At the bottom center of the hot water bath, a metal platform with perforated top surface was positioned and the water level was set at 6 cm above from the perforated top surface. Using a temperature control device, the temperature of this water was maintained at 75°C, 85°C, or 95°C. Next, a fabric specimen was attached with a rubber band to the skin simulant sensor mounted on a cylindrical weight. Using a pneumatic device, this specimen-covered sensor was immersed into the hot water bath until the whole assembly of specimen and sensor rested flat on the center of the perforated surface. To compress the specimen between the sensor and perforated surface, pressure was pneumatically applied and controlled at 14 kPa (~2.0 psi), 28 kPa (~4.0 psi), or 56 kPa (~8.0 psi). For a period of 120 s, thermal energy transmitted through the compressed specimen was processed by the sensor. From the processed thermal energy, time required to generate a second-degree skin burn was calculated by the customized software, programmed according to the HBI equation. This burn time is used as the hot water immersion and compression protective performance of the fabrics.

Recently, a full-scale instrumented manikin test has also developed at PCERF for evaluating the hot water protective performance of whole clothing (Lu, Song, & Li, 2013a; Mandal et al., 2014). In this test, the protective performance of the clothing is evaluated under hot water spray hazard. For this, a fiber glass and resin-based 40 size sensor instrumented (comprising 110 uniformly distributed skin simulant sensors), upright manikin is positioned in a water chamber and surrounded by four groups of automatically controlled three nozzle cylinder spray jets. Then, the selected clothing is donned on the manikin. By using the spray jets, hot water (85 °C) is sprayed on the clothed manikin at a pressure of 250 kPa for 10 s. During and after the exposure of hot water spray, the temperature rise in the sensors is measured and used to calculate the thermal energy transmitted through the clothing. This temperature rise is also used to calculate the amount or percentage of skin surface area affected by second-degree scalds/burns. These predicted amounts of second-degree scalds/burns of a particular specimen are used as the hot water spray protective performance value of the clothing.

Ackerman and Song (2011) assessed the protective performance of layered fabrics in hot liquid (water, drilling mud, canola oil) splash hazards. They indicated that the hot liquid splash protective performance depends on the fabric properties (e.g., weight, thickness, air permeability, finishing) and heat capacities of the liquids. They found that air permeability of a fabric is negatively associated with the protective performance and the fabrics with encapsulated fiber finishing provided the best protection from all types of hot liquid. The protective performance of a fabric was the lowest for hot water; this was thought to be because the heat capacity of hot water is higher than the heat capacity of drilling mud and canola oil. In this context, Gholamreza and Song (2013) found that a multilayered fabric with an impermeable outer layer can have high performance than a multilayered fabric with a permeable outer layer.

Recently, Lu et al. (2014) investigated the protective performance of various single-layered fabrics under hot liquids splash hazards. Depending on the surface tension between a liquid molecule and fabric, the flow pattern of a liquid on the fabric varies; eventually, the wet percentage of the exposed fabric varies. Generally, a hot liquid or highly rough fabric surface can lower the surface tension; in turn, this increases the wettability of the fabric. In the case of a highly wet fabric, the liquid can penetrate through the fabric due to wicking and cause burns on wearers' bodies. Lu et al. (2014) also mentioned that the exposed liquid can store in or transmit through the fabric depending on fabric properties (thickness, density, air permeability). If a fabric stores more and transmits less liquid, it will show high hot liquid protective performance. It was evident that an addition of thermal liner with a single-layered shell fabric can help to store more and transmit less liquid; this can enhance the protective performance of the shell fabric (Lu et al., 2013a). In fact, incorporation of more number of permeable shell fabric layers in a fabric can increase its liquid storage capacity; this situation can ultimately help to increase the hot liquid splash protective performance of the fabric (Zhang et al., 2015a; Zhang et al., 2015b). Mandal et al. (2013) also indicated that the pressure exerted by the hot liquid splash can cause some amount of conductive heat transfer through the fabrics resulting in burn injuries to wearers. Conductive heat transfer could be more prominent in the case of horizontally aligned fabric than the inclined fabric because the horizontally aligned fabric could hold the hot water on its surface. Eventually, the hot liquid protective performance of the inclined fabric is more than the horizontally aligned fabric.

To evaluate the hot liquid protective performance of fabrics, Mandal and Song (2016a), Mandal et al. (2016), and Mandal (2016) tested a set of multilayered fabrics under hot water immersion with compression hazard. It has been found that both conductive heat and hot water mass transfer could occur through the fabric toward the wearers' bodies under this hazard; this situation can decrease the protective performance of the fabric. Nevertheless, the heat and mass transfer through a fabric mainly depends on its thickness and air permeability. If the thickness is high and air permeability is low, it will resist the transmission of heat and mass through the fabric and can increase the protective performance of the fabric. Notably, the heat and mass transfer could increase with the increase in the water temperature and compression pressure, which could lower the protective performance of the fabric. Contextually, Mandal (2016) further identified that fabric thickness is less significant in comparison to the fabric evaporative resistance and air permeability when analyzing the protective performance at different water temperatures and compression pressures. Actually, the evaporative resistance and air permeability are the most significant properties to affect the performance because the mass transfer through the fabrics is the most prominent mechanism under hot water immersion and compression hazards. Finally, Mandal and Song (2016a) recommended using a multilayered fabric with moisture barrier in its outer layer; this type of fabric could effectively prevent the mass transfer through the fabric layers and that could increase the protective performance of the fabrics by resulting in fewer burn injuries on firefighters' bodies (Mandal et al., 2016).

Lu et al. (2013a) and Mandal et al. (2014) further showed that the hot water spray protective performance of air-permeable and thin fabric-based clothing is lower than the air-impermeable thick fabric-based clothing under the exposure of hot water spray. Additionally, different sizes and design (pocket, closures, vents) features of the clothing have a distinct effect on the protective performance. This is because different garment sizes can result in variation in the air gap (microclimate region) between the garment and wearers' body. This air gap controls the heat and mass (hot water) transfer through the garment toward wearers' bodies and could be critical to provide effective protection to wearers. As identified, the dimension of the air gap could be reduced depending on the pressure of the hot water spray, and this situation could lower the hot water protective performance of the clothing. Depending on the temperature and pressure of the hot water spray, the thermal conductivity and heat capacity may also change, and that could lower the protective performance of the clothing. Some essential/functional design features (e.g., reflective tape for enhancing visibility at night, closures near the collar and cuff of a garment) of clothing may also enhance its protective performance by trapping some insulative dead air inside the clothing (Mandal et al., 2014).

4.2.7 Steam Hazard

It was established that water used to extinguish fires may become hot over time and come in contact with firefighters. It was also observed that pipelines of pressurized steam may burst in a structural fire incidence; consequently, firefighters could also be exposed to the high pressurized steam while working in the fire incidence (Song et al., 2016). In this situation, steam may penetrate through the fabrics used in the clothing, depending on the diffusion resistance of the fabric. During this mass transfer (steam diffusion or penetration) process, some amount of steam may also condense into hot water, and this hot water could come in contact with firefighters' bodies. If the pressure of the steam is high, it will compress the fabric with firefighters' skin. In this situation, conductive heat transfer is prevalent from the hot fabric to the firefighters' bodies (Mandal et al., 2013; Mandal, 2016).

For evaluating the steam protective performance of fabrics, many researchers have developed customized devices since the end of the twentieth century to the beginning of the twenty-first century (Desruelle and Schmid, 2004; Le et al., 1995; Rossi et al., 2004). However, these devices were not suitable to simulate the steam pressure of more than 400 kPa, which is limited in terms of steam pressure faced by a structural firefighter. Additionally, these devices were generally instrumented with several sensors, which ultimately made these devices expensive. Considering these shortcomings, the PCERF research team has developed a state-of-the-art

Figure 4.10

Steam hazard tester. (Courtesy of Mr. Mark Ackerman, MYAC-Eng., Canada.)

device for evaluating the steam protective performance of fabrics (Ackerman et al., 2012) (Figure 4.10). In this device, a fabric specimen (150 mm diameter) is placed on the Teflon plated specimen holder that is equipped with a skin simulant sensor. The saturated and superheated steam is generated by a 3 kW boiler at a temperature of 150°C, and this steam is impinged on the specimen (from 50 to 150 mm above the fabric specimen through a nozzle having a diameter of 4.6 mm) at a pressure of 69–620 kPa for 10 s. During and after the steam hazard, the heat flux through the fabric specimen is recorded by the skin simulant sensor and the time required to generate a skin burn is calculated by a customized HBI program. Mandal et al. (2013, 2014) recently used this steam testing device with varying exposure time (according to the number of layer of the tested fabric specimen) to predict a second-degree burn injury from the sensor. Although the normal steam exposure time for this device is set at 10 s by Ackerman et al. (2012), the steam exposure time used for the thickest triple-layered fabric specimen was 30 s in the study carried out by Mandal et al. (2013, 2014). Thus, it can be concluded that the device developed by Ackerman et al. (2012) can be used for longer duration steam exposures with high pressure. Recently, this steam hazard tester of Ackerman et al. (2012) is combined with the hot liquid

4. Protective Performance of Firefighters' Clothing

hazard tester of Jalbani et al. (2012) and a standardized test method of combined steam and hot liquid hazards is described in CGSB 155.20:2017, i.e., workwear for protection against hydrocarbon flash fire and optionally steam and hot fluids (Figure 4.11).

For evaluating the steam protective performance of clothing, Desruelle and Schmid (2004) developed a thermal manikin in the early twenty-first century. This thermal manikin was divided into nine copper sheet areas and included a water cooling system to control the temperature for each area separately between 20°C and 40°C. The key features of this manikin are that it had limited internal heat storage and natural convection. The input water temperature was controlled between 20°C and 40°C and the water flow to each zone was regulated between 0.06 and 1 L/min. For the testing, clothing was put on the manikin and the clothed manikin was placed in a steam atmosphere. The steam atmosphere was created in a 7 m³ climatic chamber (that allowed the generation of a fully saturated atmosphere at 80°C) by an air conditioner (heating system and humidifier) working in a closed circuit. The humidity of the climatic chamber was increased step by step to obtain a saturated atmosphere. For each step,

Figure 4.11

A combined steam and hot liquid hazards tester. (Courtesy of Mr. Mark Ackerman, MYAC-Eng., Canada.)

the mean temperature on the surface of each area of the manikin was regulated at 33°C. Temperature and water flow were measured for each step over 7 min. From the temperature and water flow, the local and total heat fluxes were then calculated for each step, and the heat fluxes at saturation were calculated by extrapolation. These heat flux values were interpreted as the steam protective performance of the clothing. Notably, according to Sati et al. (2008), maintaining a constant manikin surface temperature of 33°C during the steam hazard test may not correspond in real-life scenarios. As the skin surface temperature generally rises during the steam hazard, it could reduce the heat transfer rates in response to reduced temperature differences in between firefighters' skin and the outer surface of the fabric.

Le et al. (1995) found that convective heat and mass (steam or condensed hot water) transfer mainly occurs through the fabric under steam hazard, and that could lower the protective performance of the fabrics. During the hazard, some amount of the steam condensation may occur within the fabric, and that could also lower the steam protective performance of the fabrics. Nevertheless, the transfer of the heat and water mass as well as the condensation of steam mainly depend on the fabric properties (Keiser et al., 2010; Keiser & Rossi, 2008; Rossi et al., 2004). For example, the heat and mass transfer is lower through a thick and water vapor–impermeable fabric, and this situation could enhance the steam protective performance of the fabric. In this context, Desruelle and Schmid (2004) found that a complex phenomenon of condensation, diffusion, and absorption of water occurs inside the water vapor–permeable fabric during the steam hazard, and this situation lowers the steam protective performance of the fabrics (Sati et al., 2008). Additionally, the condensation of steam could be high in a fabric layer with high moisture regain, and this condensed steam could increase the temperature of the other fabric layers in a multilayered fabric. This situation also lowers the protective performance of the fabric. Recently, Mandal et al. (2013, 2014) indicated that fabrics may come in direct contact with the firefighters' skin under exposure to high-pressure steam. In this situation, an air-impermeable fabric may not allow transferring the mass through it; however, it may transfer the conductive heat toward the wearers' skins and that could cause significant burn injuries on wearers' bodies.

Desruelle and Schmid (2004) found that the steam protective performance of whole clothing is highly dependent on the properties of the fabrics used in the clothing. For example, air-impermeable fabric–based clothing could provide a better protection than an air-permeable fabric–based clothing. As indicated before, depending on the size of the clothing, an optimum air gap could be created in between firefighters' bodies and the clothing. If this air gap provides proper thermal insulation, the steam protective performance of firefighters' clothing could be increased. Nevertheless, designing a garment with optimum air gap could be difficult; this may not always give a reliable protection to firefighters against a steam hazard.

4.3 Evaluation and Assessment of the Protective Performance of Fabrics and Clothing under Nonthermal Hazards

Along with the thermal hazards, firefighters also have the risk of exposing to various nonthermal hazards such as physical (e.g., sharp objects, falling debris) and chemical (e.g., toxic liquids and air vapors). Additionally, firefighters could be exposed to biological (e.g., blood-borne pathogens) and radiological (e.g., aerosols) hazards, especially when they fight a fire hazard generated through a terrorist incident. Considering this, NFPA 1971:2018 standard recommends that fabrics, seams, and/or closures (e.g., zippers, cuff, collars) used in developing structural and proximity firefighters' clothing should also provide the protection from various nonthermal hazards recommended by NFPA 1994:2018 standard for their Class 2 clothing (NFPA 1971, 2018; NFPA 1994, 2018). By following the recommended standard for Class 2 clothing, firefighters' clothing can provide limited protection to them at terrorist incidents that may involve vapor or liquid chemical hazards having the concentrations at or above immediately dangerous to life and health. Based on NFPA 1994:2018 guidelines for Class 2 clothing, the standard test methods for evaluating the protective performance of fabrics used in the clothing under various nonthermal hazards are shown in Table 4.2. In this context, it is notable that NFPA recently recommended evaluating and assessing the protective performance of firefighters' clothing under these nonthermal hazards. Therefore, limited research exactly focuses on the assessment of the protective performance of firefighters' clothing under nonthermal hazards. Nevertheless, many researchers focused on the assessment of other types of protective clothing (e.g., military protective clothing, industrial workers' chemical protective clothing) under these nonthermal hazards; the findings from their study could be partially or fully applicable in the case of firefighters' protective clothing. In the following section, various test methods for evaluating and assessing protective performance under physical, biological, and chemical/radiological hazards are briefly described.

4.3.1 Physical Hazard

Under the physical hazards (e.g., sharp objects, falling debris), a lot of force acts on the fabrics used in firefighters' clothing. Due to this force, the fabrics could break and firefighters can come in direct contact with the hazards. Eventually, the fabric should hold its strength at contact with the physical hazard.

To evaluate the physical hazard protective performance, the burst strength, puncture propagation tear resistance, and cold temperature performance of the fabrics used in firefighters' clothing should be evaluated according to the ASTM D 751:2011, ASTM D 2582:2016, and ASTM D 747:2010 standards, respectively

Table 4.2 Standardized or Customized Test Methods for Evaluating the Protective Performance under Nonthermal Hazards

Nonthermal Hazards	Tests	Test Standards	Applicable for
Physical	Burst Strength	ASTM D 751:2011	Fabrics
	Puncture Propagation Tear Resistance	ASTM D 2582:2016	Fabrics
	Cold Temperature Performance	ASTM D 747:2010	Fabrics
	Breaking Strength	ASTM D 751:2011	Seam, Closures Assembly
Biological	Viral Penetration Resistance	ASTM F 1671:2013	Fabrics, Seams
Chemical/Radiological	Chemical Permeation Resistance	NFPA 1994:2018 (Section 8.7)	Fabrics, Seam

(ASTM D 751, 2011; ASTM D 2582, 2016; ASTM D 747, 2010). Additionally, the seam and closure breaking strength should be evaluated using the ASTM D 751:2011 standard (ASTM D 751, 2011).

To evaluate the bursting strength, according to ASTM D 751:2011 standard, a fabric specimen is securely held by a ring-clamp mechanism (the internal diameter of the ring is 44.4±0.05 mm) (ASTM D 751, 2011). Then, the center of the specimen is pressed against a polished steel ball (having a diameter of 25.40±0.050 mm) until a burst is produced on the specimen. The amount of load, in Newtons (N), required for the bursting of the specimen is indicated as a bursting strength of the specimen. As per NFPA 1994:2018 standard, the bursting strength of the fabrics used in firefighters' clothing should be ≥156 N (NFPA 1994, 2018).

For evaluating the puncture propagation tear resistance of the fabric using ASTM D 2582:2016 standard, a weighted (W in kg) sharp probe (0.4 mm diameter) is released from a certain height (H in mm) to strike a 200 mm long fabric specimen (warp or weft direction) (ASTM D 2582, 2016). During this strike, the length of the tear (L in mm) on the fabric specimen is calculated to the nearest 0.1 mm. Then, by using Equation 4.5, the puncture propagation tear resistance (F) is calculated in N. As per NFPA 1994:2018 standard, the puncture propagation tear resistance of the fabrics used in firefighters' clothing should be ≥31 N (NFPA 1994, 2018). According to the ISO 11613:2012 standard, the tear strength (measured in accordance with the Method B of ISO 4674:2003 standard) of the outer layer of the firefighters' protective fabrics with performance level A1 and A2 should be ≥25 and ≥40 N, respectively (ISO 11613, 2012; ISO 4674, 2003).

$$F = ([(W \times H)/L] + W) \times 9.8065 \qquad (4.5)$$

Furthermore, as per ASTM D 747:2010 standard, cold temperature performance of a fabric specimen is measured by calculating its bending moment at −25°C (ASTM D 747, 2010). In this case, the fabric specimen is bent to a 60° angular deflection by means of a cantilever beam. The amount of load required as a function of the deflection is measured from a fixed scale attached with the load pendulum system of the cantilever beam, and this is called the load scale reading. The bending moment weight at the load scale reading of 100 is also calculated. Then, by using Equation 4.6, the bending moment of the fabric is calculated in N m. As per NFPA 1994:2018 standard, the cold temperature performance of the fabrics used in firefighters' clothing should be ≤0.057 N m (NFPA 1994, 2018).

$$\text{Bending Moment} = \frac{\text{Load Scale Reading} \times \text{Moment Weight}}{100} \qquad (4.6)$$

For evaluating the breaking strength of the seam or closure assembly, a 152×102 mm seam or closure specimen is clamped between the two jaws. Then, both the jaws are pulled in the opposite direction at a speed of 5±0.2 mm/s. The amount of load required to break the specimen is measured in N and called the breaking strength of the seam or closure assemble. As per NFPA 1994:2018 standard, the breaking strength of the seam and closure assembly used in firefighters' clothing should be ≥34N/25 mm (NFPA 1994, 2018). According to the ISO 11613:2012 standard, the breaking strength (measured in accordance with

ISO 13934-1:2013 standard) of the outer layer of the firefighters' protective fabrics with performance level A1 and A2 should be ≥450 and ≥800 N, respectively (ISO 11613, 2012; ISO 13934-1, 2013).

It has been found that the strength and tear resistance of the protective fabrics mainly depend on various factors, namely the fiber content, yarn characteristics, and constructional feature of the fabrics (Hossain, Datta, & Rahman, 2016; Kabir & Ferdous, 2012; Schiefer, Taft, & Porter, 1936; Zweben, 1978). For example, the strength of the aramid fiber is high as this fiber contains benzene ring within its molecule, which can eventually help to increase the strength and tear resistance of the fabrics. Highly twisted yarns can also increase the strength of the fabric made by these yarns. Additionally, if the number of yarns (warp and/or weft) and/or diameter of the yarn in a fabric are high, it will help to increase the strength of the fabric. This strength could ultimately help to provide better protection to wearers from physical hazards.

Contextually, Hristian et al. (2014) mentioned that the breaking strength of the fabric is important to manufacture a firefighter's protective clothing. If the breaking strength of a fabric is high, the strength of the clothing made from this fabric is high, and that could provide better protection to firefighters from physical hazards. Additionally, the tensile strength and the bending moment of the fabric have significant impact on the seam strength (Chen, Han, & Bai, 2016). If the strength and bending moment of the fabric increase, the strength of the seam becomes high. This situation also contributes to increase the strength of the protective clothing. In this context, it is notable to mention that the breaking strength of the fabric or seam could deteriorate over the exposure to various thermal hazards for certain duration (Arrieta et al., 2010). Due to this reduction in the breaking strength of the fabric, physical hazards protective performance of the firefighters' clothing could decrease and that can produce more injuries on firefighters' bodies.

4.3.2 Biological Hazard

Under the biological hazard, viruses may penetrate through the fabrics and come in contact with firefighters' bodies. In this situation, it is necessary that fabrics should have nonbreathable sealed structure to prevent any penetration of viruses through the fabrics.

For evaluating the biological hazard protective performance of fabrics or seams, ASTM F 1671:2013 standard is used (ASTM F 1671, 2013). According to this standard, a fabric or seam specimen of minimum 70 m² is prepared and conditioned at 21±5°C temperature and a relative humidity of 30% to 80%. The edges of the test specimen are sealed to prevent wicking of the biological agents through them. Then, the specimen is sterilized (if required) and placed horizontally in a sterilized penetration test cell (that is sterilized at 121–123°C temperature and 207–221 kPa pressure for 15 min in an autoclave and cooled down to room temperature). In this cell, the upper side of the specimen has a reservoir and the lower side has a viewing glass to see the back side of the specimen. Toward the lower side of the specimen, a draining valve is attached with the test cell.

To start the experiment, the test cell is vertically mounted. Then, a 60 mL liquid biological hazard [Phi-X174 Bacteriophage containing a total of 900–1200 plaque-forming unit (PFU)] is poured in the reservoir using a sterile syringe or funnel through a port attached with the reservoir. If the liquid appears to penetrate through the specimen and is visible at the back side of the specimen anytime during the pouring, the experiment is immediately terminated. If the liquid did not penetrate through the specimen during the pouring, the specimen with the liquid-filled reservoir is observed for 5 min. Afterward, the pressurized air is started to pass through the port and raised the air pressure to 13.8 kPa at a rate no faster than 3.5 kPa/s. The pressurized air at 13.8±1.38 kPa is passed through the port for 1 min and the back side of the specimen is monitored for the appearance of any liquid. The supply of the pressurized air is then turned off and vent of the cell is checked. If no liquid is observed in the vent at this point, the specimen is observed for an additional 54 min. After 54 min, the drain valve of penetration cell is opened to drain the liquid biological hazards of the reservoir. Then, the back side of the specimen is washed with 5.0 mL of sterile nutrient broth with 0.01% surfactant. The washed assay fluid is processed at a certain temperature by bacteria culture for 6–18 h. Then, the number of PFUs in the processed assay liquid is calculated to interpret the biological hazards protective performance of the fabrics or seams. As per NFPA 1994:2018 standard, PFU of the fabrics or seams used in firefighters' clothing should be zero (NFPA 1994, 2018).

Seed et al. (2008) mentioned that fabric properties have a significant effect on the biological protective performance of the fabrics. If the fabrics are porous and air permeable, it may contribute to transfer the biological hazards through it and this situation may lower the protective performance. Therefore, low porosity and/or impermeable fabrics are recommended to use for the clothing to improve the biological hazards protective performance. In this context, it is notable that porosity of a fabric can be increased by changing the weave design and/or by incorporating more number of yarns per area of the fabric during weaving (Elnashar, 2005). Although an impermeable fabric can be easily manufactured by applying the coating on a porous fabric, it is not recommended to use the impermeable fabrics–based clothing. This is because the thermoregulation performance of the impermeable fabrics is low (Wen, 2014). As a result, this type of fabric may not effectively transfer the metabolic heat and sweat vapor from firefighters' bodies and can cause a lot of physiological burden to firefighters.

4.3.3 Chemical/Radiological Hazard

Under the chemical hazard, harmful liquid mass and/or vapor particle may permeate or diffuse through the fabrics and come in contact with the firefighters' bodies. In this situation, it is necessary that the fabrics should be impermeable to prevent any permeation or diffusion of particles through the fabrics.

To evaluate the chemical/radiological hazard protective performance of fabrics or seam, NFPA 1994:2018 standard is followed (NFPA 1994, 2018) (Figure 4.12). According to this standard, a fabric specimen is placed in a permeation test cell; in this test cell, the upper side of the specimen is called the challenge side and the down side of the specimen is called the collection side. Thereafter, this test cell is conditioned for at least 24 h in an environmental chamber of 32±1°C temperature and 80±5% relative humidity. After conditioning, the test cell is mounted horizontally and connected to the air delivery system in another environmental chamber of the same temperature and relative humidity. The air is delivered (at a 32±1°C temperature and 80±5% relative humidity) toward the collection side of the test cell at a rate of 1±0.1 L/min and no air should be delivered across the challenge side of the test cell. After the air flow for at least 15 min, the test is started.

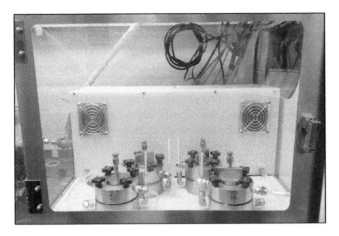

Figure 4.12

Chemical/radiological hazards tester. (Courtesy of Dr. Bryan Ormond, Textile Protection and Comfort Center, North Carolina State University, USA.)

The initiation of the test occurs when the liquid (dimethyl sulfate, tetrachloroethylene, toluene, distilled mustard, soman), gas (anhydrous ammonia, chlorine), or vapor (acrolein, acrylonitrile, diethylamine, ethyl acetate) challenge chemical is introduced into the challenge side of the test cell. In the case of liquid challenge chemicals (concentration of 10 g/m^2), 8 droplets (1 µL volume of each droplets) of a chemical are evenly applied around the perimeter of the specimen and the remaining droplets are placed at the center of the specimen (for seams the droplets at the center should be placed along the seam juncture); this liquid chemical exposed specimen is kept for 60 min. If a gas or vapor chemical (350+35/-0 ppm) challenge is used, it is exposed to the specimen for 60+1.0/-0.0 min at 32±1°C. In the collection side of the test cell, the challenged chemical permeated in the effluent air stream is collected, measured, and analyzed using either discrete or cumulative methods for the first 15+1.0/-0 min interval and overall for 60+1.0/-0 min. The average cumulative permeation (µg/m^2) for the first 15 min interval and overall for the 60 min exposure are used to determine the pass or fail protective performance of the fabric. As per NFPA 1994:2018 standard, most of the permeated chemicals (except distilled mustard and soman) for a 15 min interval and 60 min exposure should be ≤2.0 and ≤6.0 µg/m^2, respectively, to effectively pass a fabric for the use in firefighters' clothing (NFPA 1994, 2018). For distilled mustard, it should be ≤1.33 and ≤4.0 µg/m^2, respectively; it should be ≤0.43 and ≤1.25 µg/m^2, respectively, for the soman liquid. Contextually, ISO 11613:2012 standard allowed no penetration of the chemicals for the firefighters' protective fabrics with performance level A1 and A2, when measured in accordance with the ISO 6530:2005 and ISO 13994:2005 standards, respectively (ISO 11613, 2012; ISO 6530, 2005; ISO 13994, 2005).

Similar to the biological hazards protective performance, the chemical/radiological hazards protective performance is dependent on the porosity and permeability of the fabrics (Duncan, McLellan, & Dickson, 2011; Endrusick, Gonzalez, & Gonzalez, 2005; Wen, 2014; York & Grey, 1986). In general, the clothing that needs to provide protection from chemical/radiological hazards is commonly engineered with the impermeable fabrics. However, as indicated before, this impermeable fabric–based clothing could not effectively transfer the metabolic heat and sweat vapor from firefighters' bodies to the surrounding environment (Wen, 2014). Sometimes, a chemical absorbent fabric layer is also used along with the fabrics to manufacture the protective clothing (Duncan et al., 2011; Endrusick et al., 2005). This absorbent fabric layer can absorb the exposed chemical/radiological hazards and may not cause its transmission toward wearers' skin. This situation ultimately helps to increase the chemical/radiological protective performance of fabrics and clothing. Nevertheless, the absorption of chemicals in the fabric layers may restrict the mobility of the wearers by making their clothing bulky, thick, and heavy. This situation may impose physiological burden on wearers' bodies.

4.4 Key Issues Related to the Protective Performance of Firefighters' Clothing

Based on the above discussion, many standardized and customized methods are available for evaluating the protective performance of firefighters' whole clothing or fabrics used in the clothing under various thermal and nonthermal hazards. However, there are still some shortcomings that need to be addressed, so that protective performance can be evaluated effectively and accurately.

In all the standardized and customized test methods, the intensity, duration, and types of hazard are fixed. For example, the maximum intensity for flame and radiant heat hazards is ~80 kW/m^2 (ISO 9151, 2016; ISO 6942, 2015); the maximum duration for flash fire and chemical hazards are 4–8 s and 60 min, respectively (ISO 13506-1, 2017; NFPA 1994, 2018); in the case of molten metal contact and chemical hazard tests, only five to six types of metals and chemicals are used (ISO 9185, 2007; NFPA 1994, 2018). Although these standardized and customized tests are suitable for the characterization of firefighters' clothing, it is notable that the intensity, duration, and types of hazard faced by firefighters in actual firefighting scenario could be different. For example, the maximum intensity of the flash fire faced by a firefighter could be ~130 kW/m^2 (Torvi & Hadjisophocleous, 1999); in a fire incidence of a terror attack, firefighters could be exposed to different types of chemicals or biological hazards (Barker et al., 2010). Therefore, the protective performance evaluated based on the standardized and/or customized methods may not realistically provide the protection to firefighters in an actual fire scenario. Thus, it is necessary to properly detect and simulate the hazardous conditions (i.e., intensity, duration, and types of hazard) and keep evolving the test methods for evaluating the protective performance of fabrics and clothing. Another approach could be to investigate the possibilities of extrapolating and/or interpolating of the protective performance values obtained from the standardized hazards tests to address the real hazardous conditions. Furthermore, the sensors used for evaluating the protective performance of fabrics under thermal hazards are mainly suitable for short duration exposures (Torvi & Hadjisophocleous, 1999). Additionally, these sensors cannot accurately measure the steam burns because these sensors are mainly simulated to calculate the burn based on the temperature rise of the epidermis layer of the human skin. However, during the steam hazard, steam penetrates through the skin and may generate burns on the dermis layer of the human skin. The existing sensors may not effectively predict the burns or temperature rise of the dermis layer. Thus, advanced technology (e.g., optical high-power laser, fiber optic, nanotube, radiometry, black body) should be used to develop a powerful dry and wet heat flux sensor that can accurately measure the protective performance under long duration thermal hazards (Ding et al., 2004; Ding, Yamazaki, & Shiratori, 2004; Liu et al., 2004; Wang et al., 2002).

Additionally, during the exposure to thermal and nonthermal hazards, a significant amount of thermal energy (heat and mass) and hazardous substances may store inside the fabrics and clothing (Duncan et al., 2011; Endrusick et al., 2005; Song et al., 2011; Song, Cao, Gholamreza, 2011; Wen, 2014). The stored thermal energy and/or hazardous substances may provide protection to firefighters for certain duration. However, this stored thermal energy or hazardous substances could dissipate over time and/or due to compression of the fabrics/clothing toward firefighters' bodies and may generate injuries. Recently, ASTM F 2731:2011 standard has been developed to analyze the thermal energy stored inside the fabrics under low-intensity radiant heat hazard (ASTM F 2731, 2011). Nevertheless, it is necessary to develop more test methods for evaluating the protective performance considering the stored energy under other thermal hazards like flame, flash fire, and steam. Additionally, a lot of air gap can develop between the clothing and firefighters' bodies, and moisture can accumulate in the air gap or inside the clothing due to sweating of firefighters (Lawson et al., 2004; Song, 2007; Barker et al., 2006). The air gap and moisture have significant effect on the protective performance. There is a need to develop standard methods for determining the air gap and creating the active sweating while evaluating the protective performance of fabrics and clothing.

4.5 Summary and Conclusions

Firefighters could be exposed to various types of thermal and nonthermal hazards while working in a fire incidence. In this situation, a significant amount of heat and/or mass transfer occurs from the source of the hazards to the firefighters' clothing. This heat and/or mass could be transmitted through the fabrics used in the clothing and thus can cause significant injuries on firefighters' bodies.

In the last few decades, by considering the fundamentals of heat and mass transfer, many test methods have been developed for evaluating the protective performance of the fabrics used in the clothing or of whole clothing. By using these test methods, many researchers also assessed the protective performance of the fabric and clothing. Through their assessments, it has been found that various factors associated with the fabrics, clothing, and test methods could significantly affect the protective performance. Notably, although a lot of research has been carried out in the field of the protective performance under thermal and nonthermal hazards, there are still some key issues that need to be addressed. As indicated in this chapter, these key issues are mainly related to the shortcomings associated with the test methods for evaluating the protective performance.

By resolving these key issues, it is expected that the protective performance could be more realistically assessed. This kind of assessment may uncover various other factors related to fabrics and clothing that could affect the protective performance. By properly implementing this understanding, it is possible to develop the new fabrics for clothing and/or design the new clothing that possesses high protective performance. It is further notable that protective and comfort performances of the existing fabrics/clothing are generally inversely related. It means that a fabric/clothing with high protective performance possesses lower comfort performance. In the future, it is required to implement the latest technology (nanotechnology, smart textiles) for developing new fabrics for firefighters' clothing. These newly developed fabrics/clothing could possess high protective performance while maintaining high comfort performance. This kind of innovation can provide better protection and comfort to firefighters worldwide.

References

Ackerman, M. Y., Crown, E. M., Dale, J. D., Murtaza, G., Batcheller, J., & Gonzalez, J. A. (2012). Development of a test apparatus/method and material specifications for protection from steam under pressure. *Performance of Protective Clothing and Equipment: Emerging Issues and Technologies - 9th Vol., STP1544* (pp. 308–328). ASTM International: USA.

Ackerman, M. Y., & Song, G. (2011, June). *Analyzing thermal protective clothing performance against the impact of small splashes of hot liquid.* ASTM Ninth Symposium on Performance of Protective Clothing and Equipment: Emerging Issues and Technologies, California, USA.

Arpaci, V. S., & Larsen, P. S. (1984). *Convection Heat Transfer.* USA: Prentice Hall.

Arrieta, C., David, E., Dolez, P., & Toan, V. K. (2010). Thermal aging of a blend of high performance fibers. *Journal of Applied Polymer Science, 115*(5), 3031–3039.

ASTM D 2582:2016. Standard test method for puncture-propagation tear resistance of plastic film and thin sheeting. https://www.astm.org/Standards/D2582.htm

ASTM D 4108:1987. Standard test method for thermal protective performance of materials for clothing by open-flame method. https://global.ihs.com/doc_detail.cfm?gid=CQDWCAAAAAAAAAAA

ASTM D 751:2011. Standard test methods for coated fabrics. https://www.astm.org/Standards/D751.htm

ASTM D 747:2011. Standard test method for apparent bending modulus of plastics by means of a cantilever beam. https://www.astm.org/Standards/D747.htm

ASTM F 2701:2008. Standard test method for evaluating heat transfer through materials for protective clothing upon contact with a hot liquid splash. https://www.astm.org/Standards/F2701.htm

ASTM F 2731:2011. Standard test method for measuring the transmitted and stored energy of firefighter protective clothing systems. https://www.astm.org/Standards/F2731.htm

ASTM F 2700:2013. Standard test method for unsteady-state heat transfer evaluation of flame resistant materials for clothing with continuous heating. https://www.astm.org/Standards/F2700.htm

ASTM F 2703:2013. Standard test method for unsteady-state heat transfer evaluation of flame resistant materials for clothing with burn injury prediction. https://www.astm.org/Standards/F2703.htm

ASTM F 1671:2013. Standard test method for resistance of materials used in protective clothing to penetration by blood-borne pathogens using Phi-X174 bacteriophage penetration as a test system. https://www.astm.org/Standards/F1671.htm

ASTM F 1939:2015. Standard test method for radiant heat resistance of flame resistant clothing materials with continuous heating. https://www.astm.org/Standards/F1939.htm

ASTM F 1930:2015. Standard test method for evaluation of flame resistant clothing for protection against fire simulations using an instrumented manikin. https://www.astm.org/Standards/F1930.htm

ASTM F 955:2015. Standard test method for evaluating heat transfer through materials for protective clothing upon contact with molten substances. https://www.astm.org/Standards/F955.htm

ASTM F 1060:2016. Standard test method for evaluation of conductive and compressive heat resistance (CCHR).

Barker, R. L. (2005, January). A review of gaps and limitations in test methods for first responder protective clothing and equipment. USA: National Personal Protection Technology Laboratory, 1–98.

Barker, R., Deaton, S., Liston, G., & Thompson, D. (2010). A CB protective firefighter turnout suit. *International Journal of Occupational Safety and Ergonomics, 16*(2), 135–152.

Barker, R. L., Guerth-Schacher, C., Grimes, R. V., & Hamouda, H. (2006). Effects of moisture on the thermal protective performance of firefighter protective clothing in low-level radiant heat exposures. *Textile Research Journal, 76*(1), 27–31.

Barker, R. L., Hamouda, H., Shalev, I., & Johnson, J. (1999). Review and evaluation of thermal sensors for use in testing firefighters protective clothing. USA: National Institute of Standards and Technology, 1–49.

Barker, R. L., & Lee, Y. M. (1987). Analyzing the transient thermophysical properties of heat-resistant fabrics in TPP exposures. *Textile Research Journal, 57*(6), 331–338.

Barker, R., & Yener, M. (1981). Evaluating the resistance of some protective fabrics to molten iron. *Textile Research Journal, 51*(8), 533–541.

Behnke, W. P. (1984). Predicting flash fire protection of clothing from laboratory tests using second-degree burn to rate performance. *Fire and Materials, 8*(2), 57–63.

Benisek, L., & Edmonson, G. K. (1981). Protective clothing fabrics. Part I. Against molten metal hazards. *Textile Research Journal, 51*(3), 182–190.

Benisek, L., Edmondson, G. K., & Philips, W. A. (1979). Protective clothing: Evaluation of zirpro wool and other fabrics. *Fire and Materials, 3*(3), 156–166.

Benisek, L., & Phillips, A. (1979). Evaluation of flame retardant clothing assemblies for protection against convective heat flames. *Clothing and Textile Research Journal, 7*(9), 2–20.

Benisek, L., & Phillips, W. A. (1981). Protective clothing fabrics: Part II. Against convective heat (open-flame) hazards. *Textile Research Journal, 51*(3), 191–196.

Brushlinsky, N. N., Ahrens, M., Sokolov, S. V., & Wagner, P. (2017). World fire statistics. Russia: Center of Fire Statistics International Association of Fire and Rescue Services, 1–56.

Burnmeister, L. C. (1993). *Convective Heat Transfer.* USA: Wiley Publications.

Camenzind, M. A., Dale, D. J., & Rossi, R. M. (2007). Manikin test for flame engulfment evaluation of protective clothing: historical review and development of a new ISO standard. *Fire and Materials, 31*(5), 285–295.

Chen, L., Han, X. Y., & Bai, X. E. (2016, July). Correlation between fabric property and sewing quality of sportswear. 9th Textile Bioengineering and Informatics Symposium, Melbourne, Australia.

Crown, E. M., Ackerman, M. Y., Dale, D. J., & Rigakis, K. B. (1993).Thermal protective performance and instrumented mannequin evaluation of multi-layer garment systems. *Proceedings of Aerospace Medical Panel Symposium: The Support of Air Operations Under Extreme Hot and Cold Conditions,* France, pp. 14-1-14-8.

Crown, E. M., Ackerman, M. Y., Dale, J. D., Tan, Y. (1998). Design and evaluation of thermal protective flightsuits. Part 2: Instrumented mannequin evaluation. *Clothing and Textile Research Journal,* 16(2), 79–87.

Dale, J. D., Crown, E. M., Ackerman, M. Y., Leung, E., & Rigakis, K. B. (1992). Instrumented manikin evaluation of thermal protective clothing. In J. P. McBriarty & N. W. Henry (Eds.), *Performance of Protective Clothing, ASTM STP 1133* (pp. 717–733). USA: American Society for Testing and Materials.

Dale, J. D., Paskaluk, S. A., & Crown, E. M. (2016). Round-robin testing of European-weight firefighting clothing with fire engulfment instrumented manikins. In B. Shiels & K. Lehtonen (Eds.), *Performance of Protective Clothing and Equipment: 10th Volume, ASTM STP1593* (pp. 351–373). USA: American Society for Testing and Materials.

Desruelle A., & Schmid, B. (2004). The steam laboratory of the Institut de Me´decine Navale du Service de Sante´ des Arme´ es: A set of tools in the service of the French Navy. *European Journal of Applied Physiology, 92,* 630–635.

Ding, B., Yamazaki, M., & Shiratori, S. (2004). Electrospun fibrous polyacrylic acid membranebased gas sensors. *Sensor and Actuators B: Chemical, 106*(1), 477–483.

Duncan, S., McLellan, T., & Dickson, E. G. (2011). Improving comfort in military protective clothing. In G. Song (Ed.), *Improving Comfort in Clothing* (pp. 320–369). Cambridge: Woodhead Publishing Limited.

Elnashar, E. A. (2005). Volume porosity and permeability in double-layer woven fabrics. *AUTEX Research Journal, 5*(4), 207–218.

Endrusick, T. L., Gonzalez, J. A., & Gonzalez, R. R. (2005). Improved comfort of US military chemical and biological protective clothing. In Y. Tochihara & T. Ohnaka (Eds.), *Environmental Ergonomics: The Ergonomics of Human Comfort, Health and Performance in the Thermal Environment—3rd Vol. 3* (pp. 369–373). United Kingdom: Elsevier Science and Technology.

Eni, E. U. (2005). Developing test procedures for measuring stored thermal energy in firefighter protective clothing. M.Sc. Thesis. North Carolina State University, North Carolina, USA.

Ghazy, A. (2014). Influence of thermal shrinkage on protective clothing performance during fire exposure: Numerical investigation. *Mechanical Engineering Research Journal, 4*(2), 1–15.

Ghazy, A. (2017). The thermal protective performance of firefighters' clothing: The air gap between the clothing and the body. *Heat Transfer Engineering, 38*(10), 975–986.

Ghazy, A., & Bergstrom, D. J. (2010). Numerical simulation of transient heat transfer in a protective clothing system during a flash fire exposure. *Numerical Heat Transfer A, 58*(9), 702–724.

Ghazy, A., & Bergstrom, D. J. (2011). Influence of the air gap between protective clothing and skin on clothing performance during flash fire exposure. *Heat and Mass Transfer, 47,* 1275–1288.

Gholamreza, F., & Song, G. (2013). Laboratory evaluation of thermal protective clothing performance upon hot liquid splash. *The Annals of Occupational Hygiene*, *57*(6), 805–822.

Haynes, H. J. G. (2017, September). *Fire loss in the United States during 2016*. USA: National Fire Protection Association, 1–56.

Haynes, H. J. G., & Molis, J. L. (2017). United States firefighter injuries—2016. *National Fire Protection Association (NFPA)*, 1–35.

He, J., L, Y., Chen, Y., & Li, J. (2017). Investigation of the thermal hazardous effect of protective clothing caused by stored energy discharge. *Journal of Hazardous Materials*, *338*, 76–84.

Hossain, M. M., Datta, E., & Rahman, S. (2016). A review on different factors of woven fabrics' strength prediction. *Science Research*, *4*(3), 88–97.

Hristian, L., Bordeianu, D. L., Iurea, P., Sandu, I., & Earar, K. (2014). Study of the tensile properties of materials destined to manufacture protective clothing for firemen. *Materiale Plastice*, *51*(4), 405–409.

Hsu, S. T. (1963). *Engineering Heat Transfer*. Van Nostrand Company, USA: Princeton.

ISO 4674:2003. Rubber- or plastics-coated fabrics—Determination of tear resistance—Part 1: Constant rate of tear methods. https://www.iso.org/standard/65592.html

ISO 17492:2003. Clothing for protection against heat and flame—Determination of heat transmission on exposure to both flame and radiant heat.

ISO 6530:2005. Protective clothing—Protection against liquid chemicals—Test method for resistance of materials to penetration by liquids. https://www.iso.org/standard/30698.html

ISO 13994:2005. Clothing for protection against liquid chemicals—Determination of the resistance of protective clothing materials to penetration by liquids under pressure. https://www.iso.org/standard/38303.html

ISO 9185:2007. Protective clothing—Assessment of resistance of materials to molten metal splash. https://www.iso.org/standard/41479.html

ISO 11613:2012. Protective clothing for firefighters—Laboratory test methods and performance requirements. https://www.iso.org/standard/19550.html

ISO 13934-1:2013. Textiles—Tensile properties of fabrics—Part 1: determination of maximum force and elongation at maximum force using the strip method. https://www.iso.org/standard/60676.html

ISO 6942:2015. Protective clothing—Protection against heat and fire—Method of test: evaluation of materials and material assemblies when exposed to a source of radiant heat. https://www.iso.org/standard/26327.html

ISO 12127-1:2015. Clothing for protection against heat and flame—Determination of contact heat transmission through protective clothing or constituent materials. https://www.iso.org/standard/45865.html

ISO 11999-3:2015. PPE for firefighters—Test methods and requirements for PPE used by firefighters who are at risk of exposure to high levels of heat and/or flame while fighting fires occurring in structures—Part 3: Clothing. https://www.iso.org/standard/64019.html

ISO 9151:2016. Protective clothing against heat and flame—Determination of heat transmission on exposure to flame. https://www.iso.org/standard/55326.html

ISO 13506-1:2017. Protective clothing against heat and flame—Part 1: Test method for complete garments—Measurement of transferred energy using an instrumented manikin. https://www.iso.org/standard/63839.html

ISO 13506-2:2017. Protective clothing against heat and flame—Part 2: Skin burn injury prediction—Calculation requirements and test cases. https://www.iso.org/standard/63840.html

Jalbani, S. H., Ackerman, M. Y., Crown, E. M., Keulen, M., & Song, G. (2012). Apparatus for use in evaluating protection from low pressure hot water jets. *Performance of Protective Clothing and Equipment: Emerging Issues and Technologies—9th Vol., STP1544* (pp. 329–339). ASTM International: USA.

Kabir, B. R., & Ferdous, N. (2012). Kevlar—the super tough fiber. *International Journal of Textiles Science, 1*(6), 78–83.

Kahn, S. A., Patel, J. H., Lentz, C. W., & Bell, D. E. (2012). Firefighter burn injuries: Predictable patterns influenced by turnout gear. *Journal of Burn Care & Research, 33*(1), 152–156.

Karter, M. J. (2013, December). Patterns of firefighter fireground injuries. USA: National Fire Protection Association, 1–27.

Keiser, C., & Rossi, R. M. (2008). Temperature analysis for the prediction of steam formation and transfer in multilayer thermal protective clothing at low level thermal radiation. *Textile Research Journal, 78*(11), 1025–1035.

Keiser, C., Wyss, P., & Rossi, R. M. (2010). Analysis of steam formation and migration in firefighters' protective clothing using X-ray radiography. *International Journal of Occupational Safety and Ergonomics, 16*(2), 217–229.

Kemp, S. E., Annaheim, S., Rossi, R. M., & Camenzind, M. A. (2016). Test method for characterising the thermal protective performance of fabrics exposed to flammable liquid fires. *Fire and Materials, 41*(6), 750–767.

Lawson, J. R. (1996, August). Fire fighter's protective clothing and thermal environments of structural fire fighting. USA: National Institute of Standards and Technology, 1–22.

Lawson, L. K., Crown, E. M., Ackerman, M. Y., & Dale, D. J. (2004). Moisture effects in heat transfer through clothing systems for wildland firefighters. *International Journal of Occupational Safety and Ergonomics, 10*(3), 227–238.

Lawson, J. R., Twilley, W. H., & Malley, K. S. (2000, April). Development of a dynamic compression test apparatus for measuring thermal performance of fire fighters' protective clothing. USA: National Institute of Standards and Technology, 1–24.

Le, C. V., Ly, N. G., & Postle, R. (1995). Heat and mass transfer in the condensing flow of steam through an absorbing fibrous media. *International Journal of Heat and Mass Transfer, 38*(1), 81–89.

Lee, Y. M., & Barker, R. L. (1986). Effect of moisture on the thermal protective performance of heat-resistant fabrics. *Journal of Fire Sciences, 4*(5), 315–330.

Liu, H., Kameoka, J., Czaplewski, D. A., & Craighead, H. G. (2004). Polymeric nanowire chemical sensor. *Nano Letters, 4*(4), 671–675.

Lu, Y., Li, J., Li, X., & Song, G. (2013b). The effect of air gaps in moist protective clothing on protection from heat and flame. *Journal of Fire Sciences, 31*(2), 99–111.

Lu, Y., Song, G., & Li, J. (2013a). Analysing performance of protective clothing upon hot liquid exposure using instrumented spray manikin. *The Annals of Occupational Hygiene, 57*(6), 793–804.

Lu, Y., Song, G., Zeng, H., Zhang, L., & Li, J. (2014). Characterizing factors affecting the hot liquid penetration performance of fabrics for protective clothing. *Textile Research Journal, 84*(2), 174–186.

Mah, T., & Song, G. (2010). Investigation of the contribution of garment design to thermal protection. part 1: Characterizing air gaps using three-dimensional body scanning for women's protective clothing. *Textile Research Journal, 80*(13), 1317–1329.

Mah, T., & Song, G. (2010). Investigation of the contribution of garment design to thermal protection. part 2: Instrumented female mannequin flash-fire evaluation system. *Textile Research Journal, 80*(14), 1473–1487.

Makinen, H., Laiho, H., & Pajunen, P. (1997). Evaluation of the protective performance of fabrics and fabric combinations against molten iron. In J. O. Stull & A. D. Schwope (Eds.), *Performance of Protective Clothing—vol. 6, STP 1273* (pp. 225–237). ASTM International: USA.

Mandal, S. (2016). Studies of the thermal protective performance of textile fabrics used in firefighters' clothing under various thermal exposures. PhD Thesis. University of Alberta, Edmonton, Canada.

Mandal, S., Annaheim, S., Camenzind, M., & Rossi, R. (2017b, July). Characterization of thermal protective fabric materials under fire exposure. *2017 Materials Research Society Spring Meetings & Exhibit*, Arizona, USA.

Mandal, S., Annaheim, S., Pitts, T., Camenzind, M., & Rossi, R. (2017a). Studies of the thermal protective performance of fabrics used in firefighters' clothing under fire exposures: From small-scale to hexagon tests. *Textile Research Journal*. doi:10.1177/0040517517723020

Mandal, S., Camenzind, M., Annaheim, S., & Rossi, R. (May, 2017c). A new approach for evaluating the thermal protective performance of fabrics used in firefighters' clothing under flash fire exposure. *17th World Textile Conference AUTEX 2017*, Corfu, Greece.

Mandal, S., Lu, Y., Wang, F., & Song, G. (2014). Characterization of thermal protective clothing under hot water and pressurized steam exposure. *AATCC Journal of Research, 1*(5), 7–16.

Mandal, S., & Song, G. (2011, May). *Characterization of protective textile materials for various thermal hazards*. Fiber Society Spring Conference, Kowloon, Hong Kong.

Mandal, S., & Song, G. (2015). Thermal sensors for performance evaluation of protective clothing against heat and fire: a review. *Textile Research Journal, 85*(1), 101–112.

Mandal, S., & Song, G. (2016a). Characterizing fabrics in firefighters' protective clothing: Hot water immersion and compression. *AATCC Journal of Research, 3*(2), 8–15.

Mandal, S., & Song, G. (2016b). Characterizing thermal protective fabrics of firefighters' clothing in hot surface contact. *Journal of Industrial Textiles*. doi:10.1177/0123456789123456

Mandal, S., Song, G., Ackerman, M., Paskaluk, S., & Gholamreza, F. (2013). Characterization of textile fabrics under various thermal exposures. *Textile Research Journal, 83*(10), 1005–1019.

Mandal, S., Song, G., & Gholamreza, F. (2016) A novel protocol to characterize the thermal protective performance of fabrics in hot water exposures. *Journal of Industrial Textiles, 46*(1), 279–291.

Mehta, P. N., & Willerton, K. (1977). Evaluation of clothing materials for protection against molten metal. *Textile Institute & Industry, 15*, 334–340.

Mert, E., Psikuta, A., Bueno, M. A., & Rossi, R. M. (2017). The effect of body postures on the distribution of air gap thickness and contact area. *International Journal of Biometeorology, 61*(2), 363–375.

NFPA 1971:2018. Standard on protective ensembles for structural fire fighting and proximity fire fighting.

NFPA 1994:2018. Standard on protective ensembles for first responders to hazardous materials emergencies and CBRN terrorism incidents.

Pawar, M. (1995). Analyzing the thermal protective performance of single layer garment materials in bench scale and manikin tests. PhD Thesis. North Carolina State University, North Carolina, USA.

Perkins, R. M. (1979). Insulative values of single-layer fabrics for thermal protective clothing. *Textile Research Journal, 49*(4), 202–205.

Rossi, R. & Bolli, W. (2000). Assessment of radiant heat protection of firefighters' jackets with a manikin. In C. Nelson & N. Henry (Eds.), *Performance of Protective Clothing: Issues and Priorities for the 21st Century—7th Volume, STP14447S* (pp. 212-223). ASTM International: USA.

Rossi, R. M., Bruggmann, G., & Stämpfli, R. (2005). Comparison of flame spread of textiles and burn injury prediction with a manikin. *Fire and Materials, 29*(6), 395–406.

Rossi, R., Indelicato, E., & Bolli, W. (2004). Hot steam transfer through heat protective clothing layers. *International Journal of Occupational Safety and Ergonomics, 10*(3), 239–245.

Rossi, R. M., Schmid, M., & Camenzkarind, M. A. (2014). Thermal energy transfer through heat protective clothing during a flame engulfment test. *Textile Research Journal, 84*(13), 1451–1460.

Rossi, R., & Zimmerli, T. (1994, January). *Influence of humidity on the radiant, convective and contact heat transmission through protective clothing materials*. Fifth International Symposium on Performance of Protective Clothing: Improvement through Innovation, San Francisco, USA.

Saner, M. (2013). Insight on the ASTM F 1930 manikin burn test. Retrieved from http://www.workrite.com/news/iinsight-on-the-astm-f1930-manikin-burn-test/ (accessed on July 8, 2016).

Sati, R., Crown, E., Ackerman, M., Gonzalez, J., & Dale, J. D. (2008). Protection from steam at high pressure: Development of a test device and protocol. *International Journal of Occupational Safety and Ergonomics, 14*(1), 29–41.

Schmid, M., Annaheim, S., Camenzind, M., & Rossi, R. M. (2016). Determination of critical heat transfer for the prediction of materials damages during a flame engulfment test. *Fire and Materials*. doi:10.1002/fam.2362

Schiefer, H. F., Taft, D. H., & Porter, J. W. (1936). Effect of number of warp and filling yarns per inch and some other elements of construction on the properties of the cloth. *Journal of Research on the National Bureau of Standards, 16*, 139–147.

Schmid, M., Annaheim, S., Camenzind, M., & Rossi, R. M. (2016). Determination of critical heat transfer for the prediction of materials damages during a flame engulfment test. *Fire and Materials, 40*(8), 1036–1046.

Seed, M., Anand, S., Kandola, B., & Fulford, R. (2008). Chemical, biological, radiological and nuclear protection. *Technical Textile Institute, 17*, 39–46.

Shalev, I., & Barker, R. L. (1983). Analysis of heat transfer characteristics of fabrics in an open flame exposure. *Textile Research Journal, 53*(8), 475–482.

Shalev, I., & Barker, R. L. (1984). Protective fabrics: A comparison of laboratory methods for evaluating thermal protective performance in convective/radiant exposures. *Textile Research Journal*, *54*(10), 648–654.

Sharkey, R. (1997). New developments in the design and evaluation of flame-retardant clothing for protection against molten aluminum splash hazard. In J. P. Mcbriarty and N. W. Henry (Eds.), *Performance of Protective Clothing – 4th vol., STP1133-EB* (pp. 698–702). ASTM International: USA.

Siegel, R. & Howell, J. (2002). *Thermal Radiation Heat Transfer*. UK: Taylor & Francis.

Song, G. (2004). Modeling thermal protection outfit for fire exposures. PhD Thesis. North Carolina State University, Raleigh, USA.

Song, G. (2007). Clothing air gap layers and thermal protective performance in single layer garment. *Journal of Industrial Textiles*, *36*(3), 193–205.

Song, G., Barker, R. L., Hamouda, H., Kuznetsov, A. V., Chitrphiromsri, P., & Grimes, R. V. (2004). Modeling the thermal protective performance of heat resistant garment in flash fire exposures. *Textile Research Journal*, *74*(12), 1033–1040.

Song, G., Cao, W., & Gholamreza, F. (2011). Analyzing thermal stored energy and effect on protective performance. *Textile Research Journal*, *81*(11), 120–135.

Song, G., Mandal, S., & Rossi, R. (2016). *Thermal protective clothing: A critical review*. United Kingdom: Elsevier Science and Technology.

Song, G., Paskaluk, S., Sati, R., Crown, E. M., Dale, J. D., & Ackerman, M. (2011). Thermal protective performance of protective clothing used for low radiant heat protection. *Textile Research Journal*, *81*(3), 311–323.

Su, Y., Li, J., & Song, G. (2017). The effect of moisture content within multilayer protective clothing on protection from radiation and steam. *International Journal of Occupational Safety and Ergonomics*. doi:10.1080/10803548.2017.1321890

Sun, G., Yoo, H. S., Zhang, X. S., & Pan, N. (2000). Radiant protective and transport properties of fabrics used by wildland firefighters. *Textile Research Journal*, *70*(7), 567–573.

Tian, M., & Li, J. (2016). Simulating the thermal response of the flame manikin with different materials exposed to flash fire by CFD. *Fire and Materials*. doi:10.1002/fam.2363

Torvi, D. A., & Dale, J. D. (1998). Effect of variation in thermal properties on the performance of flame resistant fabrics for flash fires. *Textile Research Journal*, *68*(11), 787–796.

Torvi, D. A., & Dale, J. D. (1999). Heat transfer in thin fibrous materials under high heat flux. *Fire Technology*, *35*(3), 210–231.

Torvi, D. A., & Hadjisophocleous, G.V. (1999). Research in protective clothing for firefighters: state of the art and future directions. *Fire Technology*, *35*(2), 111–130.

Wang, M., & Li, J. (2017). Thermal protection retention of fire protective clothing after repeated flash fire exposure. *Journal of Industrial Textiles*, *46*(3), 737–755.

Wang, M., Li, X., & Li, J. (2015). Correlation of bench scale and manikin testing of fire protective clothing with thermal shrinkage effect considered. *Fibers and Polymers*, *16*(6), 1370–1377.

Wang, X., Drew, C., Lee, S., Senecal, K. J., Kumar, J., & Samuelson, L. A. (2002). Electrospun nanofibrous membranes for highly sensitive optical sensors. *Nano Letters, 2*(11), 1273–1275.

Wang, Y., Wang, Z., Zhang, X., & Li, J. (2015). CFD simulation of naked flame manikin tests of fire proof garments. *Fire Safety Journal, 71*, 187–193.

Wen, S. (2014). Physiological strain and physical burden in chemical protective coveralls. PhD Thesis, University of Alberta, Edmonton, Canada.

York, K. J., & Grey, G. L. (1986). Chemical protective clothing—Do we understand it? *Fire Engineering, 139*(2), 28–33.

Zhu, F., Zhang, W., & Chen, M. (2007). Investigation of material combinations for fire-fighter's protective clothing on radiant protective and heat-moisture transfer performance. *Fibres and Textiles in Eastern Europe, 15*(1), 72–75.

Zhang, H., McQueen, H. R., Batcheller, C. J., Paskaluk, A. S., & Murtaza, G. (2015a). Clothing in the kitchen: Evaluation of fabric performance for protection against hot surface contact, hot liquid and low-pressure steam burns. *Textile Research Journal, 85*(20), 2136–2146.

Zhang, H., McQueen, H. R., Batcheller, C. J., Ehnes, L. B., & Paskaluk, A. S. (2015b). Characterization of textiles used in chefs' uniforms for protection against thermal hazards encountered in the kitchen environment. *The Annals of Occupational Hygiene, 59*(8), 1058–1073.

Zhang, H., Song, G., Gu, Y., Ren, H., & Cao, J. (2017). Effect of moisture content on thermal protective performance of fabric assemblies by a stored energy approach under flash exposure. *Textile Research Journal.* doi:10.1177/0040517517712097

Zweben, C. (1978). The flexural strength of aramid fiber composites. *Journal of Composite Materials, 12*(4), 422–430.

5

Stored Thermal Energy and Protective Performance

Yun Su and Guowen Song

5.1 Introduction

A large number of firefighters suffer not only from fatal injuries while suppressing a fire but from a range of nonfatal injuries as well, such as skin-burn injuries, strains, bruises, and smoke or gas inhalation [1,2]. To minimize or prevent the risk of fatal or nonfatal injuries, firefighters are required to wear protective clothing; therefore, a proper evaluation of the protective clothing is critical. Realistic tests must be carried out to determine the required thermal protection in various thermal exposure conditions, such as flash fire, high-intensity thermal radiation, and hot surface contact.

During firefighting and emergency rescue, thermal energy from the fire ground is transferred to the skin through firefighting protective clothing. However, a large amount of this energy is stored in the clothing system due to the high heat capacity. Until the cool down period, the stored thermal energy can be naturally discharged to the ambient environment and human skin. The protective clothing can also be compressed to release the stored thermal energy owing to body movement or external pressure, which can cause skin burns or exacerbate stored energy burns [3–6]. It was reported that, in multilayer protective clothing exposed to flash fire testing, approximately 50% of the total heat energy to cause second-degree skin burn originated from the discharge of stored thermal energy [4]. Therefore, thermal protective clothing can be treated as a heat sink to store thermal energy for reducing heat transfer rate during exposure (positive effect), while serving as a heat source to discharge thermal energy and exacerbate skin burns during cooling (negative effect). It is extremely important to study the impact of stored thermal energy within protective clothing on skin burns under various thermal exposures.

This issue has attracted attention from many researchers in recent decades. Some assessment methods on thermal protective performance have been improved to investigate the influence of thermal stored energy on protective clothing. In this chapter, we will analyze the relationship between transmitted and stored thermal energy in protective clothing to result in an increased understanding of stored thermal energy in protective clothing. An overview of the literature on test methods and standards of thermal protective performance is presented, and the effects of various factors on transmitted and stored thermal energy of protective clothing are summarized in detail, including fabric basic properties, air gap size, moisture content, and heat exposure conditions. Finally, possible future research on thermal stored energy and protective performance is discussed, with recommendations to minimize stored thermal burns and develop advanced thermal protective clothing that can provide better occupational health and safety for firefighters.

5.2 Relationship between Transmitted and Stored Thermal Energy

When a clothed body is exposed to a fire environment, thermal energy from the external environment can be transmitted to the human skin through (protective) clothing via radiation, conduction, and convection, as shown in Figure 5.1. Protective clothing is usually composed of a porous medium so that thermal radiation can penetrate. Since convective heat transfer occurs on the surface of protective clothing, due to the special porous design of the fabric system, the heat transfer rate in multilayer protective clothing is determined by radiant and conductive heat transfer. In addition, the thermal storage capacity of protective clothing may affect the heat transfer rate during the exposure. The relationship between transmitted and stored thermal energy has been studied in previous research [7–9].

According to the variation in the rate of transmitted and stored thermal energy, heat exposure can be divided into three phases according to Figure 5.2: heat storage, thermal equilibrium, and heat release in protective clothing. Once protective clothing is exposed to flash fire, the thermal energy is first stored in the clothing.

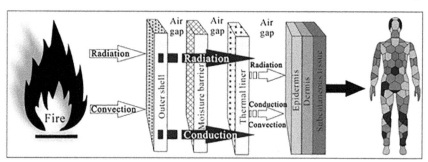

Figure 5.1

Sketch map of heat transfer process from a fire environment to the human body via multilayer protective clothing.

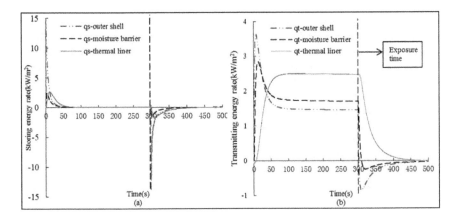

Figure 5.2

Storing energy rate (a) and transmitting energy rate (b) during and after heat exposure. (From Y. Su et al., *Applied Thermal Engineering*, 93, 1295–1303, 2016.)

With the decrease in the thermal storage capacity, the heat transfer rate in protective clothing increases. During this stage, protective clothing presents a positive effect on thermal protection from fire hazards. After a particular time under heat exposure, thermal storage will cease, since protective clothing has a limited capacity to store thermal energy. At this stage, the protective clothing system maintains a thermal equilibrium. The heat transfer rate for protective clothing keeps constant, mainly depending on the thermal conductivity and radiative properties of materials. Compared with the first stage, the clothing's protective ability decreases.

After heat exposure, the rate of transmitted thermal energy in protective clothing is not less than zero due to the existence of stored thermal energy in clothing. For this third stage, the protective clothing can be treated as a heating source, since the temperature in protective clothing is much higher than that of the external environment and skin tissues. Thus, due to the temperature difference, the thermal energy from protective clothing can be naturally released to both the ambient environment and the wearer's skin, which can amplify the extent of skin burn injuries. The heat transfer/release rate in protective clothing gradually decreases at this stage owing to the reduction of total stored thermal energy. Additionally, when firefighters move, wear the self-contained breathing apparatus, or hold a hose to extinguish a fire, the protective clothing may be compressed, increasing the heat release rate during cooling [9,10]. Generally, this compressed heat transfer in protective clothing during cooling can result in skin burn injuries.

5.3 Evaluation Method and Standard Related to Stored Thermal Energy and Protective Performance

Over the past few decades, many laboratory simulation methods have been developed to evaluate thermal protective performance of protective clothing. These assessment methods can be divided into two categories. First, thermal protective clothing is directly exposed to flash fire tests, such as the thermal protective performance (TPP) test and flame manikin test, as well as high- or low-intensity radiation, including the radiative protective performance (RPP) test, stored thermal energy test (SET), and radiant manikin test. In the second approach, researchers have developed evaluation indexes for thermal protective performance based on

Stoll's criteria and the Henriques' burn integral. These developments in testing apparatuses and evaluation indices contribute to more precise characterizations of stored thermal energy and protective performance.

5.3.1 Development of Test Apparatuses

5.3.1.1 Test Apparatuses for Heat and Flame Exposure

Originally, the burning behavior or flammability of fabrics was used to evaluate flame resistance of clothing. The common evaluation methods for flammability include limiting oxygen index (LOI), the forty-five-degree flammability test, the vertical flammability test, and the smoke concentration test, among others. Figure 5.3 shows the test apparatus of fabric LOI. According to ASTM D2863 [11], the LOI is defined as the minimum volume percent of oxygen in a mixture of oxygen and nitrogen that can support flaming combustion of a material at 23±2°C. A preconditioned test specimen is supported vertically by a specimen holder before the test. A mixture of oxygen and nitrogen with 23±2°C should be flowed upwards through a transparent chimney at a rate 40±2 mm/s. The test specimen is ignited using a flame igniter that consists of a tube with an inside diameter of 2 mm, and is then inserted into the chimney. During the test, the length of specimen burns should be recorded for evaluating the flammability of the test specimen. The LOI of fabric is calculated by gas measurements and control systems. A larger LOI value means that the test specimen is characterized by better flammability and thermal stability. Generally, the flammability of fiber can be divided into two types: fire-retardant fiber (LOI≥26%) and inflammable fiber (LOI<26%).

Although flammability tests can evaluate this performance of protective materials and determine if the tested materials are suitable for thermal protection, it is

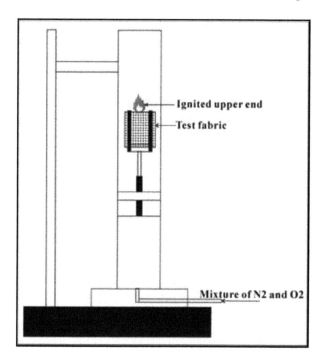

Figure 5.3

Frame design of LOI test apparatus.

hard to distinguish thermal protective performance of protective clothing when most materials have flame resistance [12]. The thermal protective performance test is used to assess the ability to resist heat transfer from flash fire and thermal radiation and determine the required time to cause skin burn injuries. To better evaluate thermal protection of firefighting protective clothing under various thermal exposures, some standardized test apparatuses were developed for simulating actual heat exposures, including flash fire and radiant heat exposure.

A test device to evaluate thermal protective performance of clothing exposed to open flame conditions was first reported in the ASTM D4108 standard [13] (see Figure 5.4). In this test standard, a Meker burner was used to produce a flame with a nominal heat flux of 84 kW/m² that consists of 30/70 radiative/convective heat transfer. The copper slug sensor was placed behind the test specimen with the introduction of an air gap to measure the thermal response of skin temperature and predict skin burn injuries based on the Stoll criterion. However, the TPP test device did not adjust the percentage of radiative/convective heat sources or consider the effect of thermal energy stored in protective clothing on thermal protective performance [14]. Thus, this test standard is not used at present.

The TPP test device was further improved in some international standards, such as NFPA 1971 [15], ASTM F2700 [16], and ISO 17492 [17]. The frame diagram of this improved test apparatus is shown in Figure 5.5. The heat source consisted of two Meker burners and a bank of nine electrically heated quartz tubes that produced a nominal heat flux of 84 kW/m², with different percentages of radiative/convective heat transfer. The angle of the Meker burners was kept at 45° to ensure that the flame converged at the center of the specimen. The testing specimen was insulated from heat sources before the test by using an automatic water-cooled shutter to ensure accurate time exposures. The temperature rise at the rear of the specimen over time was measured with a standard copper calorimeter located behind the test sample. This calorimeter consisted of a copper disk of 40 mm diameter and 1.6 mm thickness. After heat exposure, the test fabric and copper calorimeter were moved from heating position to cooling position to simulate the cooling process of

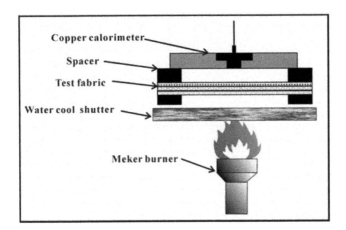

Figure 5.4

Schematic diagram of thermal protective performance tester under flame exposure from ASTM D4108. (From ASTM F1060-16, Standard Test Method for Evaluation of Conductive and Compressive Heat Resistance (CCHR), ASTM International, West Conshohocken, PA, 2016.)

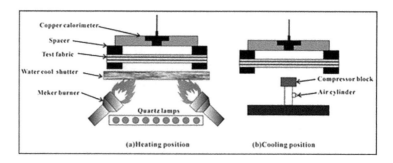

Figure 5.5

Schematic diagram of thermal protective performance tester under heat and flame exposure from ASTM D2700. (From ASTM F2700-08(2013), Standard Test Method for Unsteady-State Heat Transfer Evaluation of Flame Resistant Materials for Clothing with Continuous Heating, ASTM International, West Conshohocken, PA, 2013.)

a clothed body after exposure. The thermal response of the skin's surface during cooling was also simulated and recorded to study the release of thermal energy within protective clothing. This new test apparatus also included a compressive system at its cooling position to evaluate the effect of compression due to body movement on the release of stored thermal energy.

Although these bench-top tests can effectively evaluate the heat transfer behavior of fabrics under heat and flame exposure, they are not as effective as full-scale manikin tests for evaluating the thermal protection of clothing. This is because manikin tests can detect and record ignition and burning of the garment and heraldry, any shrinkage or sagging after flaming, as well as hole generation, smoke generation, and structural failure of seams. These tests can also be used to estimate the extent and nature of skin burns that a person would suffer if wearing the test garment under flash fire or radiant heat exposure. Thus, manikin experiments are important tools to simulate the actual wearing state of clothing and present the heat and moisture transfer behavior between the fire environment, protective clothing, and human body.

Over the years, several full-scaled manikin test systems have been established to evaluate clothing's thermal protective performance. Most manikin test systems are employed to simulate a flash fire of 84 kW/m² using 8 or 12 propane gas torches, such as PyroMan at North Carolina State University [18], the flame manikin Harry Burns of University of Alberta [19], and Thermo-Man at DuPont [20]. These manikins are tested in a standing position in initially quiescent air. Controlled air motion for simulating wind effects or body movement is not currently included in testing, but it is possible to move the manikin through a stationary flame. These manikins represent an adult human and are made of a flame-resistant, thermally stable, glass fiber-reinforced vinyl ester resin, equipped with more than 100 thermal energy sensors distributed uniformly within each area on the manikin surface. These sensor types mainly include copper, guarded copper slugs, skin simulant sensors, and embedded sensors, which can record the variation in skin temperature or heat flux over a preset time interval during and after heat exposure. These measurement data can be used to further calculate the time to cause second- or third-degree skin burn in different parts of the human body and the overall percentage of burn area based on Henriques' burn integral [21], as shown in Figure 5.6.

According to ASTM F1930 [23] and ISO 13506 [24], it is essential to calibrate the test apparatus before dressing the manikin. When the nude manikin is exposed to

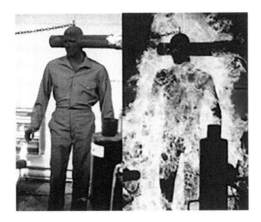

Figure 5.6

Full-scaled manikin test under heat and flame exposure. (From C. M. J. Sawcyn, Heat transfer model of horizontal air gaps in bench top testing of thermal protective fabrics, University of Saskatchewan, Saskatoon, 2003.)

flames, the average heat flux level is typically equal to 84 kW/m² with a standard deviation of 5%. An exposure duration with a range of 3 to 5 s is typically selected for the exposure of the dressed manikin, while the thermal sensors collect data for at least 60 s to ensure the discharge of thermal energy stored in the protective clothing.

5.3.1.2 Test Apparatuses for Radiant Heat Exposure

To determine the radiant protective performance of a material, or a combination of materials, used in flame-resistant clothing for workers exposed to radiant thermal hazards, many researchers use the RPP test device to assess if performance of protective materials conforms to the standards of NFPA1977 [25] and ASTM F2702 [26] (see Figure 5.7). A bank of five, 500 W, infrared tubular translucent quartz lamps is used to produce two standard sets of exposure conditions, 21 and 84 kW/m². A vertically positioned test specimen is placed between the heat source and the thermal sensor. By controlling the water cool shutter, the heat exposure duration and cooling duration of test fabrics can be adjusted to systematically evaluate a material's heat transfer properties. The sensor is a copper calorimeter that provides temperature information to the data acquisition system, which then interprets these data into a burn injury prediction using the Stoll curve.

This high-intensity radiant heat exposure is suitable to simulate a structural firefighting environment. For wildland firefighting work conditions, the reasonable maximum exposure is less than 10 kW/m². The current standard RPP test method has inherent limitations that greatly affect the results of the test at lower flux exposures, since the copper slug calorimeter used is suitable for only short-duration exposures. This is because a heat saturation effect can present in copper calorimeters for long-duration exposures, thus resulting in a low perceived heat flux [27]. The traditional RPP test also cannot assess the influence of compression on stored thermal energy. In addition, many operations in the traditional RPP test apparatus must be manually carried out, which increases the risk of error.

To cope with the above drawbacks, Lawson et al. [28] developed a new test device that could measure fabric's thermal protective performance under low-level thermal radiation, as well as the effect of heat storage, but the test apparatus could not simulate thermal release via compression. Neal et al. [29] presented an

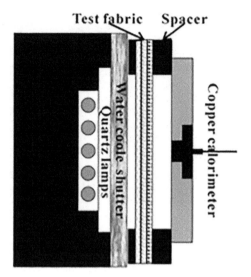

Figure 5.7

Schematic diagram of radiative protective performance tester under radiant heat exposure from ASTM F2702. (From ASTM. F2702-15, Standard Test Method for Radiant Heat Performance of Flame Resistant Clothing Materials with Burn Injury Prediction, ASTM International, West Conshohocken, PA, 2013.)

improved apparatus that included compressive equipment but did not address the air gap between the clothing and body during cooling. A new test apparatus called the stored thermal energy test (SET) apparatus from ASTM F2731 [30] was developed by North Carolina State University and Thermetrics to evaluate transmitted and stored thermal energy in firefighter turnout composites for both longer-duration radiant heat exposures and high-intensity contact with compression after radiant heat exposure [31]. The testing specimen (see Figure 5.8) is exposed to a nominal radiant heat flux of 8.5 kW/m^2 produced by a black ceramic thermal flux source. The heat flux rise versus time at the back of the specimen is measured by the sensor assembly, which consists of a water-cooled Schmidt-Boelter thermopile-type sensor suitable for long-duration exposure, and a sensor housing. The sensor assembly can be in contact with the specimen or with a spacer introducing a 6.4 mm air gap between the fabric and the sensor assembly.

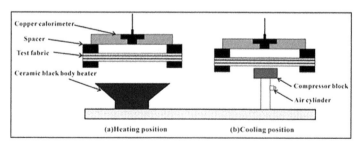

Figure 5.8

Schematic diagram of stored thermal energy test (SET) apparatus from ASTM F 2731. (From ASTM. F2731-11, Standard Test Method for Measuring the Transmitted and Stored Energy of Firefighter Protective Clothing Systems, ASTM International, West Conshohocken, PA, 2011.)

5. Stored Thermal Energy and Protective Performance

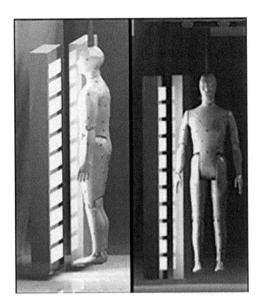

Figure 5.9

Full-scaled manikin test under radiant heat exposure. (From K. L. Watson, From Radiant Protective Performance to RadMan™: The Role of Clothing Materials in Protecting Against Radiant Heat Exposures in Wildland Forest Fires, 2014.)

The compressor assembly consists of a compressor block, air cylinder, air regulator, and a framework that rigidly holds the system in place. The compressor block is made from Marinite with thermal conductivity (0.12 W/m/K). When activated, the regulated air can activate the piston and force the circular heat-resistant block against the sample and data collection sensor with a pressure of 13.8 ± 0.7 kPa. The transfer tray connected with the sensor assembly and specimen holder can be moved between the compressor assembly and the heating source. As soon as the tray is moved over the heating source, the data acquisition system begins collecting data. At the end of 300 s exposure, the transfer tray is moved away from the heating source and over the compressor, while the data acquisition system continues to collect data for additional 300 s.

Similar to flame manikins, the RadMan test system was developed by North Carolina State University to evaluate clothing's radiant protective performance in a realistic radiant heat exposure similar to a wildland fire [32]. As shown in Figure 5.9, a radiant heat panel employed in the RadMan system can produce a heat flux of 5 to 21 kW/m². The manikin itself is constructed of a metallic and fiber layered structure to mimic human skin, and a water-cooled system is incorporated on the RadMan skin surface to simulate heat saturation and blood flow effects. Seventy RdF foil sensors are employed in the RadMan system. This type of sensor can reduce measurement errors for long duration exposures. By combining the skin heat transfer model and Henriques' burn injury model, second- and third-degree burn times are obtained.

5.3.2 Evaluation Index Development

5.3.2.1 Stoll Criterion

The Stoll criterion, developed by Stoll and Chianta [33], is extensively employed to characterize performance of materials used in firefighting protective clothing, such as TPP and RPP test. The Stoll criterion is defined as the relationship between heat flux and tolerance time to second-degree burns, which is based on Stoll's research data from a series of human or animal tests. According to the standard ASTM D4108, the relationship between the heat flux and exposure time is given as

$$q = 1.1991t^{0.2901} \qquad (5.1)$$

where q is the heat flux and t is the tolerance time. However, for a more efficient predictor of burn time using the Stoll criterion, most researchers employed the relationship between temperature rise and time to evaluate thermal protective performance. This is because temperature changes on the skin surface are obtained easily compared with the variation in heat flux or thermal energy, for example, using a copper slug sensor. The conversion equation for ASTM E457 is given as

$$T = 8.871t^{0.2905} + T_0 \qquad (5.2)$$

where T_0 is the initial temperature of sensor. The purpose of this equation is simply to determine second-degree burn injuries, but it is confined to the copper slug sensor because different sensors have different calculative formulas between heat flux and temperature. Regarding the previous TPP and RPP values, the burn time is calculated based on the relationship between the temperature rise and exposure time. The calculative equation is written as

$$TPP/RPP = burn\ time \times calibrated\ burner\ or\ radiant\ heat\ flux\ value(cal/cm^2s)$$

$$(5.3)$$

The TPP and RPP tests have been widely employed to rate thermal protective performance for flame-retardant fabrics. To meet the performance requirement of standard NFPA1971, a TPP rating of at least 35 is required for a fabric assembly used in firefighting protective clothing. This rating means that firefighters wearing this protective clothing can withstand a flash fire exposure ($84\ kW/m^2$) for 17.5 s without causing second-degree burns. The RPP rating of fabric for thermal protection specified by NFPA 1977 is more than 20; at this threshold, fabric can provide 20 s of protection time for firefighters under an approximately $42\ kW/m^2$ radiant exposure.

Both the TPP and RPP indexes were respectively replaced by HTP (heat transfer performance) in ASTM F2700-08 [16] and RHR (radiant heat resistance) in ASTM F1939-08 [34]. The HTP and RHR indexes both test the relationship between thermal energy transferred to human tissue and second-degree burn time, since cumulative thermal energy is more precise to predict burn times. Meanwhile, the HTP and RHR have a larger range of applications for other sensors in addition to the copper disk calorimeter, for example, the skin-simulant sensor. However, these evaluation indexes only account for the energy transmitted during exposure, without considering the discharge of stored thermal energy in the fabric system after exposure. Because of this, new evaluation indexes called TPE (thermal performance estimate) in standard ASTM 2703-11 [35] and RHP (radiant heat performance) in standard ASTM F2702-11 [26] have been put forward to effectively assess the stored thermal energy of clothing systems [12].

The TPE and RHP can provide comprehensive skin burn data, including the heat transfer during heat exposure and stored heat transfer during cool down.

According to ASTM F2703-11 [35], an iterative method is required to determine the certification exposure time. At this certification exposure time, a second-degree burn injury can be predicted after considering the stored energy effect. Although iteration techniques for determining the certification exposure time are central to these previous testing methods, as also suggested by other studies [9], they are also time consuming, and sometimes only approximate outcomes can be derived. To obtain a faster and more convenient way to calculate the TPE value, a new evaluation index, called the differentiating factor of second-degree burns, was developed by He and Li. This new index was also developed on the basis of Stoll criteria and defined as

$$f = \frac{\min(S(t) - Q(t))}{S(t)_{@\min}} \tag{5.4}$$

where $\min(S(t) - Q(t))$ stands for the minimum difference between the Stoll criteria and the sensor response throughout the test (J/cm²), as shown in Figure 5.10, and $S(t)_{@\min}$ stands for the value in the Stoll criteria at the time when the minimum difference is detected (J/cm²). When the calculated f is less than zero, the cumulative thermal energy from the sensor is larger than the Stoll curve, indicating the occurrence of a skin burn. On the contrary, there is no burn if f is greater than zero. This new TPE index can be used to evaluate thermal protection of fabrics for any exposure time. A larger f value indicates that the tested fabric can provide better thermal protective performance. When f is equal to zero, the TPE value is obtained. The difference of the TPE value calculated by new and previous methods has been compared, and the results demonstrated by the new method confirm the results obtained by the iterative method.

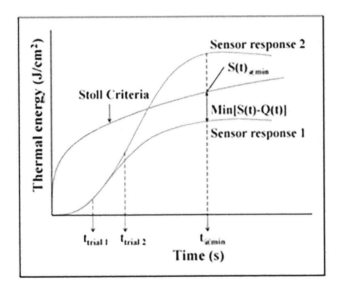

Figure 5.10

The calculative method for the differentiating factor of second-degree burns (f).

5.3.2.2 Skin Heat Transfer and Henriques' Burn Model

The required times to cause second- and third-degree skin burns, which can be calculated for skin temperature at different depths using Henriques' burn integral, are widely employed to evaluate the thermal protective performance of protective clothing systems [36]. Pennes' bio-heat transfer equation, which considers the effect of blood perfusion and metabolic heat on heat transition, has been widely used to simulate skin heat transfer [37]. It is assumed that skin heat transfer is a one-dimensional heat transfer along skin's thickness, that is, the thermal properties of each layer of skin tissue keep constant while each layer is different. Also, the body's metabolic heat can be ignored for short-term heat exposure. The blood temperature is equal to the body's core temperature, and the flow rate of partial blood remains stable. The skin heat transfer model is expressed as [38]

$$(\rho c_p)_{\text{skin}} \frac{\partial T}{\partial t} = \frac{\partial}{\partial x}\left(k_{\text{skin}} \frac{\partial T}{\partial x}\right) + \omega_b (\rho c_p)_b (T_b - T) \qquad (5.5)$$

where ρ_{skin} and $(c_p)_{\text{skin}}$ are, respectively, the density and specific heat of each layer of skin tissue, ρ_b and $(c_p)_b$ are, respectively, the density and specific heat of blood, w_b is the rate of blood perfusion at the dermis layer and subcutaneous tissue, and T_b is the blood temperature.

The skin burn damage at different skin depths is determined by entering the basal temperature of skin surface below 80 μm (or 200 μm) into Henriques' burn integral model, given as

$$\Omega = \int_0^t P \exp\left(-\frac{\Delta E}{RT}\right) dt \qquad (5.6)$$

where Ω is a quantitative measure of burn damage, R is the universal gas constant (8.31 J/mol°C), P and ΔE are the pre-exponential term and the activation energy of the skin, respectively, T is the absolute temperature at the basal layer or at any depth in the dermis, and t is the total time for which T is above 317.15 K. When Ω reaches a value of 1 at the epidermis–dermis interface and the dermis–subcutaneous tissue interface, the corresponding times are treated as second- and third-degree burn time, respectively. Full-scaled manikin tests, such as ISO 13506-2008, ASTM F1930-13, employ this method to assess second- or third-degree burn injuries in different body locations. Henriques' burn model is also used in bench top tests including ASTM F2731-11, which can predict the minimum heat exposure time of firefighters during the thermal exposure phase and cooling phase so that it can provide more precise and comprehensive thermal protective data.

Theoretically, Henriques' burn model is suitable for evaluating different degrees of skin burn injuries under all heat exposure conditions, including transition of heat sources and during the cooling phase [39], but there are actually some limitations owing to different parameters related to Henriques' burn model under different heat exposures, for example, pre-exponential term and the activation energy. These values were suggested by Weaver and Stoll [40] for the basal layer and by Takata [41] for the dermal base. The obtained burn time calculates the transmitted thermal energy during exposure and the discharged thermal energy from protective clothing during cooling. In addition, a new index to measure minimum exposure time (MET) to produce second- or third-degree skin burns was introduced to evaluate skin burns caused by transferred and stored thermal energy [9]. An iterative method similar to that used to calculate TPE was used to obtain the MET value and more precisely predict skin burn times.

5.4 Influence Factors of Transmitted and Stored Thermal Energy

The thermal protective performance of clothing is defined as (a) the resistance of heat transfer from heat sources during exposure and (b) heat release stored within protective clothing during the cooling. Previous studies demonstrated that firefighting protective clothing's thermal protective performance is determined by fabrics' physical properties, such as thickness, density, air permeability, specific heat, and thermal conductivity, as well as clothing structure and moisture content within fabric systems. The air gap between protective clothing and the body surface and exposure intensity and duration can also have an important effect on thermal protection. Therefore, the effects of these parameters on transmitted and stored thermal energy in protective clothing, specifically, have also been investigated.

5.4.1 Fabric's Basic Properties

Thermal protective clothing can be a single layer or a multilayer firefighting protective suit. Multilayer protective clothing can store more thermal energy compared with single-layer protective clothing, just as thicker fabric systems have the capacity to store more thermal energy during exposure than thin fabric systems; this includes high-intensity flame [9] and low-intensity radiant heat exposure [3]. According to the equation for stored thermal energy in protective clothing, the capacity for storing thermal energy is dependent on fabric's density, specific heat, and variation rate of temperature [7]. A multilayer or thicker fabric system can increase the volume heat capacity and contain more thermal energy, which can reduce the heat transfer rate to improve thermal protective performance. He and Li [8] analyzed the effect of fabric system thickness on the release rate of stored thermal energy after heat and flame exposure. It was found that thicker fabric systems took longer to release stored thermal energy, thus decreasing the amount of stored energy discharged toward skin tissue during cooling.

Barker et al. [31] used a stored thermal energy tester to measure the influence of different fabric layers, including the outer shell, moisture barrier, thermal liner, and reflective trim and reinforcement, on overall thermal protection. The outer shell fabrics and thermal liner components were found to influence protective performance, but these effects were not as pronounced as those produced by differences in moisture barrier permeability and trim porosity. Protective clothing systems incorporating moisture barriers, nonporous reflective trim, or nonporous reinforcement generally increased transmitted thermal energy under low-level thermal radiation, owing to the reduction of air volume trapped in the system.

The effect of fabrics' optical properties, specifically, on transmitted and stored thermal energy was studied using a radiative heat transfer model for low-level thermal radiation [42]. The results indicated that self-emission in multilayer fabric systems increases not only the rate of thermal energy transferred to the human skin during thermal exposure but also the rate of thermal energy transmitted to the ambient environment during cooling. A fabric's optical properties have a complex influence on the transmitted and stored energy in multilayer protective clothing. The optical properties of fabric do not affect the capacity for stored thermal energy within a fabric system, for example, while a fabric's reflectivity can change the rate of storing energy during exposure and the rate of discharging energy during cooling.

5.4.2 Air Gap Size

When a firefighter wearing protective clothing performs a firefighting task, the air gap below the clothing will be unevenly distributed [43]. The air gap size is dependent not only on the body shape but also on fabric properties and garment size [44]. The air gap between clothing and the body plays an important role in heat transfer and storage in protective clothing, since dead air is a good insulator compared with common fire-resistant fabrics [45]. Multiple studies have shown that the air gap can decrease the heat transfer rate from protective clothing to the body surface during an exposure [22,46].

However, when the air gap size exceeds a critical value, the air gap can have a negative effect on thermal protective performance, and the heat transfer between protective clothing and body may increase due to the occurrence of convective heat transfer in the air gap [47]. Flow visualization experiments were conducted by Torvi et al. [47] to investigate the occurrence of natural convection, and the results indicated that the optimum air gap size was about 6–7 mm. A fabric system with air gaps of 9–12 mm at different moisture content levels seemed to provide maximum thermal protection from flash fires [48]. In addition, air gaps can affect stored thermal energy in protective clothing during exposure as well as the discharge of thermal energy from protective clothing during cooling. Song et al. [4] performed a thermal protective test with and without an air gap (6.4 mm) based on an improved TPP test device to measure the effect of stored thermal energy. The results showed that an air gap can increase heat storage within protective clothing and aggravate skin burns if the heat energy is discharged to the skin. A study by He and Li [8] also indicated that the air gap can significantly increase the amount of stored thermal energy in protective clothing during heat and flame exposure, impede the discharge of thermal energy stored in protective clothing during cooling, and increase the time needed to release thermal energy.

For low-level radiant heat exposure, Su et al. [7] developed a heat transfer model of transmitted and stored heat through multilayer protective clothing to study the effect of air gap size on total stored energy. For the selected air gap ranges of 0–6.4 mm, the model predicts a decrease in total stored thermal energy when the air gap size decreases. However, there was no significant effect on the total stored thermal energy when the air gap size was more than 1.6 mm. As the air gap size increased, there was an upward trend in the percentage of stored energy transferred to the skin tissue. It is evident, then, that stored energy plays a more important role in larger air gap size.

5.4.3 Moisture Content in Protective Clothing

Moisture contained in protective clothing has a complicated effect on clothing's heat transfer and heat storage under various thermal exposures. Barker et al. [49]

studied the impact of moisture content on thermal protective performance under radiant heat flux at 6.3 kW/m^2 and found that multilayer fabrics with 15% moisture content showed the worst thermal protective performance. The effect of skin sweat on stored energy was evaluated using a simple pre-wet method in their experiment, assuming that the skin perspiration was totally absorbed by the clothing before exposure [50].

The effect of initial moisture content on transmitted and stored energy in protective clothing was also investigated by Song et al. [9] using TPP/RPP test and stored energy approaches. The traditional method tested heat transfer during exposure but did not address the effect of stored thermal energy on thermal protective performance. When the discharge process for stored thermal energy was considered using a stored energy approach, the addition of moisture increased the minimum exposure time for a fabric system without an air gap, while decreasing the minimum exposure time for a fabric system with an air gap. For the traditional TPP/RPP test, the addition of moisture content at 15% to protective clothing showed a decrease in predicted skin burn time.

The complex effect of moisture is attributed to two portions of thermal energy causing skin burn injuries: transferred heat during exposure and discharged heat during cooling. The addition of moisture at high thermal capacity can increase the capacity to store more thermal energy during exposure, while the moisture within protective clothing can evaporate with the rising of fabric temperature and condensate on the fabric and skin surface during the cooling. These changes result in the complex effect of moisture on heat transfer in protective clothing, depending on factors such as moisture content, moisture distribution in multilayer protective clothing, and exposure time and level [51,52].

5.4.4 Exposure Intensity and Duration

Firefighters are subjected to heat exposures of varying intensities during a fire event. Several studies investigated the effect of heat intensity on clothing's thermal protection and heat storage. Song et al. [3] studied thermal protection at three levels of exposure intensity (6.3, 7.5, and 8.3 kW/m^2), showing that the capacity of stored thermal energy during exposure increased with rising heat flux. This is because higher intensity heat exposure can increase the temperature difference between the front and back of protective clothing. However, it was also reported that thermal energy causing skin burn injuries increased during stored energy discharge during cooling when heat flux was measured as less than 2.4 kW/m^2 [6]. When heat flux increased to 20.9 kW/m^2, the burn mainly depended on heat transfer during the exposure itself [6].

In addition, the duration of exposure and cooling times has been showed to have an important influence on heat transfer in protective clothing. Barker et al. [31] studied the effect of exposure time (60, 90, 120 s) on protective clothing's thermal protection when exposed to intensity of 8.4 kW/m^2. It was found that skin burns were determined by heat release stored in clothing during cooling for shorter heat exposures (60 s). For longer heat exposures (90 and 120 s), the burn mainly depended on heat transfer during exposure, while discharge of stored energy could reduce skin burn time. He and Li selected heat and flame exposure of 84 kW/m^2, and a heat exposure duration of 6 to 27 s with increments of 3 s, to investigate the effect of exposure time on transmitted and stored energy. Their findings showed that the total stored energy in a selected fabric system ranged from 8.5 to 13.7 J/cm^2 over this exposure time. For a short exposure time, both transmitted energy and stored energy presented increased with the rising of exposure duration. For a longer exposure time, the amount of transmitted energy continued to increase, while the amount of stored energy decreased.

The reason for this shift might be that a fabric system has a limited capacity to store thermal energy during a specific heat exposure. After reaching the full thermal energy storage capacity, any additional thermal energy from external environment is fully transferred to skin tissue. The effect of cooling time on the discharge of stored energy was analyzed in their study. The effective cooling time to discharge stored thermal energy was 9–40 s for the selected fabric system. A longer effective cooling time was observed for longer heat exposures and thicker fabric systems, meaning the lesser heat discharging rate during the cooling. The safe time range for heat exposure duration and cooling duration was obtained to provide effective instruction to the firefighter during fire extinguishing procedures.

5.4.5 Compressive Effect During the Cooling

The protective clothing worn by firefighters while extinguishing a fire can be compressed due to body movement or contact with an object. The effect of compression on heat transfer and heat storage in protective clothing has been studied by several researchers in recent years. Applied compression reduces the air volume existing in the fabric system and decreases the thermal insulation provided by the air and the thickness of each layer's fabric, thus affecting the transmitted and stored energy. In experimental tests measuring the air gap between protective clothing and the body, compression was found to increase quickly the heat discharging rate from thermal energy stored in protective clothing.

The method described in ASTM F2731-11 [30] can be used to investigate the effect of fabrics' stored energy transferring to the body due to subsequent compression (13.8 kPa). Song et al. employed a stored energy approach to investigate the effect of moisture on thermal protection with/without the compression. When compression was applied to a fabric system with a spacer after the exposure, the minimum exposure time showed an increase with the addition of moisture as compression significantly reduced thermal and contact resistances. On the contrary, the addition of moisture reduced the minimum exposure time without compression.

Compression brings surfaces closer together and displaces air. This results in the transfer of heat between outside surfaces and clothing layers. Therefore, compressed protective clothing increases the risk of being burned by the conduction of heat and the displacement of insulating air between and within the layers of clothing. Su et al. [53] developed a heat transfer model in protective clothing during the compression. It was concluded that fabric thickness and thermal conductivity are both important factors to affect compressive heat transfer under various applied pressures, and that thermal protective performance of multilayer clothing systems decreases as the test temperature and contact pressure increase.

5.5 Future Trends

Transmitted heat during exposure, and discharged heat after exposure, determine skin burn injuries and protective clothing performance. Stored thermal energy is a potential hazard that can exacerbate burn injury, especially when the protective clothing is compressed with body movement or other contact.

Recent preliminary studies investigated the relationship between transmitted and stored thermal energy and its effect on skin burn injuries. The results demonstrated than the stored thermal energy could contribute to burn injuries. For future research, a stored thermal predictive model with compression and considering moisture effect would be helpful in protective clothing engineering.

Numerical simulation methods will help us to understand the mechanisms associated with heat and moisture transfer among multiple layers of clothing system. Increased storage of thermal energy capacity within protective clothing can potentially improve thermal protective performance, but it requires proper management of heat discharge. Phase change materials act as a heat sink and can absorb additional heat during exposure, and the incorporation of phase change materials into textiles can increase the capacity of latent heat storage. As a result, the discharge, either from nature or compression, becomes critical in thermal protection.

In addition, firefighters, in battling fire and during emergency rescue, usually get wet either from sweat due to the high-intensity activity and high-temperature environments or firefighting [54]. Continuous perspiration during exposure should be simulated to evaluate moisture effect on stored thermal energy. This perspiration, characterized by high thermal conductivity and specific heat, has the potential to increase the capacity of thermal energy storage and the rate of heat release in protective clothing systems. Finally, the safety operating time (safety exposure time and safety cooling time) should be defined to avoid skin burn injuries fighting fires.

References

1. R. F. Fahy, P. R. LeBlanc, J. L. Molis, N. F. P. Association, Firefighter fatalities in the United States—2005, National Fire Protection Association. Fire Analysis and Research Division, 2006.
2. M. J. Karter, N. F. P. Association, Patterns of firefighter fireground injuries, National Fire Protection Association Quincy, 2009.
3. G. Song, S. Paskaluk, R. Sati, E. M. Crown, J. D. Dale, M. Ackerman, Thermal protective performance of protective clothing used for low radiant heat protection, *Textile Research Journal*, 81(3) (2010) 311–323.
4. G. Song, F. Gholamreza, W. Cao, Analyzing thermal stored energy and effect on protective performance, *Textile Research Journal*, 81(11) (2011) 1124–1138.
5. R. Barker, C. Guerth, W. Behnke, M. Bender, Measuring the thermal energy stored in firefighter protective clothing, ASTM Special Technical Publication, 1386 (2000) 33–44.
6. E. U. Eni, Developing Test Procedures for Measuring Stored Thermal Energy in Firefighter Protective Clothing (2005).
7. Y. Su, J. He, and J. Li, Modeling the transmitted and stored energy in multilayer protective clothing under low-level radiant exposure, *Applied Thermal Engineering*, 93 (2016) 1295–1303.
8. J. He, J. Li, Analyzing the transmitted and stored energy through multilayer protective fabric systems with various heat exposure time, *Textile Research Journal* (2015).
9. G. Song, W. Cao, F. Gholamreza, Analyzing stored thermal energy and thermal protective performance of clothing, *Textile Research Journal*, 81(11) (2011) 1124–1138.

10. J. R. Lawson, Fire fighter's protective clothing and thermal environments of structural fire fighting, US Department of Commerce, Technology Administration, National Institute of Standards and Technology, 1996.
11. ASTM F1939-15, Standard Test Method for Radiant Heat Resistance of Flame Resistant Clothing Materials with Continuous Heating, ASTM International, West Conshohocken, PA, 2015.
12. W. P. Behnke, Thermal protective performance test for clothing, *Fire Technology*, *13*(1) (1977) 6–12.
13. ASTM F1060-16, Standard Test Method for Evaluation of Conductive and Compressive Heat Resistance (CCHR), ASTM International, West Conshohocken, PA, 2016.
14. H. Zhang, Human thermal sensation and comfort in transient and non-uniform thermal environments, Center for the Built Environment (2003).
15. NFPA Association, Standard on Protective Ensembles for Structural Fire Fighting and Proximity Fire Fighting, National Fire Protection Association, 2006.
16. ASTM F2700-08(2013), Standard Test Method for Unsteady-State Heat Transfer Evaluation of Flame Resistant Materials for Clothing with Continuous Heating, ASTM International, West Conshohocken, PA, 2013.
17. ISO (2003), Clothing for protection against heat and flame determination of heat transmission on exposure to both flame and radiant heat, ISO/TC 94/SC13 ISO 17492, ISO: Geneva, Switzerland, 2004.
18. G. Song, Modeling thermal protection outfits for fire exposures (2003).
19. J. E. Sipe, *Development of an instrumented dynamic mannequin test to rate the thermal protection provided by protective clothing*, Worcester Polytechnic Institute, 2004.
20. W. P. Behnke, A. J. Geshury, R. L. Barker, Thermo-Man® and Thermo-Leg: Large Scale Test Methods for Evaluating Thermal Protective Performance, in: *Performance of Protective Clothing*: Fourth Volume, ASTM International, 1992.
21. F. Henriques Jr., Studies of thermal injury; the predictability and the significance of thermally induced rate processes leading to irreversible epidermal injury, *Archives of Pathology*, *43*(5) (1947) 489–502.
22. C. M. J. Sawcyn, Heat transfer model of horizontal air gaps in bench top testing of thermal protective fabrics, University of Saskatchewan Saskatoon, 2003.
23. ASTM. F1930-15, Standard Test Method for Evaluation of Flame Resistant Clothing for Protection Against Fire Simulations Using an Instrumented Manikin, ASTM International, West Conshohocken, PA, 2015.
24. ISO (2004), Protective clothing against heat and flame test method for complete garments prediction of burn injury using an instrumented manikin, ISO/TC 94 ISO/DIS 13506, ISO: Geneva, Switzerland, 2004.
25. NFPA Association, Standard on protective clothing and equipment for wildland fire fighting, National Fire Protection Association, 1998.
26. ASTM. F2702-15, Standard Test Method for Radiant Heat Performance of Flame Resistant Clothing Materials with Burn Injury Prediction, ASTM International, West Conshohocken, PA, 2013.
27. R. Barker, H. Hamouda, I. Shalev, J. Johnson, Review and Evaluation of Thermal Sensors for Use in Testing Firefighters Protective Clothing. Annual Report (1999).

28. J. R. Lawson, W. H. Twilley, Development of an Apparatus for Measuring the Thermal Performance of Fire Fighters' Protective Clothing, US Department of Commerce, Technology Administration, National Institute of Standards and Technology, 1999.

29. T. Neal, R. Swain, W. Behnke, Firefighter's protective clothing stored energy test development, TPP Task Group Report, NFPA (1971).

30. ASTM. F2731-11, Standard Test Method for Measuring the Transmitted and Stored Energy of Firefighter Protective Clothing Systems, ASTM International, West Conshohocken, PA, 2011.

31. R. L. Barker, A. S. Deaton, K. A. Ross, Heat Transmission and Thermal Energy Storage in Firefighter Turnout Suit Materials, *Fire Technology*, *47*(3) (2011) 549–563.

32. K. L. Watson, From Radiant Protective Performance to RadMan™: The Role of Clothing Materials in Protecting against Radiant Heat Exposures in Wildland Forest Fires (2014).

33. A. M. Stoll, M. A. Chianta, A method and rating system for evaluation of thermal protection, DTIC Document, 1968.

34. ASTM. F1939-15, Standard Test Method for Radiant Heat Resistance of Flame Resistant Clothing Materials with Continuous Heating, ASTM International, West Conshohocken, PA, 2015.

35. ASTM. F2703-13, Standard Test Method for Unsteady-State Heat Transfer Evaluation of Flame Resistant Materials for Clothing with Burn Injury Prediction, ASTM International, West Conshohocken, PA, 2013.

36. F. C. Henriques Jr., Studies of thermal injury; the predictability and the significance of thermally induced rate processes leading to irreversible epidermal injury, *Archives of Pathology*, *43*(5) (1947) 489–502.

37. D. A. Torvi, J. D. Dale, A finite element model of skin subjected to a flash fire, *J Biomech Eng*, *116*(3) (1994) 250–255.

38. D. A. Hodson, G. Eason, J. C. Barbenel, Modeling transient heat transfer through the skin and superficial tissues—1: Surface insulation, *Journal of Biomechanical Engineering*, *108*(2) (1986) 183–188.

39. L. N. Zhai, J. Li, Prediction methods of skin burn for performance evaluation of thermal protective clothing, *Burns: Journal of the International Society for Burn Injuries* (2015).

40. J. A. Weaver, A. M. Stoll, Mathematical model of skin exposed to thermal radiation, *Aerospace Medicine*, *40*(1) (1969) 24–30.

41. A. Takata, Development of criterion for skin burns, *Aerospace Medicine*, *45*(6) (1974) 634–637.

42. Y. Su, J. He, J. Li, An improved model to analyze radiative heat transfer in flame-resistant fabrics exposed to low-level radiation, *Textile Research Journal*, *87*(16) (2016) 1953–1967.

43. G. Song, Clothing air gap layers and thermal protective performance in single layer garment, *Journal of Industrial Textiles*, *36*(3) (2007) 193–205.

44. Y. Lu, G. Song, J. Li, A novel approach for fit analysis of thermal protective clothing using three-dimensional body scanning, *Applied Ergonomics*, *45*(6) (2014) 1439–1446.

45. Y. A. Cengel, *Heat transfer: A practical approach*, McGraw-Hill, 2003.

46. D. A. Torvi, J. Douglas Dale, B. Faulkner, Influence of Air Gaps On Bench-Top Test Results of Flame Resistant Fabrics, *Journal of Fire Protection Engineering*, *10*(1) (1999) 1–12.

47. D. A. Torvi, J. D. Dale, Heat transfer in thin fibrous materials under high heat flux, *Fire Technology, 35*(3) (1999) 210–231.

48. Y. Lu, J. Li, X. Li, G. Song, The effect of air gaps in moist protective clothing on protection from heat and flame, *Journal of Fire Sciences, 31*(2) (2013) 99–111.

49. R. L. Barker, C. Guerth-Schacher, R. Grimes, H. Hamouda, Effects of moisture on the thermal protective performance of firefighter protective clothing in low-level radiant heat exposures, *Textile Research Journal, 76*(1) (2006) 27–31.

50. R. L. Barker, Effects of Moisture on the Thermal Protective Performance of Firefighter Protective Clothing in Low-level Radiant Heat Exposures, *Textile Research Journal, 76*(1) (2006) 27–31.

51. M. Fu, M. Q. Yuan, W. G. Weng, Modeling of heat and moisture transfer within firefighter protective clothing with the moisture absorption of thermal radiation, *International Journal of Thermal Sciences, 96* (2015) 201–210.

52. S. Nazaré, D. Madrzykowski, A Review of Test Methods for Determining Protective Capabilities of Fire Fighter Protective Clothing from Steam (2014).

53. Y. Su, J. He, J. Li. A model of heat transfer in firefighting protective clothing during compression after radiant heat exposure, *Journal of Industrial Textiles.* 2016: 1528083716644289.

54. J. R. Lawson, Fire fighters' protective clothing and thermal environments of structural fire fighting, ASTM Special Technical Publication, 1273 (1997) 334–335.

6

Functionality and Thermophysiological Comfort of Firefighter Protective Clothing

A Case Study

Olga Troynikov and Nazia Nawaz

6.1 Introduction

Millions of years of evolution have resulted in the human body developing a complex and dynamic physiological response to changing environmental conditions. The body aims to maintain its optimum state of thermophysiological comfort by altering the thermal balance between itself and the environment. To avoid heat stress in hot surroundings, the proper balance between body heat production and heat loss to the environment is crucial. Transportation of heat and moisture through the human–clothing–environment system achieves this balance (Li 2001; Li & Holcombe 1998).

Clothing that maintains proper thermal balance is obviously crucial for firefighters, both in terms of their work performance and survival. In recent years, there has been a growing recognition of the importance of incorporating physiological and ergonomic comfort factors in the design of clothing for firefighters and others who must work or perform in high temperatures. Firefighters' clothing is designed primarily to protect the wearer from fire and extremely hot environments. These garments enable firefighters to perform tasks quickly, effectively, and with minimum energy expenditure, and further increase safety by maximizing their visibility. In addition, firefighters' apparel must provide protection from mechanical and chemical hazards.

Due to these many and varied performance demands, firefighters' protective garments are typically bulky, heavy, and uncomfortable, imposing a physiological burden on the wearer and in many cases exacerbating heat stress (Chung & Lee 2005; Cheung, Petersen, & McLellan 2010; Rossi 2005). The same applies to other protective systems and garments worn during physical activity performed in extremely hot conditions, such as military operations or high-energy sport. The engineering challenge is to produce protective materials, material structures, and clothing systems that protect against the hazard while maximizing the wearer's thermal comfort (Song, Barker, Hamouda, Kuznetsov, Chitrphiromsri, & Grimes 2004; Torvi 1997); in other words, materials and clothing that provide the best balance between the physiological burden and protection, enabling optimal performance. Moreover, optimal interaction between the human body, clothing, and the microclimate between skin and material is vital; ill-fitting protective apparel can impose additional thermophysiological burden.

The shielding attributes of firefighters' protective garments depend on the relevant hazards; as already noted, they have to be waterproof, fireproof, chemically impermeable, impact-resistant, and so on. The protective performance attributes of firefighters' clothing are currently well defined and highly regulated (for example, ISO 6942-2002, ISO 9151-1995). The heat transfer index enables the ability of clothing structures to delay the transfer of heat from an open flame to be assessed. Indices must be between 3 and 6 for a single-layer material and 13–20 for triple-layered material structures. However, attributes of thermophysiological comfort or the thermophysiological performance of materials are not so clearly defined and regulated. For example, practice in some industrial sectors indicates that for any single layer of a material structure, thermal resistance (ISO 11092-1993) should be <0.055 m^2KW^{-1} and water vapor resistance <10 m^2PaW^{-1}. However, this is far from universally accepted, and researchers continue to develop and evaluate new materials and standards.

The extremely hazardous environments in which firefighters work preclude the use of human subjects in developing and testing protective clothing. Instead, researchers employ laboratory methods such as flame test manikins (Kim, Kim, Lee, & Coca 2017; Li, Shen, & Guo 2016) and flame and radiant test devices (Tessier 2018). Modeling of human physiological responses to different protective systems has become highly sophisticated, including "virtual" computational thermal manikins, physiological models, and thermal comfort analysis (Cheng, Niu, & Gao 2012; Foda, Almesri, Awbi, & Sirén 2011). These models permit safe, rapid evaluation of environments, materials, and protective systems and their structures before prototyping begins.

This chapter addresses a case of application of new materials to firefighting protective systems, with a focus on thermophysiological human comfort.

Objective laboratory testing and mathematical modelling were used to test the protective materials and their structures. In addition, we assessed the fit and ergonomics of protective garments and their effect on the thermal status of the wearer using 3D body scanning technology.

6.2 Materials and Material Structures

As described above, firefighters' apparel is intended to protect the wearer against thermal hazards such as high-temperature radiant sources, flame impingement, hot liquids and gases, molten substances, and hot solids and surfaces (Ajayi 1992). Firefighting is exhausting physical work that generates considerable body heat, and the extremely hot working environment elevates core body temperature substantially. Losing body heat through evaporation of sensible liquid sweat or through insensible vapor is vital to avoid heat stress in such circumstances (Stapleton, Wright, Hardcastle, & Kenny 2012).

Firefighters' protective garments involve layered material structures typically composed of a flame-resistant outer shell and several inner layers (Figures 6.1 and 6.2). Inner layers are usually a moisture barrier, a thermal barrier, and an

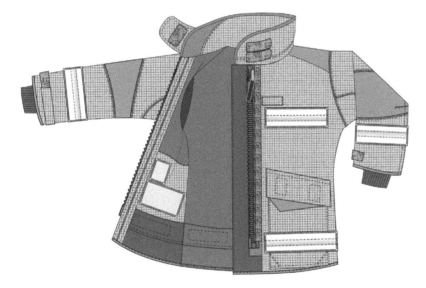

Figure 6.1

A firefighter's protective jacket.

Figure 6.2

Material layered structures: Generation I (left), Generation IV (right).

inner liner. When firefighters are at work, their body heat and perspiration are partially trapped inside the protective clothing. To balance thermal protection and the physiological comfort of the wearer and avoid heat stress, the system of protective garments and material layered structures must be designed carefully (Chung & Lee 2005).

6.3 Material Attributes Relevant to Thermal Comfort

Here we consider a comparison of textile materials and their structures (Figure 6.2) from two generations of firefighters' protective clothing (Figure 6.1): "Generation I" and "Generation IV." We begin with an evaluation of their thermal, vapor, and liquid moisture transfer properties using traditional objective methods.

6.3.1 Generation I and Generation IV: Thermal, Vapor, and Liquid Moisture Transfer Properties

Both protective garments are constructed of three-layered textile material structures, but the textile components of each layer are different. Both structures' outer layers, which provide protection from fire and radiative heat, are a blend of polybenzimidazole (PBI) fibers and Kevlar® fibers. PBI is a synthetic fiber with a high melting point and low heat-deflection temperature; its thermal and chemical stability results in its low flammability.

The Generation I structure's middle layer is made from a blend of Nomex® and Kevlar® fibers using the Sontara® spunlace process. It incorporates a barrier membrane to protect the inner layer from environmental moisture. Two layers of Sontara® quilted to Nomex®/Flame-retardant (FR) Viscose woven fabric (an artificial cellulosic fiber incorporating phosphorous flame retardant in the viscose fiber matrix) form the inner layer.

The Generation IV structure's middle layer is a Crosstech® moisture barrier membrane with added heat-stable and chemically inert spacer Airlock® technology. This creates extra air spaces between the outer and middle layers, lightening the overall structure while improving thermal protection, and avoiding the need for heavy additional thermal protective layers as in the inner layers of Generation I structures. A thin and light 100% Nomex® woven fabric constitutes the inner layer of this structure.

We collated the physical parameters and the thermal, vapor, and liquid moisture transfer properties of the textile layers of Generation I and IV clothing and evaluated their thermophysiological comfort attributes. We estimated the areal weight of fabrics by calculating the mean mass per unit area of five specimens (AS 2001.2.13—Standards Australia 1987). We measured the thickness of material samples and their structures using the distance between the reference plate and parallel presser foot of the thickness tester (AS 2001.2.15—Standards Australia 1989). We measured component fabric density as the number of warp yarns and weft yarns per centimeter in an accurately measured length of fabric (AS 2001.2.5—Standards Australia 1991). We tested the air permeability of fabrics by measuring the air flow rate through them (ISO 9237-1995). We evaluated the fabrics' moisture management properties with a Moisture Management Tester used according to AATCC Test Method 195–2009 (AATCC 195-2011, 2009). A series of indices were defined and calculated: top absorption rate (ARt), bottom absorption rate (ARb), top maximum wetted radius (MWRt), bottom maximum wetted radius (MWRb), top spreading speed (SSt), bottom spreading speed (SSb), accumulative one-way transport index (AOTI), and overall moisture management capacity (OMMC) (Hu, Li, Yeung, Wong, & Xu 2005). The reported OMMC is an index indicating the overall capacity of the fabric to manage the transport of liquid moisture, which includes three aspects: ARb, AOTI, and SSb (Yao, Li, Hu, Kwok, & Yeung 2006).

Fabric thermal resistance (Rct) and water vapor resistance (Ret) were tested using a sweating guarded hot plate (ISO 11092-1993) at air temperature 20°C, relative humidity 65%, and air speed 1 m/s. We calculated the thermal resistance of the fabric after the system reached a steady state using Equation 6.1:

$$Rct = A(Ts - Ta)/H - RcT0 \qquad (6.1)$$

where Rct is the thermal resistance of the fabrics, m^2KW^{-1}; $Rct0$ is the thermal resistance of boundary air layer, m^2KW^{-1}; A is the test section area, m^2; Ts is the plate's surface temperature, K; Ta is the ambient air temperature, K; and H is the electrical power, W, the instrument uses.

The fabric's water vapor resistance was calculated using Equation 6.2:

$$Ret = A(Ps - Pa)/H - Ret0 \qquad (6.2)$$

where Ret is the vapor resistance of the fabrics, m^2Pa/W; $Ret0$ is the vapor resistance of boundary air layer, m^2Pa/W; A is the area of specimen test section, m^2; Ps is the water vapor pressure at the plate surface, Pa; Pa is the water vapor pressure of the air, Pa; and H is the electrical power, W, the instrument uses.

As the outer layers of the Generations I and IV fabrics have similar mass/unit area and thickness (Table 6.1), their comparable thermal and vapor resistance contributions were similar. Both outer layers have zero OMMC, showing that they prevent environmental moisture from reaching the inner layers but cannot transport liquid moisture from the inner microclimate and skin to the environment.

The middle layers differ substantially from the outer layers in their parameters and performance attributes. Generation IV's middle layer is heavier and thicker due to the spacer configuration, but it provides better physiological comfort

Table 6.1 Physical Parameters and Attributes Relevant to Physiological Comfort of Materials

Materials Structure	Functional Layer	Mean Mass/Unit Area (g/m²)	Mean Thickness (mm)	Mean Air Permeability (mm/sec)	Mean Rct (m²K/W)	Mean Ret (m²Pa/W)	Mean OMMC
Gen. I	Outer	204	0.45	58.79	0.014	3.099	0.000
	Middle	96	0.49	n/a	0.048	9.169	0.000
	Inner	276	0.36	82.44	0.067	5.697	0.003
	Layered structure	Σ = 576	Σ = 1.30	n/a	0.203	23.105	0.000
Gen. IV	Outer	206	0.40	75.84	0.008	3.145	0.000
	Middle	250	1.58	n/a	0.013	4.158	0.000
	Inner	132	0.27	234.14	0.003	2.840	0.576
	Layered structure	Σ = 588	Σ = 2.25	n/a	0.081	12.241	0.000

with its *Rct* and *Ret* being less than 50% of those in Generation I's middle layer. Similarly, the inner layer of Generation IV, due to its relative lightness, has a fraction of the relevant values of *Rct* and *Ret* of Generation I's inner layer. However, these values are still substantially higher than those for everyday clothing, meaning protective apparel incorporating Generation IV fabric will still impose a substantial thermophysiological burden on wearers in hot conditions. The liquid moisture transfer properties of the inner lining material of Generation IV are significantly better than those of the inner lining of Generation I. This means the inner layer can more effectively transfer liquid moisture from the wearer's skin surface to the middle layer, thereby increasing the wearer's comfort.

Our results illustrate that Generation IV has higher heat transfer capacity, and greater ability to transfer liquid and vapor from the body to the environment, than Generation I. We conclude that Generation IV's new material structures would improve the thermophysiological comfort of firefighters operating in extremely hot conditions. We follow with the case study of assessment of materials through the use of mathematical modeling and simulation.

6.3.2 Generation I and Generation IV: Mathematical Modeling and Simulation

Mathematical modeling at the garment system level can be used to evaluate the performance of different material structures of firefighters' protective clothing (Yermakova & Candas 2008). To accurately predict the thermophysiological performance of firefighters' clothing, they should be tested in both steady and transient states. Several mathematical models and simulation tools that simulate or predict human thermal physiological status have been developed (Fiala, Lomas, & Stohrer 1999; Wong, Li, & Yeung 2006). However, few researchers have used them to study dynamic changes in the thermal responses of firefighters wearing ensembles constructed of different textile materials and their combinations, especially in hot and humid environments.

Two protective ensembles made of Generation I and Generation IV materials (Table 6.1) were used for modeling; they consisted of a turnout helmet, turnout jacket, turnout overtrousers, turnout boots (fully lined leather structural firefighters' boots), and firefighters' gloves made of leather and a moisture barrier. Undergarments consisted of 100% cotton single jersey T-shirt and briefs. Helmet, boots, gloves, and undergarments remained the same for Generation I and Generation IV ensembles, while the textile materials structures of the jacket and overtrousers varied.

We derived the mathematical model of human heat exchange and thermoregulation utilized here (Troynikov, Nawaz, & Yermakova 2013) from the class of multicompartmental models realized for the prediction of human thermoregulatory responses while clothed and working in different environments (Candas & Yermakova 2008; Yermakova & Candas 2007). The mathematical models are built from differential equations that describe metabolic heat, heat transfer by conduction, convection by blood flow, radiation and convection heat exchange with the environment, and heat loss by evaporation from skin and through respiration and allow for modeling transient and steady-state changes of physiological characteristics such as specific data points and the mean values for temperatures, sweat rate, dripping rate, total water loss, evaporation, maximal evaporation, blood flow, heart rate, and heat flow by convection and radiation. The models include steps designed to evaluate the efficacy of bodies sweating in Generation I and Generation IV protective ensembles in hot environments: evaluation of each ensemble's effect on the thermal responses of firefighters working in a dry, hot environment, evaluation of their effect on thermal responses of firefighters working in a humid, hot environment, and comparative evaluation of these effects. The modeling experiments were carried out under the following conditions: air temperature 30°C, wind velocity 1 m/s, relative humidity 20% and 85%, and work intensity 160 W/m².

Sweat evaporation is the main thermoregulatory response that impedes overheating of the human body in a hot environment. This dominant physiological process can enable the human body to lose sufficient heat to maintain temperature homeostasis. Figure 6.3a and b shows the dynamics of sweating for a man wearing Generation I and IV protective clothing in dry and humid environments. In a dry, hot environment (30°C, RH = 20%), both ensembles are acceptable to the wearer; all sweat is evaporated, and no dripping occurs. The sweat rate

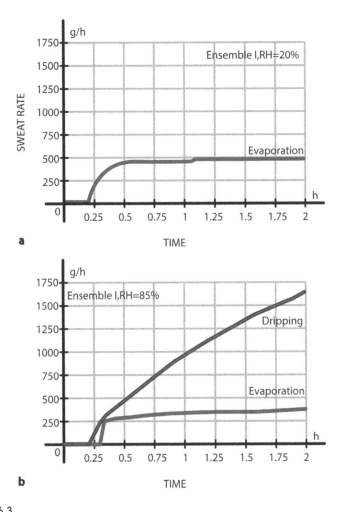

Figure 6.3

Dynamics of evaporation and dripping sweat rate for firefighters in two kinds of protective ensembles in dry (a,c) and humid (b,d) environments. (From Troynikov, O. et al., *Advanced Materials Research*, Vol. 633, Trans Tech Publications, 2013.)

(Continued)

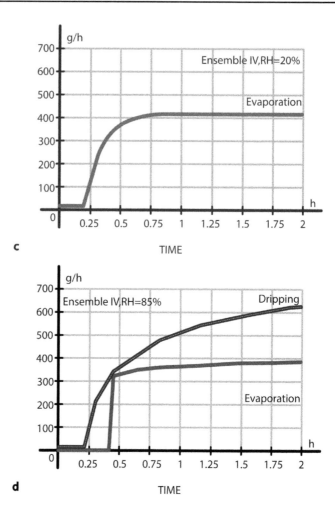

c

d

Figure 6.3 (Continued)

Dynamics of evaporation and dripping sweat rate for firefighters in two kinds of protective ensembles in dry (a,c) and humid (b,d) environments. (From Troynikov, O. et al., *Advanced Materials Research*, Vol. 633, Trans Tech Publications, 2013.)

for the Generation I ensemble is 500 g/h and for Generation IV 400 g/h, proving that a person wearing the more modern ensemble accumulates less heat than a person wearing the older ensemble and can attain the required thermal balance by exuding less sweat.

The results related to work in a humid hot (30°C, RH = 85%) environment (Figure 6.3a,c) are of greatest interest. Neither ensemble is physiologically comfortable during physical activities in a humid environment, but the Generation IV ensemble has a clear advantage over Generation I. Two simultaneous processes occur (Figure 6.3c,d): evaporation of a portion of sweat and dripping of the remainder. The amount of dripping sweat for the Generation I ensemble increases from 0 to 1600 g/h in 2 h, in which time only 260 g/h is evaporated. From a physiological perspective, the ratio of these two outputs for the Generation IV ensemble is preferable, because the rate of sweat evaporation increases from 0 to 400 g/h in 2 h, while dripping increases from 0 to only 600 g/h. Therefore, after

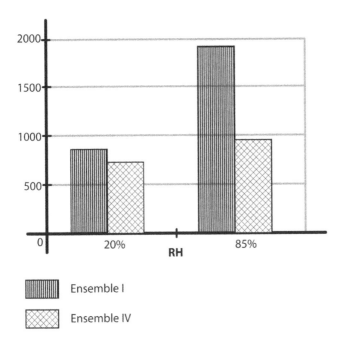

Figure 6.4

Total water loss, g, in 2 h of work in dry and humid environments.

2 h, the Generation IV ensemble generated only 38% as much dripping sweat as the Generation I ensemble.

Mean skin temperature at the end of 2 h of exposure was also substantially different for the two fabric ensembles. Humid conditions increased mean skin temperature to 38.2°C for Generation I and 36.4°C for Generation IV. Heart rate also increased, to 150 and 120 beats/min after 2 h for Generation I and IV ensembles, respectively. There is no doubt that the ensembles' performances during work in the dry and humid environments differ significantly (Figure 6.4). The Generation I ensemble generated total body water loss of 1700 g (more than 2% of body weight) and thus presents a considerable dehydration risk. In contrast, the Generation IV ensemble generated total body water loss of 900 g after 2 h of exposure, presenting no formal risk of dehydration; nonetheless, wearers should still be cautious and be prepared to cease working to avoid it.

The results outlined in this section reinforce our conclusion that the Generation IV ensemble produces better thermophysiological comfort during physical work, especially in hot humid environments, than the Generation I ensemble.

6.4 Use of Absorbent Materials for Improved Comfort

The use of absorbent materials provides another interesting case study in the quest for improved comfort in firefighters' protective clothing. Firefighting conditions almost invariably involve the ambient air temperature being significantly higher than human body temperature; therefore, a negative temperature gradient exists between the skin and the environment. Hence, the human body is unable to lose metabolic heat and its temperature rises (Bartkowiak 2010; Guidotti 1992), causing the body to retaliate with intensive sweating.

As we described in Section 6.1, the moisture management capacity of the middle layer of the assemblies used in firefighters' protective garments is extremely low. When combined with intensive sweating, wearer comfort becomes similarly low. Bartkowiak (2006) demonstrated that using super-absorbent nonwoven material as an undergarment or as a lining in protective clothing improves its comfort characteristics. Hence, our hypothesis in this case study was that "incorporating materials with high moisture absorption capacity improves comfort during working conditions, as they absorb liquid and ... keep the skin dry."

We studied the comfort-related properties of material assemblies with an inner lining of highly absorbent material, such as super-absorbent fiber (SAF), and Coolmax®, a high-wicking polyester fiber (Houshyar, Padhye, Troynikov, Nayak, & Ranjan 2015). SAF not only absorbs sweat but also contains it, even under the pressure of a heavy outer garment. To prevent this material assembly contacting the wearer's skin, a hydrophobic layer was added to the skin side of the assembly, ensuring the fabric surface next to the skin remained dry.

Various compositions of SAF (Technical Absorbents Ltd.) and Coolmax® (Invista Pty Ltd) were used as nonwoven and woven middle layers in material assemblies. The inner layer of the material structures we evaluated was a thin and light 100% Nomex® woven fabric, the same as that used in the Generation IV material assembly considered in our first case study. This SAF consists of a cross-linked acrylate copolymer partially neutralized by the sodium (Na) salt (http://chimianet.zefat .ac.il/download/Super-absorbant_polymers.pdf). We stitched the nonwoven SAF-containing material to the inner layer of the material assemblies and tested their physical parameters (Table 6.2) using the methods applied in our first case study.

The maximum moisture pickup and the drying rate for all five assemblies were measured. After conditioning at standard environmental conditions for 24 h, material assemblies were immersed in distilled water at 20°C for 5 min and weighed after removal from the water and dripping had stopped. The moisture pickup was calculated by subtracting the weight of the conditioned sample from that of the moistened sample. The evaporation rate was measured on samples saturated with distilled water and padded to remove the excess water; the weight loss was measured at 30, 60, 90, 120, 180, 240, and 300 min, and the results from three fabric samples were averaged. Materials thermal resistance (Rct) and water vapor resistance (Ret) were measured utilizing the same methods as described in our first case study.

Table 6.2 Physical Parameters of the Material Assemblies

Material Assembly	Inner Layer	Middle Layer, Fiber Content	Total Thickness, mm
IL	Generation IV (Table 1)	No middle layer	0.27
MA1 Nonwoven		Coolmax/SAF/polyolefin (75/15/10%)	1.07
MA2 Nonwoven		Coolmax/SAF/polyolefin (65/25/10%)	1.67
MA3 Woven		Warp: polyester (100%); Weft: polyester/SAF (75/25%)	1.27
MA4 Nonwoven		Nomex® (DuPont TM Nomex® E89, 100%)	0.50

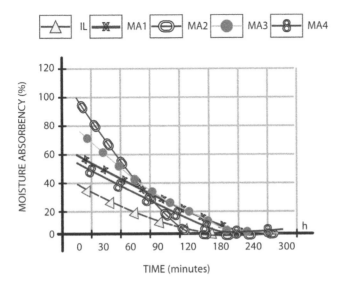

Figure 6.5

The absorbency of the material assemblies.

As Figure 6.5 shows, the SAF and SAF-containing materials in the material assemblies retain significant amounts of moisture despite high rates of evaporation in standard conditions. Evaporation rates will be significantly higher in hot conditions, and we expect the materials containing SAF will hold higher amounts of moisture without any reduction in their evaporative abilities.

Thermal and water vapor resistance values of the material assemblies are summarized in Table 6.3. The MA2 assembly returned the highest R_{ct} value, whereas the IL inner layer returned the lowest R_{ct} value. The highest value of R_{ct} can be attributed to the greatest thickness of that assembly (Bartkowiak 2010).

The effect of the fiber composition on the thermal property of the sample is modified by the SAF materials' liquid transport properties. SAF absorbs more sweat from the skin than the unlined fabric, resulting in less sweat accumulating on the skin surface and thus less wet clinginess. SAF retains liquid inside by absorbing water and water vapor, improving the hygienic properties of the ensemble. Li, Wang, and Ren (2005) stated that wearers' perception of comfort was positively related to skin temperature, but nonlinearly and negatively related to RH in the clothing microclimate. Therefore, introducing highly absorbent materials will reduce the microclimate's humidity within the clothing ensemble and improve the wearer's comfort. However, this property can change over time; SAF can eventually become saturated due to high sweat generation and low evaporation. This would raise the humidity of the clothing microclimate, leading to sensations of clinginess and discomfort. Nevertheless, incorporating optimized

Table 6.3 Thermal and Water Vapor Resistance Values of the Material Assemblies

Fabric/ Parameters	IL	MA1	MA2	MA3	MA4
Ret (m²Pa/W)	2.36 ± 0.07	5.99 ± 0.13	7.51 ± 0.29	5.13	4.99 ± 0.36
Rct (m²K/W)	0.03 ± 0.001	0.05 ± 0.01	0.08 ± 0.001	0.04	0.05 ± 0.001

amounts of super-absorbent materials into garment structure can improve the level and duration of comfort without compromising garment performance.

6.5 Effect of Surface Characteristics of Inner Layer Fabrics on Comfort

As noted already, appreciation of the need for greater consideration of physiological comfort factors in the design of effective firefighters' clothing has grown in recent years. Sensorial or tactile comfort, usually described as "handle" or "hand," is an important part of overall physiological comfort and can be evaluated in multiple ways. The most commonly used objective method of evaluation of comfort in fabrics and materials involves the four modules of the Kawabata System (KES), used in place of the evaluation of the panel of trained experts (Kawabata 1973, 1980; Barker 2002). Research shows that it is possible to predict and evaluate the sensorial comfort of clothing and materials by reducing the many sensory and mechanical properties that can be measured on fabrics to a few independent components (Cardello 2008). Note that the assessment of fabrics using KES and other instrumental methods is comparative and requires a large amount of comparative data to categorize the textiles being assessed (Du & Yu 2007).

Friction between fabric and skin and surface roughness are important elements in evaluating the sensorial comfort of materials. The frictional properties of fibrous assemblies are generally assessed using a modified adhesion model, in which friction between textile substrates depends mostly on the area of contact, the surface geometry, the applied load, and (sometimes) the duration of contact. However, ploughing or shearing of surfaces in contact is also involved in sliding friction; for the sliding process to continue, the asperities must be sheared (Ramkumar 1998). Briscoe and Tabor (1978) correlated the physical properties of textiles to their frictional characteristics, and Briscoe (1996) showed that the apparent area of contact and the ploughing component contribute independently to net frictional force. As fabric moves along the skin, the dynamic contact of textile material with the skin increases and roughness or smoothness is perceived. In most cases, the frictional force is positively correlated with the roughness of the fabric's surface (Bhupender & Yehia 1991). Moisture in fabric–skin contact increases the frictional force and thus the perception of roughness: a fabric that is comfortable in low humidity may become uncomfortable in high humidity or during sweating (Li 2001). Kenins (1994) evaluated the effects of fiber type and moisture on fabric-to-skin friction, finding that moisture on the skin is more critical than fiber type or fabric construction parameters. In addition, Hes, Marmarali, Ozdil, Oglakcioglu, and Lima (2008) stated that a fabric produces an uncomfortable feeling to the underlying skin due to the high moisture absorption leading to the higher friction between them. In other words, skin moisture and friction strongly enhance unpleasant perceptions of clothing in hot and damp environments, and therefore clothing designed for hot and moist environments should minimize skin wetness and the sensation of surface roughness (Kenins 1994).

Ajayi (1992) found that increased density of weft yarn sett with the yarn count being constant was associated with heightened frictional force, and speculated that this was due to increased yarn crimp, producing a "knuckle" effect. Ajayi and Elder (1997) later studied the effect of yarn surface properties on fabric hand, concluding that yarn structure and fabric friction are positively correlated; that is, for a given fabric structure and finishing treatment, yarns with a higher coefficient of friction will yield fabrics with a higher coefficient of friction. In addition, Ajayi and Elder (1994) also asserted that yarns' frictional properties are related to their fiber content, and yarn structure and surface hairiness affect fabric friction.

Kondo (2002) studied the frictional properties of plain weave fabrics and their relationship to human comfort using various textile materials. He found that the coefficients of friction (MIUs) of fabrics made of spun yarn were higher than those made of fabrics made of filament yarns. A study of the handle properties of microdenier polyester filament yarns concluded that surface friction did not show any fixed trend with the change from coarser filament to finer microdenier filament; however, the results showed clearly that the finest deniers gave the lowest surface roughness (Behera & Sobti 1998).

Fabric friction is reportedly influenced by many factors other than fabric structure, such as fiber type, fiber blend ratio, fabric compressibility, and yarn structure. Thus, if the frictional properties of different component materials differ, their use in fabrics and blend ratios influences the resultant fabrics' frictional properties (Apurba & Nagaraju 2005). Moreover, fabric treatments such as bleaching, dyeing, and finishing have been observed to increase disturbance in surface fibers and fabric surface irregularities in cotton knitted fabrics, amplifying their friction and roughness (Hassani 2010). Laundering also changes the surface properties of materials, depending more on fabric structural parameters than on laundering temperature (Quaynor & Nakajim 2000).

The conclusion to draw from the evidence presented above is that a fabric's surface characteristics and physical attributes are crucial to the wearer's overall tactile comfort. For comfortable wear, a fabric has to allow garments to slide easily on themselves or on the surface inside them—most importantly, the skin. Clothing that does not slide easily during donning, doffing, use, or wear causes an uncomfortable sensation and increases the energy involved in flexing joints, movement, and performing normal human functions (Gwosdow, Stevens, Berglund, & Stolwijk 1986).

6.6 Effect of Surface Characteristics of Lightweight Fiber-Blended Fabrics on Comfort

Our third case study consists of an evaluation of the surface characteristics of lightweight Nomex® and Nomex® fiber-blended fabrics employed as the inner layer of firefighters' apparel, such as in the Generation IV example previously (Figure 6.2). The effects of fabric composition, construction, and structural parameters on the fabric surface with respect to the practical wear and maintenance of garments were investigated. In addition, the influence of moisture, specifically sweat absorption during strenuous physical activity, on the surface characteristics of the fabrics was examined.

We investigated the surface properties and characteristics of the fabric samples under various conditions: in virgin state, after washing and drying, after abrasion, and under moist conditions. (Virgin fabrics are materials that have received no commercial treatment since final production, as available to a garment manufacturer; they have not been washed, dried, rubbed, or had any further chemicals applied.) The standard deviation for the MIU and SMD of individual samples of each fabric group was calculated to demonstrate the variation in their surface

Table 6.4 Experimental Fabric Details

Fabric	Fabric Description	Fiber Composition,%	Yarn Structure	Weave/Knit
F1	Nomex® INSULTEX™ Runner	100% Nomex®	Filament	Plain weave
F2	Nomex® Delta C™ Scrim	100% Nomex®	Long-staple yarns	Plain weave, Rip-stop
F3	TenCate Aralite®	100% Nomex®	Staple	Plain weave
F4	Nomex®/FR Viscose	50% Nomex®, 50% FR Viscose	Staple	Plain weave, Rip-stop
F5	Nomex®/FR Viscose liner Black	50% Nomex®, 50% FR Viscose	Staple	Plain weave, Rip-stop
F6	Nomex®/FR viscose liner Blue	50% Nomex®, 50% FR Viscose/aramid	Staple	Plain weave, Rip-stop
F7	Nomex® Delta C™ liner	93% Nomex®, 5% Kevlar®, 2% P140	Long-staple yarns**	Plain weave, Rip-stop

properties. Student's t-test was used to identify statistically significant differences between fabrics.

We selected seven commercial woven fabrics suitable for an inner liner layer of protective garment assemblies of different weaves, yarn constructions, and fiber compositions (F1–F7) (Troynikov & Nawaz 2010), as shown in Table 6.4.

Each fabric was treated as follows prior to evaluation of its influence on wear and care:

- Group 1 (F1–F7) were virgin commercially produced fabrics in their state (we applied no further treatment).
- Group 2 (2F1–2F7) were samples F1–F7, washed at 40°C and tumble-dried at low temperature five times according to AS 2001.5.4 Procedure 2A (Standards Australia 2005).
- Group 3 (3F1–3F7) were samples F1–F7, subjected to abrasion treatment in accordance with Martindale Abrasion Method (AS 2001.2.25.4—Standards Australia 2001) on the side intended for contact with the wearer's skin. The samples were also rated for appearance change after 200 rubbing cycles, according to the visual standards prescribed in AS 2001.2.25.4 (Standards Australia 2001).
- Group 4 (4F1–4F7) were samples F1–F7, conditioned using the following procedure. Samples were dried completely to determine their dry weight, then immersed in water. We then passed the samples through the squeezing rollers of the padder mechanism twice, rendering the fabric wet but not dripping with moisture before testing. The fabrics were weighed again to determine their moisture pickup as a percentage of dry weight. After weighing, we enclosed each sample in a polyethylene bag to prevent moisture loss and tested its surface properties, simulating the fabric lying next to the skin while saturated with sweat. (The natural and maximum moisture contents of the samples during practical use depend on environmental conditions and were not determined in this research).

All fabrics (excluding Group 4) were conditioned for 24 h to eliminate the influence of atmospheric moisture content according to AS 2001.1-1995 (Standards Australia 2001).

Fabrics' physical parameters were tested using the relevant methods described in Section 6.1. The fabric surface properties corresponding to the human senses of "numeri" (smoothness) and "zaratsuki" (roughness) (Barker 2002; Kawabata 1980) using the KESFB4-A (Kato Tech Co. Ltd) evaluation system were measured. The system uses two sensors simultaneously to measure a fabric's frictional coefficient and roughness in three samples of area of 0.002 m × 0.002 m sample.

The frictional coefficient MIU was calculated by averaging the output over the distance between 0 and 20 mm, and is defined as

$$MIU = F/P \tag{6.3}$$

where F is frictional force, N, and P is sensor load, N.

The geometrical SMD is the value obtained by eliminating the low-frequency harmonic wave. This is achieved by filtering the measurement curve through the low-cut filter, extracting the frequency components higher than 1 mm/s, and integrating over the absolute value of the distance moved from the standard position along the sensor's path. Geometrical SMD is expressed as follows:

$$SMD = 1/L_{max} \int^{Lmax} 0 |Z_0 - Z| dL, \tag{6.4}$$

where L_{max} is the distance the sensor moves over the fabric, and Z_0 is the standard sensor position.

The friction coefficient (MIU), mean deviation of the coefficient of friction (MMD), and roughness signals (SMD in μm) were recorded individually and simultaneously for the "go and return" stroke. We took three measurements on each fabric sample in both the warp and weft directions.

All fabrics were tested on the side next to the wearer's skin. We evaluated fabrics with the assumptions that the value of MIU ranges from 0 to 1, and values approaching 1 imply increasing friction and decreasing smoothness; SMD ranges between 0 and 20, and values approaching 20 denote increased surface roughness and surface irregularities.

The fabrics' physical and structural parameters (Table 6.5) as well as their appearance after the abrasion test (Table 6.6) demonstrate that yarn structure and fabric density strongly influence fabric appearance after abrasion.

Ratings of 4 for F1 and F7 and 2–3 for F2 indicate that abrasion did not damage the fabrics' surface greatly. These fabric samples are made of filament or

Table 6.5 Physical and Structural Properties of Experimental Fabrics in Virgin State

Fabric Code	Mean Thickness (mm)	Mean Mass/ Unit Area (g/m²)	Mean No. Ends/cm	Mean No. of Picks/cm	Mean Thread Count/cm²
F1	0.28	133	13	13	26
F2	0.37	130	16	9	25
F3	0.39	120	11	11	22
F4	0.48	150	10	10	20
F5	0.49	171	9	9	18
F6	0.48	167	8	8	16
F7	0.32	128	12	12	24

Table 6.6 Visual Appearance Rating after Abrasion Test
(200 Cycles)

Fabric Code	Fabric Description	Rating
F1	Nomex® Insul-Tex Runner	4
F2	Nomex® Delta C™ Scrim	2–3
F3	TenCate Aralite®	2
F4	Nomex®/FR Viscose	3
F5	Nomex®/FR Viscose Liner Black	1
F6	Nomex®/FR Viscose liner Blue	1–2
F7	Nomex® Delta C™ Liner	4

combinations of long-staple yarns, but differ in their densities: F2 is an unbalanced fabric with lower weft density, while F1 and F7 are square-sett fabrics. Fabrics F3–F6 consist of spun yarns with lower thread counts and densities, and abrasion process changed their surfaces markedly. Our findings demonstrate that compact woven structures, such as F1, F2, and F7, containing filament or long-staple yarns are more durable, with surface characteristics less susceptible to damage during clothing use and care. In addition, visual evaluation of F1 revealed that the virgin fabric surface has a smooth and lustrous appearance, with only a few protruding fiber ends.

The virgin fabric sample F2 is quite smooth and regular, with few single fiber ends protruding from the surface. Abrasion created soft, malleable pill balls, unlike those formed on the surface of F1. F2 contains a combination of long-staple yarns, meaning that more fiber ends protrude from the surface after abrasion, making it fuzzy and yielding pill balls. F1 and F2 are made of 100% Nomex® fiber; their different yarn structures are the only explanation for the divergence in surface appearance resulting from abrasion.

Fabric sample F3's appearance in virgin state is smooth and regular; it is constructed from the same fiber as F1 and F2 but in staple-spun yarns, which explains its large change in surface characteristics compared to F1 and F2. F4 is a blended fabric made of thicker spun yarns with lower and uniform warp and weft density (Table 6.5); its virgin surface is quite fuzzy, with abrasion causing many protruding fibers, prominent and soft pill balls, and surface matting. The formation of prominent pill balls is most likely due to the use of staple yarns with coarser counts. The major surface change in F4 is likely to increase its surface roughness during practical wear and garment care.

Visual evaluation of sample F5 after abrasion revealed numerous small pill balls of greater softness than the pill balls generated from fabric F4. F5 has a somewhat matted surface appearance even in virgin state, caused by long, intertwined protruding fibers that are certain to increase surface roughness and the coefficient of friction. The physical parameters of fabric sample F6 approximate those of F5, as does its surface appearance after abrasion; thus, the surface properties of F5 and F6 were expected to be similar.

The appearance of the virgin-state fabric sample F7, which has similar physical parameters to F1 and F2, is smooth and lacks any significant irregularities. Abrasion produced only a few broken filaments on the surface, resulting in F7 receiving the same rating as F1 and F2.

The surface properties of the seven experimental fabrics were assessed not only in virgin state but also after washing and tumble drying (Figures 6.6 and 6.7)

6. Functionality and Thermophysiological Comfort of Firefighter Protective Clothing

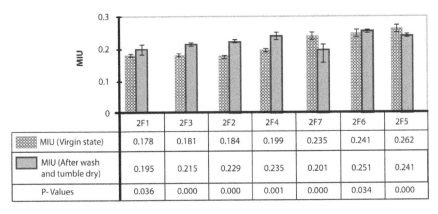

Figure 6.6

MIU of sample fabrics in virgin state and after washing and tumble drying, Groups 1 and 2.

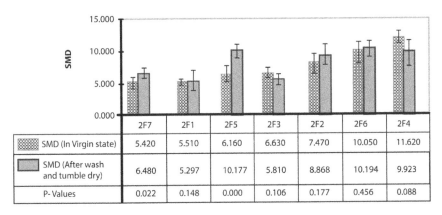

Figure 6.7

SMD of sample fabrics in virgin state and after washing and tumble drying, Groups 1 and 2.

and after abrasion (Figures 6.8 and 6.9). Mean values of MIU and SMD for warp and weft and corresponding p-values are presented in the following figures. In the bar charts, the respective mean values of MIU and SMD of different fabrics are plotted with error bars on both sides showing the standard deviation for each fabric. The standard deviation (SD) for each fabric was calculated with 95% confidence intervals (CIs). For MIU and for the sample fabrics in Group I, SDs range between 0.003 and 0.011; for the sample fabrics of Group II, SDs range between 0.003 and 0.015. The SDs of both groups demonstrate that washing increases the variability of the MIUs of the sample fabrics. The SD and 95% CI for the SMD of each fabric were also calculated. In virgin state, the SD for each fabric ranged between 0.496 and 1.813, but after washing and tumble drying, it ranged between 0.998 and 1.841, indicating that washing increased their roughness. Significant

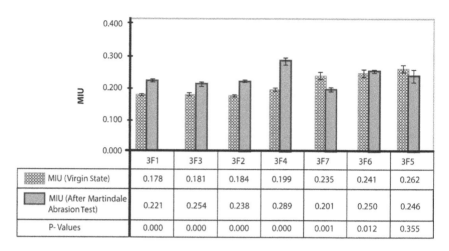

Figure 6.8

MIU of sample fabrics in virgin state and after abrasion, Groups 1 and 3.

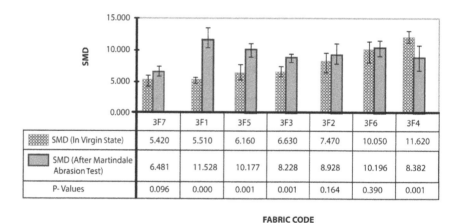

Figure 6.9

SMD of sample fabrics in virgin state and after abrasion, Groups 1 and 3.

differences exist between the MIU values for all fabrics, and between the SMD values for fabrics 2F7, 2F5, 2F4 at a 10% significance level, indicating that washing and tumble drying influenced these characteristics of these fabrics substantially.

Washing and tumble drying raised the coefficient of friction for samples 2F1, 2F2, 2F3, 2F4, and 2F5 (Figure 6.6), but decreased it for 2F5 and 2F7. These two fabrics have higher MIU values than the other fabrics in virgin state, and washing did not change their values markedly. Samples 2F1, 2F2, and 2F3 have similar values of MIU in virgin state, lower than other fabrics in this group; washing and tumble drying increased 2F2's MIU by the largest amount of the three (24.4% of virgin MIU), and 2F1 by the smallest—9.5% of virgin MIU. Sample 2F1 consists of filament yarns with a high weave density and a square sett, so it has a smoother surface giving a smaller contact area, and thus has the lowest MIU.

Sample 2F3's spun yarn structure results in a large contact area due to its surface hairiness; this, and lower weave density, produced higher MIUs in virgin state and after washing and tumble drying than for the fabrics made of filament yarns. Sample 2F2 is made of long-staple yarn with the same fiber composition as 2F1 and 2F3, but it has rip-stop weave of high density and raised "ridges" on its surface, whereas the plain weave structure of 2F1 and 2F3 provide a smooth surface. It is probable that these differences caused the higher MIU values of 2F1 and 2F3 in virgin state and after washing and drying.

Sample 2F4 is made of spun yarns and is fiber-blended with FR Viscose. Viscose fiber is a soft raw material that creates a large contact area in a fabric surface, resulting in higher MIU than in fabrics made of both filament and staple Nomex® yarns. Washing and drying decreased MIU by 14%, possibly by reducing surface hairiness or increasing fabric softness.

Samples 2F5 and 2F6 are constructed from the same spun yarns with the same fiber composition; a slight difference in weave density gives 2F5 higher density and weight. The minor difference in the MIUs of 2F5 and 2F6 (Figure 6.6) could be a result of fiber blending ratios.

After washing and tumble drying, the MIU of sample 2F5 increased 4% and the MIU of sample 2F6 decreased by the same percentage. These fabrics are similar in virgin state and after washing and tumble drying, but in virgin state they have higher MIUs than 100% Nomex® fabrics. Sample 2F7 is similar to 2F6, but its MIU is slightly lower due to a higher weave density and long-staple yarns in its warp. SMDs of 2F1, 2F2, and 2F6 did not change significantly (at 10% level of significance) as a result of washing and drying (Figure 6.7). Samples 2F1's and 2F3's roughness SMDs are the lowest of all the fabrics tested; washing and tumble drying produced minimal change in their SMDs, as was the case for samples 2F6 and 2F2. Washing and tumble drying significantly decreased SMD for 2F4, which already had the highest SMD in virgin state of all the fabrics tested (Figure 6.7). Sample 2F5 recorded the highest increase in SMD (65% of the virgin state) after washing and tumble drying. Therefore, in this group, both 2F1 and 2F3 perform well in terms of visual appearance after abrasion and also after washing and tumble drying, with low virgin MIU and SMD values. We conclude that 100% Nomex® fiber in filament form in plain weave fabric construction gives the best surface comfort characteristics, and is also highly acceptable in spun yarn form for next-to-skin applications.

We now turn to the effect of abrasion on fabric surface properties. Mean MIU values increased after abrasion for all fabrics other than 3F5 and 3F7 (Figures 6.8 and 6.9). For all experimental fabrics other than 3F5, abrasion alters MIU significantly (at the 5% level).

The SDs of the MIUs for the sample fabrics of Group 3 range between 0.004 and 0.023, demonstrating that abrasion increases the variation in MIU from virgin state to a larger extent than washing and tumble drying. Increased MIU is due to the mechanical action of abrasion disturbing the yarn and fibers on the fabric surface. The greatest percentage increase of MIU among all the fabrics of this group occurred in sample 3F4 (45.2% of virgin MIU), which also has the highest virgin MIU. The decrease in MIU in samples 3F5 and 3F7 could be due to abrasion causing a smaller surface contact area (this outcome requires further investigation).

The SDs of SMDs for sample fabrics of Group III range between 0.645 and 2.236, demonstrating a greater variation in roughness after abrasion. SMDs for all fabric samples of this group (other than 3F4) increased after abrasion. No statistically significant difference in SMD due to abrasion was found in samples 3F2 and 3F6. SMD increased most (109.5% of the virgin state) in sample 3F1, which has a smooth surface in its virgin state, but abrasion disturbs the surface significantly, creating pills and broken yarns. The pill balls formed through abrasion of sample 3F1 are substantially harder than those formed from the staple filaments of the spun yarns of the other fabrics, due to 3F1's continuous and rigid filament fiber structure. Hence, 3F1's MIU increased significantly through abrasion, signaling that this fabric may not respond well to the repeated rubbing involved in donning and doffing of garments in practical use.

We measured the effect of moisture on fabric surface properties: the fabric samples' varying fiber composition and structural attributes translate into different moisture pickup characteristics (Figure 6.10).

Fabrics containing Nomex® and aramid fibers, both in filament and staple form, had lower moisture pickup, with the lowest pickup identified in 4F1 (Figure 6.11). Moisture also influenced the surface characteristics of most fabric samples (Figures 6.11 and 6.12). The SDs of MIU values for the sample fabrics of Group 4 range between 0.005 and 0.019, demonstrating that moisture also increased the MIU of most fabrics—to a lesser extent than abrasion, but more than washing.

Fabric MIUs increased significantly due to the presence of moisture for fabric samples 4F3, 4F4, and 4F6; sample 4F1 recorded a significant decrease. Moisture in fabric increases the adhesive forces between contact surfaces, creating greater resistance to sliding. Sample 4F1's filament yarn construction gives it a smooth, lustrous surface; due to the fabric's low moisture pickup water molecules do not penetrate but roll on the surface, lowering friction. MIU increased most for samples 4F4 and 4F6 (45.2% and 35.6% of virgin state, respectively) due to the high moisture content of the absorbent fiber, signifying that these are unsuitable fabrics for firefighting garments. We measured minor increases in MIU in 4F2 and 4F3 due to moisture, meaning these fabrics and 4F1 could be considered to be most suitable for firefighting applications.

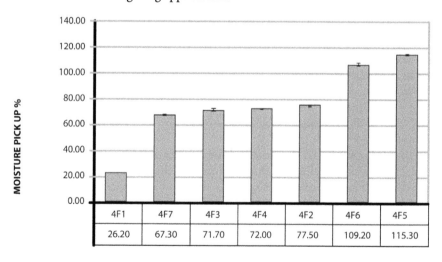

Figure 6.10

Moisture pickup, Group 4.

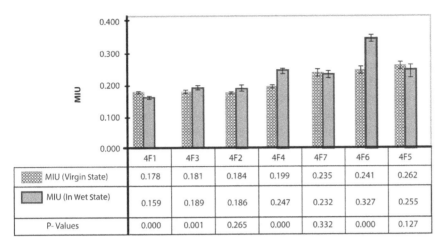

FABRIC CODE	4F1	4F3	4F2	4F4	4F7	4F6	4F5
MIU (Virgin State)	0.178	0.181	0.184	0.199	0.235	0.241	0.262
MIU (In Wet State)	0.159	0.189	0.186	0.247	0.232	0.327	0.255
P- Values	0.000	0.001	0.265	0.000	0.332	0.000	0.127

Figure 6.11

MIUs of sample fabrics in virgin and wet states, Groups 1 and 4.

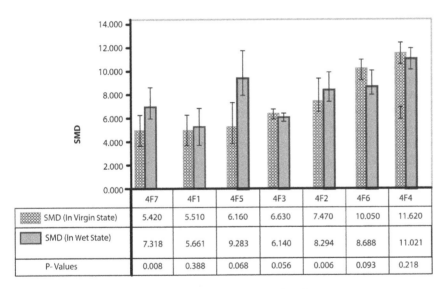

FABRIC CODE	4F7	4F1	4F5	4F3	4F2	4F6	4F4
SMD (In Virgin State)	5.420	5.510	6.160	6.630	7.470	10.050	11.620
SMD (In Wet State)	7.318	5.661	9.283	6.140	8.294	8.688	11.021
P- Values	0.008	0.388	0.068	0.056	0.006	0.093	0.218

Figure 6.12

SMDs of sample fabrics in virgin and wet states, Groups 1 and 4.

The SDs of the SMDs for the sample fabrics of Group 4 ranged between 0.417 and 2.936. The presence of moisture was not associated with statistically significant differences (at 10% level of significance) in SMD in samples 4F1 and 4F4. Moisture content increased the roughness (SMD) of fabrics 4F2, 4F5, and 4F7 significantly, but decreased it significantly for 4F3 and 4F6. Sample 4F4 had the highest roughness of all tested fabrics in virgin state, possibly due to moisture and swelling of FR viscose yarns that increases the irregularity of surfaces.

In conclusion, this case study indicates that fabric friction (MIU) and fabric surface roughness (SMD) are affected by many factors, including fabric structure, fiber type, fiber blend ratio, yarn structure, yarn density, moisture content, and mechanical treatments. These factors influence SMD to a greater extent than MIU, and fabric roughness varies significantly as a result of treatments involving moisture. We have shown that yarn structure (spun or filament) and blending ratio (material composition) are the two most critical factors that influence surface properties in fabrics containing specialized performance yarns used in firefighters' apparel.

6.7 Garment Fit and Comfort

The case studies considered above demonstrate that materials can significantly add to or detract from human physiological comfort and performance. However, evaluating protective clothing's thermal comfort and protection at the level of material structures and physiological modeling alone is insufficient. Garment style and (ultimately) fit are also important, because the fit of the garment determines the air gaps between human skin and clothing, and thus powerfully influences the protective efficiency of the garment as well as the dynamics of heat loss (Torvi, Dale, & Faulkner 1999).

Air is an efficient insulator due to its low thermal conductivity, reducing energy transfer between the skin and the environment (Krasny 1986). The topology of air layers between the skin and a garment depends on garments' style, size, fit, ease, and fabric drape, in addition to features such as collar folds, pockets, and cuff and neck openings. Loose apparel tends to trap more air and have larger neck, waist, and other openings, reducing their thermal insulation and moisture vapor resistance during windy conditions and body movement (Chen, Fan, Qian, & Zhang 2004). Fabric properties such as stiffness, bending rigidity, and drapeability also affect the fit of apparel (Figure 6.13).

In our final case study, we compared the air volumes between the body of a male thermal manikin and two different sizes of a protective jacket at multiple body zones.

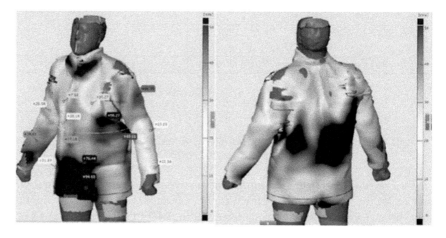

Figure 6.13

Size and distribution of air gaps between a male subject and a protective jacket, Generation IV.

6. Functionality and Thermophysiological Comfort of Firefighter Protective Clothing

The firefighters' protective jacket (made of Generation IV material, as described above) is a three-layered garment consisting of a structural fire coat with removable inner liner, as shown in Figure 6.1. We used small (S) and medium (M) sizes for this case study.

An NX-16 [TC]² three-dimensional (3D) white light body scanner ([TC]² 2010) was used to generate a 3D point cloud from a Newton thermal manikin (Measurement Technology Northwest, Seattle) (Figure 6.14). The Newton enables objective thermal testing of apparel; its surface temperature can be controlled in 20 segments independently, and the total heat input required to achieve set surface temperatures is measured accurately (Figure 6.15).

The thermal manikin Newton in male form was scanned nude and dressed in jackets in sizes S and M (Figure 6.14). Each configuration was scanned five times, and the means of the air volumes measured in the five scans were calculated for each configuration. We calculated the volume (cm³) of each zone covered by the jacket and subtracted from it the volume of the corresponding zone of the nude manikin using Rapidform software, giving the volume of air entrapped in each zone (Figure 6.16).

Figure 6.16 shows that the volume of entrapped air at each zone increased when the jacket size was changed from S to M, except at the shoulders and back. Size M is quite loosely fitted, causing it to shift forward and reduce the air volume in the shoulder and back zones while increasing the air volume at the stomach zone (29% greater) and chest zone (26%). In practical use, these two zones will

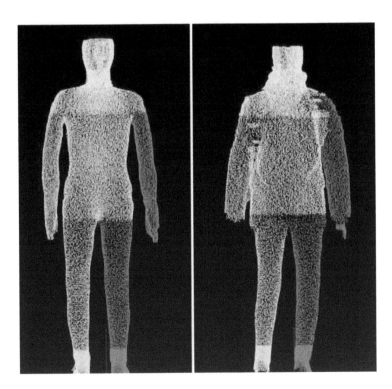

Figure 6.14

3D point cloud data: (a) thermal manikin, (b) thermal manikin dressed in protective jacket.

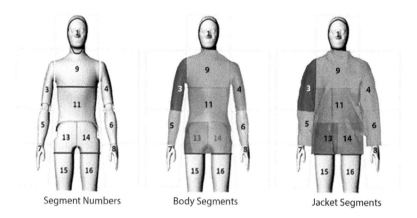

Segment Numbers   Body Segments   Jacket Segments

Figure 6.15

Manikin segments used for air volume calculations.

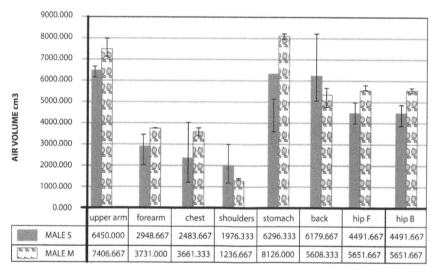

	upper arm	forearm	chest	shoulders	stomach	back	hip F	hip B
MALE S	6450.000	2948.667	2483.667	1976.333	6296.333	6179.667	4491.667	4491.667
MALE M	7406.667	3731.000	3661.333	1236.667	8126.000	5608.333	5651.667	5651.667

MANIKIN ZONES

Figure 6.16

Air volume (cm³) between the body of a thermal manikin and jackets of different sizes.

accumulate hot air, causing discomfort or even burns to the wearer. The total air volume is 16% higher for the M jacket than the S jacket.

These preliminary results indicate a correlation between jacket fit and the air volume between the jacket and the body. Hence, the fit of apparel is likely to affect the overall garment attributes that determine thermophysiological comfort in extremely hot conditions.

6.8 Conclusions

The primary engineering challenge in the design of material structures for firefighters' protective apparel is to attain the best balance between the physiological burden these materials impose and the protection they offer, and thereby permit optimal performance. In this chapter, holistic approach to engineering and evaluation of new protective materials and garments through a combination of objective laboratory testing, multicompartmental mathematical modeling and 3D body imaging, analysis, and design. It is important to note that evaluation of clothing on sweating thermal manikins, as well as human trials, are also critical and should be taken into consideration.

The attributes of materials used in protective garments, such as thermal and water vapor resistance and overall moisture management capacity, are critical to the thermophysiological comfort of the wearers. Our case studies revealed that the new structure of Generation IV material has greater ability than Generation I material to transfer heat and moisture in liquid and vapor form from the human body to the environment. Mathematical modeling confirmed that these Generation IV materials impose much lower physiological burden on the wearer.

These engineering approaches to the performance of firefighters' protective clothing could be applied in other fields where the balance between clothing performance and wearer comfort is crucial, such as sporting apparel and work wear.

References

AATCC, *Test method 195-2011, Liquid Moisture Management Properties of Textile Fabrics.* 2009.

Ajayi, J. O., Effect of fabric structure on frictional properties. *Textile Research Journal,* 62 (1992) 87–93.

Ajayi J. O., and Elder, H. M., Comparative studies of yarn and fabric friction. *Journal of Testing and Evaluation,* 22 (1994) 463–467.

Ajayi, J. O., and Elder, H. M., Fabric friction, handle, and compression. *Journal of the Textile Institute,* 88(3) (1997) 232–241.

Apurba, D., and Nagaraju, V., A study on frictional properties of woven fabrics. *AUTEX Research Journal,* 5 (2005) 133–140.

Barker, R. L., From fabric hand to thermal comfort: The evolving role of objective measurements in explaining human comfort responses to textiles. *International Journal of Clothing Science and Technology,* 14 (2002) 181–200.

Bartkowiak, G., Liquid sorption by nonwovens containing superabsorbent fibers. *Fibres and Textiles in Eastern Europe,* 14 (2006) 57–61.

Bartkowiak, G., Influence of undergarment structure on the parameters of the microclimate under hermetic protective clothing. *Fibres & Textiles in Eastern Europe,* 18 (2010) 82–86.

Behera, B. K., and Sobti, M., Studies on handle of microdenier polyester filament dress materials. *International Journal of Clothing Science and Technology,* 10 (1998) 104–113.

Bhupender, S. G., and Yehia, E. E., Friction in fibrous materials. *Textile Research Journal,* 61 (1991) 547–554.

Briscoe B., and Tabor, D., Friction and wear of polymers: The role of mechanical properties. *British Polymer Journal,* 10 (1978) 74–78.

Briscoe, T., *Interfacial friction of solids in solid-solid interactions.* Imperial College Press, 1996.

Candas, V., and Yermakova, I., Computer simulation of human physical activity in moderate heat. In M. Silva (Ed.), *Manikins and Modelling,* Coimbra, Portugal, 2008.

Cardello, A. V., The sensory properties and comfort of military fabrics and clothing. In E. Wilusz (Ed.), *Military Textiles*, Sawston, Woodhead Publishing, 2008, pp. 71–106.

Chen, Y. S., Fan, J., Qian, X., and Zhang, W., Effect of garment fit on thermal insulation and evaporative resistance, *Textile Research Journal*, 74(8) (2004) 742–748.

Cheng, Y., Niu, J., and Gao, N., Thermal comfort models: A review and numerical investigation, *Building and Environment*, 47 (2012) 13–22.

Cheung, S. S., Petersen, S. R., and McLellan, T. M., Physiological strain and countermeasure with firefighting, *Scandinavian J. of Medicine & Science in Sports*, Suppl. 3 (2010) 103–116.

Chung, G.-S., and Lee, D. H., A study on comfort of protective clothing for firefighters. In T. Yutaka & O. Tadakatsu (Eds.), *Environmental Ergonomics— The Ergonomics of Human Comfort, Health, and Performance in the Thermal Environment*, Vol. 3, Elsevier, 2005, 375–378.

Du, Z., and Yu, W., A comprehensive handle evaluation system for fabrics: II. Theoretical analyses and modeling of friction and tensile properties. *Measurement Science and Technology*, 18 (2007) 3555–3564.

Fiala, D., Lomas, K., and Stohrer, M., A Computer model of human thermoregulation for a wide range of environmental conditions: Passive system. *Journal of Applied Physiology*, 87(5) (1999) 1957–1972.

Foda, E., Almesri, I., Awbi, H. B., and Sirén, K., Models of human thermoregulation and the prediction of local and overall thermal sensations, *Building and Environment*, 46 (2011) 2023–2032.

Guidotti, T. L., Human factors in firefighting: Ergonomic-, cardiopulmonary-, and psychogenic stress-related issues. *International Archives of Occupational and Environmental Health*, 64 (1992) 1–12.

Gwosdow, A. R., Stevens, J. C., Berglund, L. J., and Stolwijk, J. A. J., Skin friction and fabric sensations in neutral and warm environments. *Textile Research Journal*, 56(9) (1986) 574–580.

Hassani, H. Effect of different processing stages on mechanical and surface properties of cotton knitted fabrics. *Indian Journal of Fiber & Textile Research*, 35 (2010) 139–144.

Hes, L., Marmarali, M., Ozdil, N., Oglakcioglu, N., and Lima, M., The effect of moisture on friction coefficient of elastic knitted fabrics. *Tekstil ve Konfeksiyon*, 3 (2008) 206–210.

Houshyar, S., Padhye, R., Troynikov, O., Nayak, R., and Ranjan, S., Evaluation and improvement of thermo-physiological comfort properties of firefighters' protective clothing containing super absorbent materials. *The Journal of the Textile Institute*, 106(12) (2015) 1394–1402.

Hu, J., Li, Y., Yeung, K. W., Wong, A. S. W., and Xu, W., Moisture Management Tester: A method to characterize fabric liquid moisture management properties. *Textile Research Journal*, 75(1) (2005) 57–62.

ISO 11092-1993, *Textiles—Physiological effects—Measurements of thermal and vapour resistance under steady-state conditions sweating guarded hot plate.* © ISO, 1993.

ISO 9237-1995, *Textiles—Determination of the permeability of fabrics to air.* © ISO, 1995.

ISO 9151-1995, *Protective clothing against heat and flame—Determination of heat transmission on exposure to flame.* © ISO, 1995.

ISO 6942-2002, *Protective clothing—Protection against heat and fire—Method of test: Evaluation of materials and material assemblies when exposed to a source of radiant heat.* © ISO, 2002.

Kawabata, S., Characterization method of physical property of fabrics and the measuring system for hand feel evaluation. *Journal of Textile Machinery Society of Japan*, 26 (1973) 721–728.

Kawabata, S., *The Standardization and Analysis of Hand Evaluation* (2nd ed.). Osaka, Japan: The Textile Machinery Society of Japan, 1980.

Kenins, P. Influence of fiber type and moisture on measured fabric-to-skin friction. *Textile Research Journal*, 64 (1994) 722–788.

Kim, J. H., Kim, D. H., Lee, J. Y., and Coca, A., 2017, September. Relationship between Total Heat Loss and Thermal Protective Performance of Firefighter Protective Clothing and Consequent Influence on Burn Injury Prediction via Flame-Engulfment Manikin Test. In Proceedings of the Human Factors and Ergonomics Society Annual Meeting (Vol. 61, No. 1, pp. 1468–1471). Sage CA: Los Angeles, CA: SAGE Publications.

Kondo, S. The frictional properties between fabrics and human skin. *Journal of the Japan Research Association for Textile End Uses*, 43(3) (2002) 264–275.

Krasny, J. F., Some characteristics of fabrics for heat protective garments. In R. L. Barker & G. C. Coletta (Eds.), *Performance of protective clothing*, American Society for Testing and Materials, Philadelphia (1986) 463–474.

Li, C., Shen, Y., and Guo, Y., Thermal Protective Performance Tests of Firefighter Clothing Based on the Flame-mannequin Method. *Manikin and Modelling (11i3m)*, (2016) 42.

Li, C., Wang, L.-P., and Ren, R.-M., Research on evaluating the indices of dynamic fabric heat and moisture comfort, *Journal of Industrial Textiles*, 34 (2005), 255–272.

Li, Y., The science of clothing comfort, *Textile Progress*, 31(1) (2001).

Li, Y., and Holcombe, B. V., Mathematical simulation of heat and moisture transfer in human-clothing-environment system, *Textile Research Journal*, 68(6) (1998) 389–397.

Quaynor, M. L., and Nakajim, M., Effects of laundering on the surface properties and dimensional stability of plain knitted fabrics. *Textile Research Journal*, 70 (2000) 128–135.

Ramkumar, S., *A study on handle characteristics of 1-1 rib cotton weft knitted fabrics.* PhD thesis, Leeds: University of Leeds, 1998.

Rossi, R., Interaction between protection and thermal comfort. In: R. A. Scott (Ed.), *Textiles for protection*, Woodhead Publishing Ltd., Cambridge, UK, 2005, 233–240.

Song, G., Barker, R. L., Hamouda, H., Kuznetsov, A. V., Chitrphiromsri, P., and Grimes, R. V., Modeling the thermal protective performance of heat resistant garments in flash fire exposures, *Textile Research Journal*, 74(12) (2004) 1033–1040.

Standards Australia, *AS 2001.1-1995, Conditioning of Textile Materials.* Standards Australia. 2001.

Standards Australia, *AS 2001.2.13-1987, Methods of test for textiles—Physical tests—Determination of mass per unit area and mass per unit length of fabric.* Standards Australia, 1987.

Standards Australia, *AS 2001.2.15-1989, Methods of test for textiles—Physical tests—Determination of thickness of textile fabrics.* Standards Australia, 1989.

Standards Australia, *AS 2001.2.5-1991, Methods of test for textiles—Physical tests—Determination of the number of threads per unit length in woven fabric.* Standards Australia, 1991.

Standards Australia, *AS 2001.2.25.4-2001, Determination of the abrasion resistance of fabrics by Martindale method—Assessment of appearance change.* Standards Australia. 2001.

Standards Australia, *AS 2001.5.4-2005, Methods of test for textiles—Dimensional change- Domestic washing and drying procedures for textile testing.* Standards Australia, 2001.

Stapleton, J. M., Wright, H. E., Hardcastle, S. G., and Kenny, G. P., Body heat storage during intermittent work in hot–dry and warm–wet environments, *Applied Physiology, Nutrition, and Metabolism,* 37(5) (2012) 840–849.

Tessier, D., 2018. Testing thermal properties of textiles. In *Advanced Characterization and Testing of Textiles* (pp. 71–92).

[TC]² 2010, *3D Body Scanning & Technology Development.* Available from http://www.tc2.com/index_3dbodyscan.htm

Torvi, D. A., *Heat transfer in thin fibrous materials under high heat flux conditions,* Department of Mechanical Engineering, Doctoral Thesis, University of Alberta Canada (1997).

Torvi, D. A., Dale, J. D., and Faulkner, B., Influence of air gaps on bench-top test results of flame resistant fabrics, *Journal of Fire Protection Engineering,* 10(1) (1999) 1–12.

Troynikov, O., and Nawaz, N., *Firefighter survey results to evaluate the physiological comfort of structural firefighter's ensembles.* Unpublished results, RMIT University 2010.

Troynikov, O., Nawaz, N., and Yermakova, I. Materials and engineering design for human performance and protection in extreme hot conditions, *Advanced Materials Research.* Vol. 633. Trans Tech Publications, 2013.

Wong, A. S., Li, Y., and Yeung, K. W., Performances of artificial intelligence hybrid models in prediction of clothing comfort from fabric physical properties, *Sen-i Gakkaishi,* The Society of Fiber Science and Technology, Japan, 59(11) (2003) 429–436.

Yao, B-g., Li, Y., Hu, J., Kwok, Y., and Yeung, K., An improved test method for characterizing the dynamic liquid moisture transfer in porous polymeric materials, *Polymer Testing,* 25(5) (2006) 677–689.

Yermakova, I. Mathematical modelling of thermal processes in man for development of protective clothing, *J. of the Korean society of living environmental system,* 8(2) (2001) 127–133.

Yermakova, I., and Candas, V., Practical use of thermal model for evaluation of human state in hot environment: Summary of French-Ukrainian project, *ICEE X11,* Ljubliana (2007) 468–471.

Hot Plates and Thermal Manikins for Evaluating Clothing Thermal Comfort

Ying Ke, Udayraj, Ziqi Li, and Faming Wang

7.1 Introduction

Thermal comfort is conveniently defined as "that condition of mind that expresses satisfaction with the thermal environment" (ASHRAE, 2013). There are mainly four essential conditions for an individual person to reach a thermal comfort status (Fanger, 1970): (1) the whole body is in thermal equilibrium (which means the body heat gain equals the heat loss); (2) the body sweating is within comfortable limits; (3) the mean skin temperature shall be within comfortable limits (i.e., 32–34°C); and (4) there is no local discomfort at any body part. From a physical viewpoint, thermal comfort is mainly affected by six variables, namely, air temperature, relative humidity, radiation, air velocity, metabolic rate, and clothing thermal insulation (Parsons, 2014). Clothing plays an important role in maintaining body heat balance as it serves as a prominent heat and moisture barrier between the human body and its surrounding thermal environment. Obviously, the clothing thermal comfort of the wearer is one of the most important aspects contributing to human health and well-being (Kilinc-Balci, 2011). Numerous instruments were designed to objectively quantify thermal comfort properties of clothing ensembles. Sweating guarded hot plates and thermal manikins are the two most representative testing systems to determine clothing/fabric thermal insulation and evaporative resistance, which are widely accepted as the most important parameters to characterize clothing thermal comfort.

The sweating guarded hot plate (SGHP), often referred to as the "skin model," is designed to simulate the heat and moisture transfer from the human skin, through the fabric, to the environment (Huang, 2006). It is a complete unit that has the hotplate and controls integrated into a chamber that maintains specified conditions regarding temperature, relative humidity, and airspeed. The SGHP measures the thermal properties and water vapor resistance (or evaporative resistance) of fabrics and other materials for use in clothing systems such as films, coatings, foams, leathers, and multilayer assemblies under steady-state conditions. It is commonly recognized as one of the most accurate techniques for determining the thermal conductivity of insulating materials (Huang, 2006; Salmon, 2001).

Given the fact that SGHPs can only estimate the heat and moisture transfer properties of fabric samples, they fail to take into consideration the effects of other important clothing factors such as pattern design, human body configurations, air gaps between clothing and human body, and body movement/postures on thermal comfort of a clothed person. Thermal manikins were therefore designed to mimic the actual heat and moisture transfer through the human body, clothing, and environment system. Generally, a thermal manikin is a human-shaped heated dummy equipped with heating wires, sweating systems (if any), temperature measuring sensors, and/or heat flux sensors (Wang, 2017a). They are mainly used to measure thermal and water vapor resistances of clothing ensembles to characterize the ability of clothing to impede the heat and moisture flow from the skin to the environment. The first thermal manikin was developed in the 1940s, and since then many manikins have been developed and designed (Clarke, 2015). Presently thermal manikins are widely used in the research area of clothing comfort and to interpret the actual performance of different clothing. Compared with human wear trials, manikin tests are usually faster, more cost-effective, and repeatable (Holmér, 2004). Nevertheless, most of the existing manikins are unable to evaluate the dynamic thermal properties of clothing. More importantly, traditional manikins fail to examine how the wearers respond psychologically and physiologically to clothing they wear as well as thermal environments at their workplaces. For these reasons, a more advanced test method based on the thermoregulatory model controlled thermal manikin (sometimes referred to as adaptive manikin) has been proposed to investigate the thermal interaction in the human body–clothing–environment system (Psikuta et al., 2017a; Wang, 2017b). In this chapter, working principles and the development history of the sweating guarded hot plate and thermal manikins are discussed. Recent development on the advanced thermoregulatory models controlled thermal manikins will also be addressed. Finally, applications of the hot plate and the manikin in the area of clothing research will be presented.

7.2 Hot Plate

7.2.1 Development of Hot Plate

Development of the guarded hot plate (GHP) apparatus started in the early twentieth century based on the Lees' disk method (Lees, 1892). Lees disk method is a simple and effective way to calculate the thermal conductivity of materials with low thermal conductivities, such as glass, wood, or polymer, and it is a steady-state method. The experimental setup involves the use of two metal disks, a steam chamber, the sample (thermal conductivity of which is to be measured), and two thermometers to measure the temperature (see Figure 7.1).

7. Hot Plates and Thermal Manikins for Evaluating Clothing Thermal Comfort

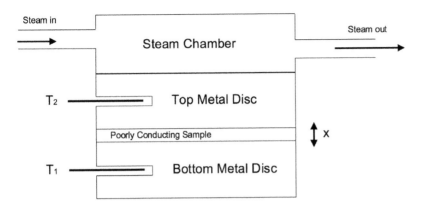

Figure 7.1

Experimental setup of the Lees' disk steady-state method for thermal conductivity measurement.

Based on the Lees' disk method, American scientists Hobart Cutler Dickinson of the National Institute for Standards and Technology (NIST) and Milton S. van Dusen of the National Bureau of Standards (NBS) built the first GHP in 1912 (Zarr, 1997). Around the same time in Germany, Ricard Poensgen developed his own GHP apparatus, but it started to measure thermal conductivity in 1912 (Flynn et al., 2005). Several different guarded hot plate apparatuses were built at NIST between 1912 and 1928. In 1929, Van Dusen built what was to be the final version of this type of guarded hot plate apparatus. This particular apparatus operated consistently for NIST for more than 50 years until 1983. The GHP uses a steady-state method to determine the thermal conductivity of an insulating material. The apparatus setup (Figure 7.2) involves a heating plate (guarded on all lateral sides) sandwiched in

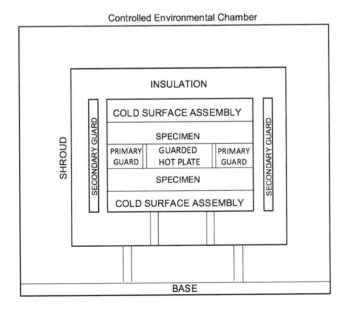

Figure 7.2

Experimental setup of an early GHP apparatus.

between two samples (of the same dimensions) of the insulator, the thermal conductivity of which is to be measured. An unheated "cold plate" is then placed on either side of the insulating material, and once the system has reached thermal equilibrium, the thermal conductivity is calculated from the temperature rise of the cold plates (which is representative of heat conduction through the insulating material).

In 1964, Robinson presented the basic design of the line-heat-source GHP. Compared with the conventional GHP, the line-heat-source GHP used circular line-heat sources at precisely specified locations. It has simpler construction, improved accuracy, simplified mathematical analyses for calculating the mean surface temperature of the plate, as well as determining the errors resulting from heat gains or losses at the edges of the specimens, and operates under vacuum conditions. Near the end of 1980, the second line-heat-source guarded-hot-plate apparatus was completed from the efforts of Hahn and Peavy of NIST and Ober. From 1981 to 1996, more than 75 measurements were provided. In 1996, the American Society for Testing and Materials (ASTM) formally adopted the line-heat-source concept as a standard practice based, in part, on NIST's design (Zarr, 1997, 2001).

Most GHPs have been constructed using a metal-surface laminated design, in which an electrical heater on a central electrically insulating plate is sandwiched between two thin electrically insulating plates, which in turn are sandwiched between two metal surface plates. While such designs can work quite well in 240 to 340 K air temperature range, problems are often encountered at higher or lower temperatures or if it is desired to measure the thermal transmission of specimens under vacuum conditions (Flynn et al., 2002). Considering the above problems, NIST proposed a new GHP for use at temperatures from 90 to 900 K (Flynn et al., 2002).

Though different systems were developed, the working principles of the GHPs remain almost the same. Previous GHPs could only evaluate the thermal conductivities of materials. In 1993, ISO 11092 proposed a standard to measure both the heat and moisture transfer properties of materials using a sweating guarded hot plate (SGHP). The first SGHP was operational in 1985. Compared with the dry GHP, the SGHP added a system for feeding water to the surface of the test plate and guard section (Cŭbrić et al., 2011; Huang, 2006; Kar et al., 2007) as shown in Figure 7.3. In the SGHP, a PTFE microporous membrane is placed between the

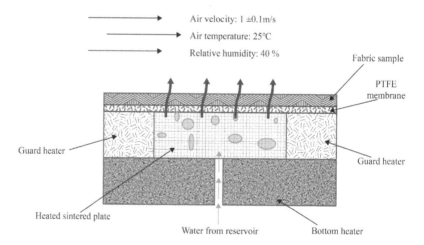

Figure 7.3

Schematic of a sweating guarded hot plate (SGHP) apparatus.

hot plate and the fabric sample to prevent water penetration and wicking into the tested fabric while allowing water vapor transfer through the fabric sample.

7.2.2 Measurements Using Sweating Guarded Hot Plates

Sweating guarded hot plate (SGHP) is well known and one of the most widely used test instruments for measuring dry and wet heat transfer through textile materials. Thermal resistance and the water vapor resistance of textile materials can be determined using a sweating hot plate (ISO 11092, 2014; ASTM F1868, 2017). It simulates the conditions that prevail in the human body (heat and water vapor transfer from the surface of the body to the environment through the fabric) in a better way (Huang, 2006). To measure the thermal resistance, the hot plate temperature is maintained at 35±0.5°C and environmental air temperatures are maintained in the range 4±0.1°C to 25±0.1°C (ASTM F1868, 2017). Air velocity in the environmental chamber is maintained at 1.0±0.1 m/s and the relative humidity at 40±4% as shown in Figure 7.3. A heater is provided to heat the plate, and there are two other heaters, guard and bucking heaters, to minimize heat transfer in other directions. Amount of power supplied by the heater to the plate to maintain it at a constant temperature is measured, and based on it, the total dry thermal resistance ($R_{th,total}$) of fabric samples can be calculated by

$$R_{th,total} = \frac{(T_H - T_{amb})A}{Q} \tag{7.1}$$

where T_H is the hot plate temperature (°C), T_{amb} is the ambient air temperature (°C), A is the surface area of the plate measurement section (m^2), and Q is the power required to maintain the plate at a constant temperature T_H (W). Total thermal resistance considers the thermal resistance of the fabric sample and air layer present above the fabric sample. The ability of a fabric sample to transfer heat is generally better represented by intrinsic thermal resistance ($R_{th,i}$) as it is less affected by the air layer or wind speed as compared to the total thermal resistance ($R_{th,total}$). It can be calculated as

$$R_{th,i} = R_{th,total} - R_{air-layer} \tag{7.2}$$

where $R_{air-layer}$ is the resistance of the air layer present above the fabric sample. It can be determined by conducting experiments without fabric sample and using Equation 7.1.

Evaporative resistance can be determined by measuring the amount of power required to maintain the plate at a constant temperature when the plate is saturated with water. As the saturated plate is heated, the water evaporates by absorbing heat and the water vapor diffuses through the fabric sample into the environment. To measure the evaporative resistance, the hot plate and environmental air temperatures are maintained at 35±0.5°C (ASTM F1868, 2017). Under the steady state, the total evaporative resistance ($R_{e,total}$) can be calculated as

$$R_{e,total} = \frac{(P_s - P_{amb})A}{Q_{s,plate}} \tag{7.3}$$

where P_s is the saturated water vapor pressure at the plate surface (Pa), P_{amb} is the saturated water vapor pressure of the ambient air (Pa), A is the surface area of the plate measurement section (m^2), and $Q_{s,plate}$ is the power (W) required to maintain the plate at a constant temperature when the plate is saturated with water. It is important to note that the case of measurement of the evaporative resistances at isothermal conditions is not always true. In real-life situations, dry and evaporative heat transfers simultaneously, and hence non-isothermal situations may exist (Huang, 2006). To take this into account, the non-isothermal method has been included in ASTM F1868 (2017). The measured evaporative resistance under the non-isothermal condition is normally referred to as the apparent evaporative resistance. Although GHPs and SGHPs have been used widely, these are only designed for use under steady-state conditions, and the process of measurement is time-consuming (at least 90 min). To deal with the unsteady conditions, dynamic hotplate systems (DHSs) were proposed. Thermetrics LLC (Seattle, WA) invented a DHS (Model DHS-8.2). It can measure the water vapor resistance and thermal resistance in situations of positive or negative heat flux, and in dynamic transient environments with good accuracy. The system can be used for evaluating PCM (phase change material) samples, tests of heated fabrics and pads, or textile testing under sustained solar loads and elevated ambient temperatures (up to 50°C). Also in case of GHPs and SGHPs, when the test specimens are thicker than 5 mm, corrections for water vapor losses are required. But manikin tests can make up for this limitation.

7.3 Thermal Manikins

Thermal manikins are available in a variety of sizes, body shapes, and configurations. Although most of the manikins are made in the male shape, the first female thermal manikin, "Nille," was developed in 1989 (Holmér, 2004). It was made of plastic and could also simulate breathing. Then female manikins "Maria" by Coelho and "Niana" by Konarska et al. were also developed (Konarska et al., 2006; Konarska et al., 2007; Silva and Coelho, 2002). Not all these female manikins could simulate perspiration. Besides full body size manikins, many thermal manikin body parts were designed and used for testing, such as the thermal hand, the thermal foot, and the thermal head (Fonseca, 1974; Zwolińska et al., 2014). They are cheaper, more portable, and easier to operate while not compromising with the accuracy of analysis as compared to full body size manikins. The initial reports on the thermal head (also called manikin head) were published in 1974 (Fonseca, 1974; Zwolińska et al., 2014) where the head of a thermal manikin was used to study military helmets.

In recent years, a number of new thermal manikins have been developed as shown in Table 7.1. In 2017, Thermetrics LLC (Seattle, WA) developed an 11-zone baby sweating thermal manikin. The baby manikin is of the size of a 9-month-old child and is used to evaluate the thermal properties of clothing and baby gear, diapers, indoor environments, bedding, and baby carriers, including car seats and strollers. Overall, there are two different directions for thermal manikin development. One is represented by simple, yet accurate, reliable and rather inexpensive manikins, which are meant to satisfy the needs of occupational hygienists or quality control officers of clothing manufacturers. The other direction is a more sophisticated, multifunctional approach for research and advanced testing (Rossi & Psikuta, 2010). Although different manikins were developed by

Table 7.1 Milestones in the Development of Human-Shaped Thermal Manikins

Type	Material	Measurement Method	Adjustability	Development Location and Time
One segment	Copper	Analogue	–	USA 1945
Multisegment	Aluminum	Analogue	–	UK 1964
Radiation manikin	Aluminum	Analogue	–	France 1972
Multisegment	Plastics	Analogue	Movable	Denmark 1973
Multisegment	Plastics	Analogue	Movable	Germany 1978
Multisegment	Plastics	Digital	Movable	Sweden 1980
Multisegment	Plastics	Digital	Movable	Sweden 1984
Immersion manikin	Aluminum	Digital	Movable	Canada 1988
Sweating manikin	Aluminum	Digital	–	Japan 1988
	Plastic	Digital	Movable	Finland 1988
	Aluminum	Digital	Movable	USA 1996
Female manikin	Plastics Single wire	Digital, comfort regulation mode	Movable	Denmark 1989
Breathing thermal manikin	Plastics Single wire	Digital, comfort regulation mode	Movable, breathing simulation	Denmark 1996
Sweating manikin	Plastic	Digital, 30 dry and 125 sweat zones	Realistic movements	Switzerland 2001
Self-contained, sweating field manikin	Metal	Digital, 126 zones	Articulated	USA 2003
Virtual, computer manikin	Numerical, geometric model	Heat and mass transfer simulations	Articulated	China 2000
	Numerical, geometric model	Heat and mass transfer simulations	Articulated	UK 2001
	Numerical, geometric model	Heat and mass transfer simulations	Articulated	Sweden 2001
	Numerical, geometric model	Heat and mass transfer simulations	Articulated	Japan 2002
One-segment, sweating manikin	Breathable fabric	Digital, water heated	Movable	China 2001
One-segment manikin	Windproof fabric	Digital, air heated	Movable	USA 2003

Source: Holmér, I., *Eur. J. Appl. Physiol.*, **92**(6): 614–618, 2004.

various research groups having different materials, shapes, structures, methods of temperature control, and testing conditions, the main principles are the same (Pamuk, 2008). For example, most of the manikins have 15 or more thermally independent segments where each segment has its own computer-controlled heating system. The surface temperature of the segments can be maintained constant at 34°C and the temperature can be measured by temperature sensors. These manikins are placed in a controlled climatic chamber where the air flow is either horizontal or vertical. The thermal insulation is calculated according to either the serial or parallel methods.

Sweating manikins have been used significantly not only for cold protective clothing but also for protective clothing meant for hot conditions (Ross, 2005; Richards, 2003; Richards & McCullough, 2005; Richards et al., 2003). A number of good reviews of historical development and application of thermal manikins can be found (Wyon, 1989; Holmér, 2000; Holmér, 2004). Recent advances in the area of clothing comfort and human–thermal interaction also lead to the development of various sweating and walking thermal manikins.

It is more realistic to use a sweating thermal manikin to quantify heat and moisture transfer through clothing as both heat and moisture are coupled during the testing. The evaporative resistance of clothing can be determined using a sweating thermal manikin. Dry thermal manikins (or "thermal manikins" as referred above) are modified to incorporate an integrated sweating mechanism to simulate human perspiration. In general, there are five different types of sweating thermal manikins available with the following sweating mechanisms:

- Dry thermal manikins simply covered by a tight-fitting pre-wetted underwear or fabric "skin." For example, the Tore manikin.
- Gore-Tex "skin" bonded with capillary materials on its back side (spread water). For example, the Coppelius type manikin.
- Dry thermal manikins with an arrangement to supply water to the skin surface of the manikin. Amount of water to the skin or section of the skin can be controlled as per requirement to simulate different activity levels. Skin surface contains a large number of micropores depending upon the sweating gland density at different body locations. Such manikins can simulate liquid sweating. For example, the SAM manikin.
- Manikins incorporating sweating nozzles with a piece of highly stretchable knitted fabric "skin." For example, the Newton manikin.
- Dry thermal manikins covered with waterproof but water vapor-permeable fabric materials. There is an arrangement in these manikins also to supply the desired amount of water to the skin. Since the "skin" is covered with waterproof and water vapor-permeable material, vaporous sweating can be simulated through these manikins. For example, the Walter manikin.

Actually, a systematic approach to analyze any clothing should be to determine the thermal and water vapor transmission properties of fabric materials of a different layer of the clothing. This will help in selecting proper fabric materials for different layers. Then the manikin (thermal or sweating thermal manikin) testing should be carried out, as it will give a proper understanding of how the clothing will behave in actual situations. It should be followed by the subjective testing if it is possible to perform human wear trial tests under the environments in which the clothing is to be used.

7.3.1 Development of Thermal Manikins

There are more than 150 manikins in use worldwide. The first thermal manikin "Chauncy" was developed for the U.S. Army in the early 1940s, which was a one-segment copper manikin. It was designed for the measurement of thermal insulating properties of military garments and headgear for the U.S. military (Gagge et al., 1941; Holmér, 2004; Clarke, 2015). Since then, many thermal manikins have been developed. Table 7.1 shows the milestones in the development of human-shaped thermal manikins as summarized by Holmér (2004).

7. Hot Plates and Thermal Manikins for Evaluating Clothing Thermal Comfort

Overall, thermal manikins could be grouped into four generations (Sun & Fan, 2017). The first-generation manikins were standing and nonperspiring (Fonseca, 1975; McCullough et al., 1989; Mecheels & Umbach, 1977). The second-generation manikins were moveable but non-perspiring ones such as "Charlie" in Germany (Hanada, 1979; Olesen et al., 1982; Wu, 2010). The third-generation manikins were able to simulate perspiration but were unable to move such as the "Taro" in Japan (Yasuhiko, 1991). The latest-generation manikins could simulate both human perspiration and body motion. Moreover, they are now widely used in clothing comfort and ergonomics research. They include the "Newton" type manikin from the United States, "Coppelius" in Finland, "SAM" in Switzerland, "KEM" manikin in Japan, and "Walter" manikin from Hong Kong. Table 7.2 shows the details of the

Table 7.2 Sweating Thermal Manikins Used in Clothing Comfort Area

Name	Body Section	Sweating Mechanism	Shape	Movement	Skin
Coppelius (Meinander, 1992a,b; Meinander & Hellsten, 2004; Meinander et al., 2004; Wu, 2010)	18	A computer-controlled sweating system; 187 sweat glands.	A standard Swedish man, size 40	Permit movements and different body postures	Two layers: a nonwoven inner layer and a microporous outer layer
SAM (Richards & Fiala, 2004; Richards & Mattle, 2001; Richards et al., 2006)	26, 30	Both vapor sweating (when at rest) and combined vapor and liquid sweating (at heavy workloads); 125 sweat glands	Female	Repetitive body movements such as walking and climbing can be performed during an active test phase	–
Walter (Chen, 2003; Fan & Chen, 2002)	1	Pumping system is used to supply water to the extremities of the manikin	Male (height 172 cm, surface area 1.79 m²)	Can simulate walking	High-strength breathable fabric with waterproof but moisture-permeable (PTFE Gore-tex membrane)
ADAM (Rugh et al., 2006)	126	Arrangement to provide variable lateral (across segments) sweat distribution and flow regulation	American male (175 cm height, 61 kg weight)	Permit movements and different body postures	–
KEM (Fukazawa et al., 2004)	17	17 individually controlled sweat glands to obtain 0–1500 g/m²/h sweat rate	–	Can realize any desired posture	Thick and strong surface material with good water-vapor permeability
Newton (Wang et al., 2012; Wang, 2017a)	20, 26, 35	Uniformly distributed 134 sweating holes; individual control of sweat rate is possible	Western or Asian male	Simulate walking motion (up to 6 km/h) and allows different body postures	–

sweating manikins (Fan & Chen, 2002; Fan & Qian, 2004; Fukazawa et al., 2004; Koelblen et al., 2017; Mattle, 1999; Meinander, 1999).

7.3.2 Controlling Modes of Thermal Manikins

7.3.2.1 Basic Controlling Modes of Thermal Manikins

Normally, thermal manikins can be operated in three controlling modes: the constant heat flux (CHF) mode, the constant skin surface temperature (CST) mode, and the comfort equation (CE) mode. The CHF mode is that when the manikin supplies a constant level of power, set by the user, and the skin temperature of the different segments is measured. In the CST mode, the surface temperature of the manikin is maintained constant at a user-specified value (normally around 34.0°C), while the power increases or decreases depending on the environmental conditions. This may arguably be considered another method as well, as one can set the entire manikin to maintain the same temperature in all zones, or choose specific temperatures for each zone. The CE mode uses the comfort equation proposed by Fanger in the 1970s to control the manikin (Malina, 1973). Among these modes, the CE mode is considered to be the most accurate representation of the actual heat distribution across the human body, while the CHF mode is primarily used in high-temperature settings (when room temperatures are likely to be above 34°C) (Melikov, 2004). Also, when using Fanger's model, the setting of the skin surface temperature of the manikin is according to neutral conditions using a linear correlation with the sensible heat loss.

7.3.2.2 Thermal Manikins Controlled by Thermoregulatory Models

To provide a reliable prediction of the human thermoregulatory response under steady and transient conditions, a number of thermophysiological models have been developed, and they have been coupled with the thermal manikin system to form a new testing system, which may be called the thermoregulatory model controlled manikin or the adaptive manikin (Foda & Sirén, 2012; Psikuta et al., 2008; Wang et al., 2014). This control model may be referred to as the fourth control mode of a manikin (in addition to the aforementioned three commonly used control modes). For a thermoregulatory model controlled manikin, the manikin measures the thermal exchanges between environments and the clothed body. Thus, there is no need to measure complicated environmental parameters, which are normally served as key inputting parameters for the virtual clothing-thermoregulatory models prior to being used to predict human thermophysiological responses. In this mode, the manikin simulates the response of the human being exposed to a particular environment. Thermal manikin and the thermoregulatory model are coupled with a feedback loop.

Generally, two methods may be used to couple a human thermoregulatory model with a manikin (Psikuta et al., 2017b): the Neumann approach (or the boundary condition type 2 method) and the Dirichlet approach (or the boundary condition type 1 method). For the Neumann approach, the skin heat flux of the manikin is used as the inputting parameter for the thermoregulatory model. The predicted skin temperature and sweat rate based on the thermoregulatory model are used to control the manikin explicitly. In contrast, the Dirichlet coupling method uses the manikin skin surface temperature as the input to the thermoregulatory model, whereas the predicted skin heat flux and sweat rate are used as the feedback parameter for the manikin. Table 7.3 presents a list of the up-to-date development of thermoregulatory model controlled manikins.

Table 7.3 Summary of Developed Thermoregulatory Model Controlled Manikins

Manikin	Thermophysiological Model	Coupling Method (Feedback Parameters)	Number of Sectors	Primary Application	Lab
Advanced Automotive Manikin ADAM	Thermoregulation model by Rugh et al. (2004)		126	Automotive engineering	National Renewable Energy Lab (USA)
Sweating thermal cylinder Torso	Thermoregulation model by Fiala et al. (2012)	Boundary condition type 2 (heat flux)	1	Fabric assemblies	Empa (Switzerland)
Sweating Agile thermal Manikin SAM	Thermoregulation model by Fiala et al. (2012)	Boundary condition type 2 (heat flux)	22	Clothing and PPE	Empa (Switzerland)
Thermal sweating manikin Newton	Manikin PC2 (based on Fiala thermoregulatory model)	Boundary conditions type 1 (surfaces temperature)	26/34	Clothing, PPE, automotive and building engineering	Thermetrics LLC and ThermoAnalytics (USA)
Thermal sweating manikin Newton	Improved Xu and Werner's thermoregulation model	Boundary condition type 2 (heat flux)	38	Clothing	Decathlon (France)
Thermal sweating manikin Newton	Improved Tanabe thermoregulation model	Boundary condition type 2 (heat flux)	20	Clothing and PPE	Tsinghua University (China)
Therminator	Pierce's thermoregulation model	Boundary condition type 2 (heat flux)	24	Building engineering	Aalto University (Finland)
Sweating thermal head manikin	Fiala thermoregulation model	Boundary condition type 2 (heat flux)	4	Headgear	Empa (Switzerland)

Source: Psikuta, A. et al., *Renew. Sust. Energ. Rev.,* **78**: 1315–1330, 2017b.

Rugh's team from the National Renewable Energy Laboratory (NREL) in the United States developed an Advanced Automotive Manikin (ADAM) to predict human thermal comfort in transient, non-uniform thermal environments (see Table 7.3). The system includes a 126-segment sweating manikin and a human thermal physiological model. The model is a three-dimensional transient finite-element model incorporating a detailed simulation of human internal thermal physiological systems and thermoregulatory responses. The model consists of two interactive systems: a human tissue system and a thermoregulatory system. The thermoregulatory system controls physiological responses such as vasomotor control, sweating, and shivering. The human tissue system represents the human body, including the physiological and thermal properties of the tissues (Rugh et al., 2004). The manikin measured the heat loss from each body segment, while the thermophysiological finite-element model calculated the surface temperatures and sweat rates for each segment. The ADAM was proved to be better at predicting skin temperature distribution similar to the actual human body; it results in significant difference when predicting the core temperature and local skin temperature as compared to the real human body (Rugh & Bharathan, 2005; Rugh & Lustbader, 2006).

In 2005, a thermal sweating cylinder torso combined with the Fiala thermoregulation model was developed at Empa (Psikuta et al., 2008; Richards et al., 2006).

Fiala's model (Fiala et al., 2010), which could predict human thermo-physiological and thermal sensation responses under both steady and transient boundary conditions, was used. The model consists of two interacting systems: the controlling active system and the controlled passive system. The passive system simulates the physical human body and the heat and mass transfer within the body, and at its surface, the active system mimics the behavior of the human temperature regulation system predicting responses of the central nervous system. The passive system is a multisegmental, multilayered representation of the human body. To simulate the overall physiological response, the torso was taken to represent the entire of the human body (i.e., trunk, head, and extremities). The coupling method was based on the real-time iterative exchange of the relevant data between the torso and Fiala's model as shown in Figure 7.4.

Area-weighted averages of local skin temperatures and local sweat rates from the physiological model were used as control parameters for the torso. The average heat flux from the cylindrical surface of the torso was then used as the feedback signal, which represented the mean of heat exchanged with the environment for the clothing worn during a set time interval. Other physiological and perceptual parameters, such as the core temperature, skin blood flow, and heart rate, were calculated by the physiological model and stored in data files for post-processing (Psikuta et al., 2008). For the simulator, Fiala's model was modified to use a single

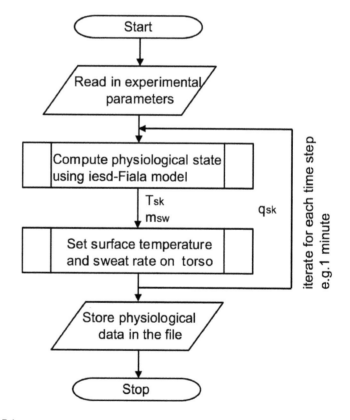

Figure 7.4

Flowchart showing data exchange in the thermophysiological human simulator. T_{sk}—average body skin temperature, °C; m_{sw}—average body sweat rate, g/min; q_{sk}—average body heat flux, W/m².

7. Hot Plates and Thermal Manikins for Evaluating Clothing Thermal Comfort

integral input value of the measured surface heat flux, which was used as the boundary condition at the body's skin surface in the model.

Besides the above sweating torso, the sweating agile thermal manikin SAM was also coupled with Fiala's model (Psikuta, 2009). It was also assumed to produce homogenous skin temperatures and sweat rate over the entire body, which are changing with time. The area-weighted averages of the skin temperature and the sweat rate from the physiological model were used as input parameters for the shell parts of the manikin. The feedback signal from the manikin was the average heat flux from all shell parts. The single average heat flux was applied to all body parts of the physiological model (Psikuta et al., 2008).

In 2009, an active physiological and comfort model was integrated with the sweating manikin Newton, and it is now a commercial product. The coupling was done for the first time with a manikin using the skin temperature as an input to the model. This method demands adjustment of the model code to extract the heat flux values at the skin level (to be provided to the manikin) and to accept resultant manikin surface temperature as a single-parameter boundary condition at the skin. On the other hand, the characteristics of the manikin's thermal mass need to be carefully evaluated and accounted for in the model-manikin control system (Psikuta et al., 2017a).

In addition, the manikin Newton was also coupled with Xu's thermoregulation model (Xu & Werner, 1997). In this case, the human body is subdivided into six segments, and each segment consists of four layers: core, muscle, fat, and skin layers. The afferent signal of the controlling system is composed of the weighted temperatures measured by thermal receptors at sites distributed in the body. Also, the difference between the signal and its threshold activates the thermoregulatory actions (Xu & Werner, 1997). This improvement of the simulator of using heat flux from the thermal manikin surface provides more reliable operations of the system in transient conditions.

Further, the Newton manikin has also been coupled with Tanabe's 65-node thermoregulation (65MN) model (see Figure 7.5) based on the Stolwijk model (Yang et al., 2014). The model has 16 body segments, and each segment consists of four layers: core, muscle, fat, and skin. The 65th node in the model is the central blood compartment (Tanabe et al., 2002). Convective and radiant heat transfer

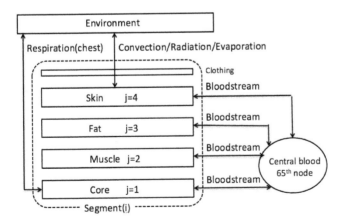

Figure 7.5

Conceptual figure of 65MN Tanabe's thermoregulatory model. (From Tanabe, S. et al., *Energ. Build.*, **34**(6): 637–646, 2002.)

coefficients and clothing insulation were derived from the thermal manikin experiments (Stolwijk, 1970, 1971). Yang et al. (2014) from the Tsinghua University of China coupled the Newton manikin with the modified Tanabe's model to create a dynamic system to simulate human thermal response. The human body is divided into 20 segments, and the central blood compartment is the 81st node. The model uses parameters including air temperature, mean radiant temperature, relative humidity, wind speed, and heat production, and measures heat exchange with the environment to predict skin temperature and sweating rate of each body node. The manikin skin temperature and sweating rate were predicted by the thermal model. The measured convective, radiant, and evaporation heat exchange will give a feedback signal to the thermal model to predict thermal response in the next time interval (Yang et al., 2014).

The manikin Therminator was also coupled with the improved multisegmental Pierce (MSP) model based on the Pierce model (Foda & Siren, 2011, 2012). In addition, Martínez et al. (2016) proposed a multisector thermophysiological head simulator by coupling a sweating thermal head manikin with Fiala's model. It is the first partial coupling system where heat flux at the head site is directly measured by the manikin, whereas the rest of the body is virtually simulated.

Although adaptive manikins have many advantages, such as it is less time consuming, low operational cost, operational readiness, no ethical issues, good repeatability, and easy operation as compared to human trials, it has some limitations as well. Adaptive manikins lack adequate validation and evaluation of the thermophysiological model and the thermal manikin individually before coupling (Psikuta et al., 2017b). The thermophysiological model is usually based on a simplified clothing model, which may lead to significant error in case of protective clothing or non-uniform environmental conditions. Manikin thermal characteristics and control system may also pose limitations on heating and cooling rate or skin temperature change rate. Besides, the simulation of human postures and body movements is often neglected in adaptive manikins, which may introduce big errors on the predicted thermophysiological responses (Wang, 2017b). It is therefore important to analyze the performance of both the thermal manikins and thermoregulatory model individually. Numerous validation studies shall also be performed even after the coupling.

7.3.3 Calculation of Thermal and Evaporative Resistances

Thermal manikins are widely used nowadays to study the performance of clothing. The thermal and evaporative resistances are the two main parameters of interest (Holmér & Nilsson, 1995). This section is devoted to describing the calculation procedure of these two parameters.

7.3.3.1 Thermal Insulation (Thermal Resistance)

Total thermal insulation, that is, insulation of clothing and boundary air layer near clothing surface, is defined as

$$R_t = \frac{(T_{surf} - T_{air})A}{Q_{dry}} \tag{7.4}$$

where T_{surf} is the manikin's surface temperature, which is usually maintained at 33–35°C, T_{air} is the ambient air temperature (°C) as shown in Figure 7.6, A is the

7. Hot Plates and Thermal Manikins for Evaluating Clothing Thermal Comfort

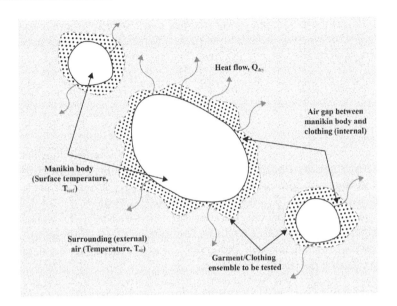

Figure 7.6

Heat transfer through clothing and air layer with a manikin at a cross section of the upper torso of the manikin body.

heated surface area of the manikin (m²), and Q_{dry} is the total input power to a thermal manikin (W). Figure 7.6 shows a manikin body with clothing at a cross section near the torso of a manikin. Air layer (air gap) between the body and the clothing can also be observed from the figure. At steady state, Q_{dry} amount of heat transfer takes place from the manikin surface to the ambient through clothing. Thermal insulation is frequently expressed as "clo" value.

$$I_t = \frac{6.45(T_{surf} - T_{air})A}{Q_{dry}} \qquad (7.5)$$

One "clo" is the thermal insulation of an ensemble required to keep a man at rest, exposed to an environment with 0.1 m/s wind speed and 21°C temperature in thermal comfort (Babus' Haq et al., 1996). It is a practical unit and 1 "clo" approximately represents the value of the insulation of one's everyday clothing (Gagge et al., 1941). Usually, R is used to represent clothing insulation in terms of the SI units (m²K/W) and I is used to indicate the clothing thermal insulation in the clo units (ASHRAE, 2013). There are other practical units to represent thermal insulation also like "Met," "Tog," and "Com" (Huang, 2006b; Huang & Xu, 2006). In case of the thermal manikin Walter, part of the total input power (Q_{sweat}) supplied to the manikin is used to evaporate water to simulate sweating (Zhou et al., 2010). Total thermal resistance, in that case, is calculated using the following relation:

$$R_{t,sweat} = \frac{(T_{surf} - T_{air})A}{Q_{sweat} - Q_{evap}} \qquad (7.6)$$

where Q_{sweat} is the total input power supplied to the sweating thermal manikin, and Q_{evap} is the fraction of total input power supplied to the manikin, which is used to evaporate the water. It is known as evaporative heat loss and calculated as

$$Q_{evap} = \dot{m}_{w,evap} h_{fg@T_{surf}} \tag{7.7}$$

where $\dot{m}_{w,evap} = (m_{w,total} - m_{w,cond})/t$ is the rate of water evaporated (kg/s), $m_{w,total}$ is the total amount of water supplied to the manikin (kg), $m_{w,cond}$ is the amount of water condensed in clothing (kg), t is the time, and $h_{fg@T_{surf}}$ is the heat of vaporization of water at the manikin surface temperature, T_{surf} (J/kg).

It is important to note here that even with the above-defined relations, there are different methods to calculate thermal insulation of any clothing using thermal manikins. Parallel, serial, and global calculation methods are commonly used to calculate the thermal insulation.

7.3.3.2 Parallel Calculation Method

In the parallel calculation method, before calculating thermal insulation of clothing, the summation of heat losses and area-weighted temperatures is calculated.

$$T_{surf} = \sum_{i=1}^{n} \frac{A_i}{A} T_{surf,i} \tag{7.8}$$

$$Q_{dry} = \sum_{i=1}^{n} Q_{dry,i} \tag{7.9}$$

where A_i is the surface area of the ith segment of the manikin, $T_{surf,i}$ is the surface temperature of the ith segment of the manikin, $Q_{dry,i}$ is the heat loss from the ith segment of the manikin, and n is the total number of segments in the manikin. Combining Equations 7.8 and 7.9 with Equation 7.4 results in the following definition of the parallel calculation method (Huang, 2008; Kuklane et al., 2012):

$$I_t = \frac{\left(\sum_{i=1}^{n} \frac{A_i}{A} T_{surf,i} - T_{air} \right) A}{\sum_{i=1}^{n} Q_{dry,i}} \tag{7.10}$$

However, some scientists (Oliveira et al., 2008) suggested that instead of Equation 7.10, the following represents the calculation of the parallel calculation method of thermal insulation better:

$$\frac{1}{I_t} = \sum_{i=1}^{n} \frac{A_i}{A} \frac{Q_{dry,i}}{(T_{surf,i} - T_{air})A_i} = \sum_{i=1}^{n} \frac{A_i}{A} \times \frac{1}{I_{t,i}} \tag{7.11}$$

7.3.3.3 Serial Calculation Method

In serial calculation method, the individual area-weighted thermal insulation of all body segments is calculated and then summed (Huang, 2008; Kuklane et al., 2012).

$$I_t = \sum_{i=1}^{n} \frac{A_i}{A} \frac{(T_{surf,i} - T_{air})A_i}{Q_{dry,i}} = \sum_{i=1}^{n} \frac{A_i}{A} \times I_{t,i} \qquad (7.12)$$

7.3.3.4 Global Calculation Method

Apart from the parallel and the serial calculation methods, standard ISO 9920 (2007) suggests another method to calculate the thermal insulation of clothing using thermal manikin known as the global method. It calculates thermal insulation using

$$I_t = \frac{\left(\sum_{i=1}^{n} \frac{A_i}{A} T_{surf,i} - T_{air} \right) A}{\sum_{i=1}^{n} \frac{A_i}{A} Q_{dry,i}} \qquad (7.13)$$

In fact, it is a general method applicable for wider conditions, whereas the parallel and the serial calculation methods are specific cases and can be obtained from the global method (Oliveira et al., 2008). In case of constant skin temperature over the manikin body, the global calculation method in Equation 7.13 becomes equivalent to the parallel method of Equation 7.10. Similarly, in case of uniform heat flux over the manikin body, the global calculation method of Equation 7.13 becomes equivalent to the serial method of Equation 7.12 (Huang, 2012).

It can be observed that different standards suggest using different calculation methods for thermal insulation. ISO 15831 (2004) specified the parallel and the serial calculation methods. EN 342 (2008) specified the need to calculate thermal insulation using the parallel, serial, and mean of both the parallel and serial methods. On the other hand, ISO 9920 (2007) specified the parallel, serial, and global calculation methods. It has been found that the thermal insulation values calculated by these three methods may differ significantly, especially in case of clothing of higher thermal insulation (Xu et al., 2008) and unevenly distributed clothing (up to 24%) (Kuklane et al., 2004; Lee et al., 2011). In case of constant skin temperature distribution over the manikin body, the serial method gives higher thermal insulation value than the parallel or the global method (Huang, 2008; Lee et al., 2011). Lee et al. (2011) found that this difference between the serial and the global is higher for garments/ensembles with a larger covering area per unit mass, that is, winter garments. In case of non-uniform temperature distribution over the manikin surface, the serial method gives lower thermal insulation than the parallel method (Huang, 2008).

It was found that during the thermal comfort (CE) regulation mode, the serial, parallel, and global methods give different values of thermal insulation (Oliveira et al., 2008). For daily wear garments, the mean relative difference between the serial and global methods is found to be 33.6%, and that between the global and the parallel methods is found to be −8.6%. This difference due to calculation methods further increased in case of cold protective garments. For cold protective garments, the mean relative difference between the serial and the global methods is found to be 78.4%, and between the global and the parallel methods is found to be −15.2%. Hence, cold protective garments are more sensitive to calculation methods. This study was later extended to include all three different modes: constant skin temperature (CST), constant heat flux (CHF), and thermal comfort (CE) regulation modes of using thermal manikins (Oliveira et al., 2008b). They (Oliveira et al., 2008b) analyzed thermal insulation of cold protective ensemble, summer ensemble, and business suit under these three modes. Significant differences between the thermal insulations calculated by the serial and the parallel methods were observed especially in case of the CE and CST modes and with the cold protective ensemble. Lee et al. (2011) and Kuklane et al. (2007) compared thermal insulation values obtained by using the series and the global (or conventionally "parallel") methods with the subjective wear-trial experiments. Lee et al. (2011) used thermal manikin under the constant surface temperature (CST) mode to obtain thermal insulation of 150 single garments and 38 clothing ensembles. Wear-trial results were obtained using 26 clothing ensembles with five female participants as the subject. It was found that the calculation method is one of the sources of differences in the thermal insulation results obtained by thermal manikins and wear trial. Global methods result in lower thermal insulation than wear-trial results for all seasonal ensembles, whereas the serial method results in higher thermal insulation than wear-trial results for only winter wear and lower for spring and summer ensembles. These results are in agreement with the findings of Kuklane et al. (2007), who also reported higher thermal insulation values in case of the serial calculation method compared to the parallel method for cold protective clothing.

An accurate value of thermal insulation is essential as it is an important input of models related to cold stress estimation. It is suggested that the parallel method of calculation of the thermal insulation using manikins should be used for clothing having non-uniform insulation distribution (Kuklane et al., 2012), whereas the series method is better to analyze local insulation or thermal sensation as it calculates thermal insulation of each segment (Huang, 2008). Moreover, it is recommended (Huang, 2012; Xu et al., 2008) to use the serial model in case of the constant heat flux (CHF) mode, whereas the parallel method should be used in case of the constant skin temperature (CST) mode. The global method can be used in all cases irrespective of the regulation modes (Huang, 2012; ISO 9920, 2007).

As mentioned earlier and shown in Figure 7.7, the total thermal insulation, I_t, includes both clothing thermal insulation and the air layer thermal insulation. Air layer thermal insulation depends on climatic conditions such as wind velocity, ambient air temperature, etc., and hence thermal insulation values change significantly depending on these conditions even for the same clothing ensemble (Havenith et al., 1990). Hence, to segregate other effects from the total thermal insulation and to determine thermal insulation of only clothing ensemble, intrinsic (or basic) clothing insulation is used. The intrinsic clothing insulation (I_{cl}) is thus independent of wind effect and defined as

$$I_{cl} = I_t - \frac{I_a}{f_{cl}} \tag{7.14}$$

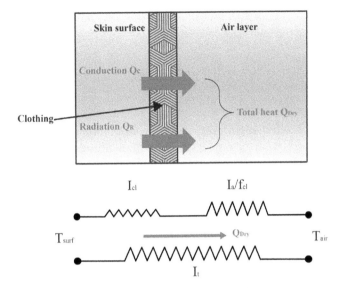

Figure 7.7

Schematic and network analogy showing the total thermal insulation, which includes intrinsic clothing and air layer thermal insulation.

where I_a is the thermal insulation offered by boundary air layer (clo), and f_{cl} is the clothing area factor (the ratio of the surface area of the clothed body to the surface area of the nude body). The value of the clothing area factor is always greater than 1.0 (1.0 corresponds to nude manikin), and it can be maximum 1.7 for protective clothing (McCullough, 2005). There are various techniques available to measure the clothing area factor of any manikin with a clothing ensemble (Alajmi et al., 2008; Fanger, 1970; Gao et al., 2005; McCullough et al., 2005; Kakitsuba, 2004). Apart from this, values provided in standards ISO 9920 (2007) and ASTM F 1291 (2016) or database (McCullough et al., 1985; McCullough et al., 1989) can be used for the clothing area factor (i.e., f_{cl}) directly.

7.3.3.5 Evaporative Resistance

There are two widely used methods to define clothing total evaporative resistance (R_{et}), that is, the overall evaporative resistance of the clothing ensemble and the boundary air layer. These are known as the heat loss method and the mass loss method (ASTM F2370, 2016; Wang et al., 2011). According to the heat loss method, total evaporative resistance can be calculated as follows:

$$R_{et,h} = \frac{(P_{surf} - P_{air})A}{Q_{evap}}$$

(7.15)

where P_{surf} and P_{air} are saturated water vapor pressure (Pa) at the manikin "skin" surface and partial water vapor pressure in the ambient air near the clothing surface, respectively. A is the total sweating surface area of the manikin (m²), and Q_{evap} is the evaporative heat loss (W) as defined in Equation 7.7.

According to the mass loss method, evaporation rate of water ($\dot{m}_{w,evap}$) from the manikin surface is first calculated by weighing sweating thermal manikin at regular intervals (Richards & McCullough, 2005). Evaporative resistance is then calculated using the following correlation:

$$R_{et,m} = \frac{(P_{surf} - P_{air})A}{\dot{m}_{w,evap}h_{fg@T_{surf}}} \qquad (7.16)$$

It has been found that air layer resistance contributes to the total resistance, and in some cases, it can be even larger than the clothing evaporative resistance itself (Fourt & Harris, 1947). Similar to the intrinsic thermal insulation defined above, the intrinsic evaporative resistance (R_{ecl}) is also defined as

$$R_{ecl} = R_{et} - \frac{R_{ea}}{f_{cl}} \qquad (7.17)$$

where R_{ea} is the boundary air layer evaporative resistance of the nude manikin ($m^2 \cdot Pa/W$), and f_{cl} is the clothing area factor as defined earlier.

Both the heat and mass loss methods result in different values of the evaporative resistance even for the same clothing. Havenith et al. (2008a) and Wang et al. (2011) found that the evaporative resistance calculated using the heat loss method was higher than the evaporative resistance calculated using the mass loss method. The difference in the evaporative resistance calculated by the two methods was found in the range 11.2–37.1%. This may further result in a significantly different prediction of heat stress (Havenith et al., 2008b). Wang et al. (2011) suggested using the mass loss method for calculating the evaporative resistance. However, even the heat loss method can be used with suitable correction in the temperature value (Wang et al., 2015). It can be observed here that the calculation of the evaporative resistance is based on the manikin surface temperature as evident from Equation 7.16. In practice, in a sweating thermal manikin, evaporating water takes away some heat and it results in a lower skin temperature than the manikin surface temperature. This leads to the error in the calculation of the evaporative resistance (Wang et al., 2011; Wang et al., 2012b). Wang et al. (2012b) observed that using the nude manikin surface temperature in the calculation of the evaporative resistance can result in 3.8% to 23.7% higher estimation than the real evaporative resistance. Empirical equations were proposed to predict the wet fabric "skin" surface temperature (Ueno & Sawada, 2012; Wang et al., 2010a; Wang et al., 2010b; Wang et al., 2017). This can be used to correct the heat loss method to obtain better values of the evaporative resistance using the heat loss method (Wang et al., 2015).

Evaporative resistance was not only affected by calculation method but also influenced by other factors such as sweating method, temperature control, fabric "skin," sweating rate, and test conditions. Evaporative resistance calculated using thermal manikins should, therefore, be corrected to take into account these factors. It is recommended that isothermal conditions should be maintained and the mass loss method should be used while calculating evaporative resistance using sweating thermal manikins (Wang, 2017a).

Over the years, various clothing ensembles were tested on thermal manikins to calculate their thermal insulation and evaporative resistance. A comprehensive database of the thermal and evaporative resistances for various clothing ensembles are available now (Kim & McCullough, 2000; McCullough et al., 1985; McCullough et al., 1989; McCullough & Hong, 1992, 1994). These values of the thermal insulation and evaporative resistance are required in thermal models to predict thermal comfort and heat transfer to or from the human body in various ambient conditions and at different metabolic heat levels. These values can be directly taken from the database.

7.4 Applications

7.4.1 Measurements of Clothing Performance under Steady-State Conditions

SGHPs and sweating thermal manikins have been widely used for simulation of heat transfer from the human body to the environment. For the SGHPs, they are mainly used to measure the thermal and water vapor resistances of clothing fabrics, under steady-state conditions (Huang, 2006; McCullough et al., 2004). Standards for measuring these have been proposed, such as ISO 11902 (2014) *Textiles—Physiological effects—Measurement of thermal and water-vapor resistance under steady-state conditions (sweating guarded-hotplate test)* and ASTM F1868 (2017)—*Standard test method for thermal and evaporative resistance of clothing materials using a sweating hot plate*. The resistance data can also be used to calculate performance indexes such as the permeability index, total heat loss, insulation per unit thickness, and insulation per unit weight for comfort evaluations. But when using the fabric data for clothing comfort evaluation, clothing variables such as the amount of body surface area covered by different fabrics, the distribution of fabric layers, air gap thickness, and volume should also be considered. Therefore, sweating thermal manikins that can simulate the actual shape of the human body are more often used for clothing comfort evaluation as compared to hotplates.

Sweating thermal manikins are usually operated at uniform steady-state surface temperatures and homogenous sweat rates in comparative measurements. Based on this, several standards such as ASTM F1291(2016): *Standard Test Method for Measuring the Thermal Insulation of Clothing Using a Heated Manikin*, ASTM F1868(2017): *Standard Test Method for Thermal and Evaporative Resistance of Clothing Materials Using a Sweating Hot Plate*, ISO 15831 (2004): *Clothing—Physiological effects—Measurement of thermal insulation by means of a thermal manikin*, ISO 9920 (2007): *Ergonomics of the thermal environment— Estimation of thermal insulation and water vapor resistance of a clothing ensemble*, ASTM F2370 (2016): *Standard Test Method for Measuring the Evaporative Resistance of Clothing Using a Sweating Manikin*, and ASTM F1720 (2014): *Standard Test Method for Measuring Thermal Insulation of Sleeping Bags Using a Heated Manikin*, were proposed.

The standards above only study and evaluate the thermal insulation in a static condition. As body posture, activity level, air velocity, accumulation of sweat, compression, thickness, number of clothing layers, and fit of clothing may change thermal insulation (or evaporative resistance) significantly, thermal manikins with different motions were used to study heat transmission properties of clothing. Besides, the European standard for protective clothing against cold EN 342 (2008) also suggests that the testing of thermal insulation should be made on a moving thermal manikin. Afanasieva et al. (2000) adopted a thermal manikin to simulate a walking speed of 0.8 m/s to study the effect of wind on the thermal insulation of cold protective clothing. Nilsson et al. (2000) studied the effects of wind, permeability, and walk on the total thermal insulation, and found that the combined effect of the body movements and air velocity can reduce the total thermal insulation by up to 70%. Oliveira et al. (2011) studied the thermal insulation of clothing ensembles both in static and walking conditions using a thermal manikin operating under the thermal comfort mode. They also compared the thermal insulation differences under various thermal insulation calculation methods.

In addition, effects of clothing microclimate on heat and mass transfer properties of the clothing were also analyzed based on thermal manikin testing. Wang et al. (2016) studied the air layer distribution and evaporative resistances of 39 sets of male Chinese ethnic clothing using a sweating thermal manikin. The relationship between the evaporative resistance and air layers (air gap thickness and air volume) was explored. They found that the clothing total evaporative resistance increases with the increasing air gap size/air volume, but the rate of increase gradually decreases as the mean air gap size or the total air volume becomes larger. The study contributes to the body knowledge on thermal properties of non-Western ethnic clothing and the relationship between clothing moisture transfer property and fit. Considering the heterogeneous distribution of air gap under real wearing conditions, Mert et al. (2015) studied the effects of air gaps on dry heat loss through garment using a vertical heated sweating cylinder.

Besides thermal and water vapor resistance measurements, thermal manikins can also be used to assess the performance of various types of clothing, especially the personal protective equipment (PPE) evaluation. PPE is designed to protect against natural or human-made hazards, such as chemical, biological, radiological, nuclear, and explosive hazards, and includes firefighting, first responder, HAZMAT, and certain occupational and combat uniforms (ASTM F2668, 2007; O'Brien et al., 2011). All the new PPE designed should be evaluated in terms of health and performance risks before use. Usually the systematic evaluation includes evaluation of insulation and permeability of textiles on a guarded hot plate and ensembles on a thermal manikin; modeling to predict thermal strain, including work limits and fluid requirements; and human testing to quantify physiological responses in a controlled laboratory setting that simulates the conditions under which the PPE will be used in the field (O'Brien et al., 2011). Testing should begin with an evaluation of fabrics according to the ISO 11092 (2014), to determine insulation and permeability values. Then the testing should be followed by the whole PPE on a thermal manikin.

PPE ensemble wearers may encounter heat stress while performing work under difficult environmental conditions. Kim et al. (2014) conducted a study to evaluate the predictive capability of fabric total heat loss values on the thermal stress. The SGHP test was conducted in accordance with ASTM F1868 (2017). The sweating thermal manikin test was conducted in accordance with ASTM F1291 (2016) and ASTM F2370 (2016). Human trials tests were also conducted. They found that the physiological benefits from wearing a more breathable PPE ensemble may not be feasible with incremental total heat loss values less than approximately 150–200 W/m^2. And the effects of thermal environments on a level of heat stress in PPE ensemble wearers are greater than ensemble thermal characteristics.

Besides the heat and cold stress evaluation of PPE using thermal manikins, cooling effects (or cooling performance) of personal cooling systems (PCS) can also be evaluated by the systems. Dionne et al. (2003) evaluated liquid cooling garments intended for use in hazardous waste management through a sweating thermal manikin. Three tube suits containing different densities of tubing were evaluated. The environmental chamber and manikin surface temperatures were set at 35°C. For fluid flow rates ranging from approximately 250 to 750 mL/min, and inlet temperatures to the tube suit ranging from 7°C to 10°C, heat removal rates between 220 and 284 W were measured, indicating the effectiveness of the tube suit at removing excessive body heat.

In 2005, ASTM F2371 (now updated to 2016 version) was proposed to specify the measurement of the average heat removal rate from a sweating heated manikin as well as the duration of cooling provided by a cooling garment. A multisegmental sweating manikin should be used to do testing. At least, the chest, back, abdomen, buttocks, arms, and legs should be sweating. The standard testing condition is isothermal, that is, 0.4 ± 0.1 m/s air velocity, $40\pm5\%$ relative humidity, 35 ± 0.5°C air temperature, and mean skin temperature during a rest. A baseline (control) test without cooling should be conducted first. The baseline test involves determining the power input to the manikin when without cooling but with sweating. When doing PCS performance test, it is required to keep the cooling system switched off before the temperature of the manikin is stabilized. As soon as the cooling system of the PCS is ON and the manikin is powered, start recoding the manikin's skin temperature, the air temperature, and the power input to the manikin. Take measurements until the effective cooling rate has decreased to 50 W, or up to a maximum of 2 h. The cooling rate is computed as the time average of the power input to the manikin from the time that the data were recorded. The time-averaged power is obtained from the numerical integration of the power input versus time until the effective cooling rate has decreased to 50 W, divided by the time it took to reach the value from when the time data were recorded. The effective cooling rate is determined by subtracting the average power input value and the baseline power value (ASTM F2371, 2016).

Yi et al. (2017) used a "Newton" thermal manikin to measure the cooling power of PCS according to ASTM F2371 (2016). The sweating rate was set to 1200 mL/h/m² to simulate the human body in heavy sweating. Segmental heat losses were recorded and the torso heat loss was calculated. Ouahrani et al. (2017) evaluated the PCM cooling vest performance using a 20-zone thermal manikin. The constant skin temperature of 35°C was used and the chamber temperature was 32°C. The total manikin torso heat flux for 200 min was recorded to evaluate the cooling effect of garments. As the heat removal rate was not constant for PCMs, Gao et al. (2010) studied the gap between the measured heat removal rate of smart clothing with PCMs obtained on a thermal manikin in a stable state, and clothing effects on local human skin and on core temperature. They suggested that thermoregulatory manikins should be used when evaluating the cooling performances of intelligent clothing and smart textile.

It is needed to mention that, according to ASTM F2371 (2016), the sweating manikins were always operated in the constant temperature mode, to determine the heat removal rate and cooling duration of PCSs. Unfortunately, the test method is proved to yield unrealistically high cooling rates for air cooling systems due to a fully saturated fabric skin surface being used during testing (Wang & Song, 2017).

7.4.2 Clothing Performance under Transient State Conditions

Performing steady-state testing is insufficient to quantify clothing characteristics because the properties and performance of clothing in real situations change with time and the physiological effects of the wearers need to be predicted. A more realistic method for analyzing clothing performance under transient conditions are human tests, but they need to consider costs, ethical restrictions, and intra- and inter subject variability. To solve the above issues, adaptive manikins were used to evaluate the clothing performance under transient conditions.

Psikuta et al. (2008) developed a single-sector thermophysiological human simulator consisting of a sweating heated cylinder "Torso" coupled with the Fiala multinode model of human physiology and thermal comfort. Further, they validated the system for various types of protective clothing (water vapor permeable and impermeable, thick multilayered and thin single-layered) and activities (reclining, sitting, walking, and running). The simulator can also reflect the moisture retention within the clothing and the additional cooling due to the heat pipe effect in impermeable clothing (Psikuta et al., 2013).

Another application of the single-sector thermophysiological human simulator was related to the evaluation of the cooling garments (a cooling shirt and an ice vest) using sweating agile thermal manikin SAM. The results of manikin testing were compared with human trials. The results show that the cooling power determined using the thermal manikin was two times higher for the cooling shirt and 1.5 times higher for the ice vest compared to that determined using human participants. For the thermophysiological human simulator, the cooling power of the cooling shirt was similar to that obtained using human participants. However, it was two times lower for the ice vest when using the thermophysiological human simulator. The thermophysiological human simulator is shown to be a useful tool to predict thermophysiological responses under mild cooling intensity. But it needs to be further improved for strong cooling intensities under heterogeneous conditions (Bogerd et al., 2010).

Except for the single-sector thermophysiological simulator, multisector adaptive manikins are also used to evaluate clothing performance. The Newton thermal manikin together with the thermophysiological models and comfort ManikinsPC² remains the only commercially available thermal human simulator system up to date (Psikuta et al., 2017). Lai et al. (2017) evaluated a hybrid personal cooling system using a manikin operated in constant temperature mode and thermoregulatory model control mode in warm condition. For the constant temperature mode, the surface temperature was set at 34°C. For the thermoregulatory model control mode, the manikin surface temperature and the sweating rate were controlled in analog to the prediction of the thermoregulatory model based on the Fiala model (Fiala et al., 1999; Wang et al., 2014). The real-time heat flux observed from the manikin was used as the feedback to the thermoregulatory model to determine physiological prediction outputs for the next time interval. Wang and Song (2017) also used the above adaptive manikin to investigate the thermophysiological responses of a human while using different personal cooling strategies during heatwaves. The documented studies on adaptive manikin show that the system can be used for examining the impact of smart clothing or protective clothing on the state of the human body and allowing a comparison between different prototypes. But it is necessary to conduct thorough qualitative and quantitative validation for a wide range of conditions including scenarios with clothing (Psikuta et al., 2017).

7.5 Conclusions

The hot plate and the thermal manikin are among the most important tools to evaluate the thermal and evaporative properties of textiles and clothing. They are being used increasingly in the clothing comfort research. Importance of accurately determining the thermal and evaporative resistances of textile or clothing is more than ever with the development of thermophysiological models. These thermophysiological models are widely used nowadays to predict human physiological responses under different environmental and clothing conditions. To predict reliable human physiological responses, it is important to input accurate values of clothing properties such as thermal and evaporative resistances. Although the SGHP and thermal manikins have been developed for many decades, there are still many issues that raise concerns regarding the accuracy of determined thermal and evaporative properties of clothing using these methods. It is therefore important to further analyze these measurement methods and involved instruments. First, the relationship between testing results of the hot plate and those of the thermal manikins needs to be studied further. The textiles are eventually made into clothing. Hence, the relationship between thermal and water vapor transmission properties of fabrics and clothing should be analyzed, although it requires consideration of many physical factors. Second, validation studies should be conducted using human trials during each stage of the adaptive manikin development to avoid any possible error.

References

Afanasieva, R. F., Bessonova, N. A., Burmistrova, O. V., Burmistrov, V. M., Holmér, I., Kuklane, K. (2000) Comparative evaluation of the methods for determining thermal insulation of clothing ensemble on a manikin and person. In: Proceedings of Nokobetef 6 and 1st European Conference on Protective Clothing (ECPC), National Institute for Working Life, Stockholm, Sweden, pp. 188–191.

Alajmi, F. F., Loveday, D. L., Bedwell, K. H., Havenith, G. (2008) Thermal insulation and clothing area factors of typical Arabian Gulf clothing ensembles for males and females: Measurements using thermal manikins. *Appl. Ergon.* **39**(3): 407–414.

ANSI/ASHRAE 55 (2013) Thermal environmental conditions for human occupancy. American Society of Heating, Refrigerating, and Air-Conditioning Engineers, Atlanta, GA.

ASTM F1291 (2016) Standard test method for measuring the thermal insulation of clothing using a heated manikin, ASTM International, West Conshohocken, PA.

ASTM F1720 (2014) Standard test method for measuring thermal insulation of sleeping bags using a heated manikin, ASTM International, West Conshohocken, PA.

ASTM F1868 (2017) Standard test method for thermal and evaporative resistance of clothing materials using a sweating hot plate. ASTM International, West Conshohocken, PA.

ASTM F2370 (2016) Standard test method for measuring the evaporative resistance of clothing using a sweating manikin. ASTM International, West Conshohocken, PA.

ASTM F2668 (2007) Standard practice for determining the physiological responses of the wearer to protective clothing ensembles, ASTM International, West Conshohocken, PA.

Babus'Haq, R. F., Hiasat, M. A. A., Probert, S. D. (1996) Thermally insulating behaviour of single and multiple layers of textiles under wind assault. *Appl. Energy* **54**(4): 375–391.

Bogerd, N., Psikuta, A., Daanen, H., Rossi, R. (2010) How to measure thermal effects of personal cooling systems: Human, thermal manikin and human simulator study. *Physiol. Meas.* **31**(9): 1161–1168.

Chen, Y. (2003) The development of a perspiring fabric manikin for the evaluation of clothing thermal comfort. Ph.D. Thesis. The Hong Kong Polytechnic University, Hong Kong.

Clarke, T., Jr. (2015) Chauncy, the copper thermal manikin. *Mil. Med.* **180**(6): 718–719.

Čubrić, I. S., Skenderi, Z., Mihelić-Bogdanić, A., Andrassyc, M. (2011) Experimental study of thermal resistance of knitted fabrics. *Exp. Therm. Fluid. Sci.* **38**: 223–228.

Dionne, K., Semeniuk, K., Makris, A., Teal, W., Laprise, B. (2003) Thermal manikin evaluation of liquid cooling garments intended for use in hazardous waste management. In: WM'03 Conference, Tucson, AZ.

EN 342 (2008) Ensembles and garments for protection against cold. European Committee for Standardization, Brussels.

Fan, J., Chen, Y. (2002) Measurement of clothing thermal insulation and moisture vapour resistance using a novel perspiring fabric thermal manikin. *Meas. Sci. Technol.* **13**(1): 1115–1123.

Fan, J., Qian, X. (2004) New functions and applications of Walter, the sweating fabric manikin. *Eur. J. Appl. Physiol.* **92**(6): 641–644.

Fanger, P. O. (1970) *Thermal comfort: Analysis and applications in environmental engineering.* New York: McGraw-Hill.

Fiala, D., Havenith, G., Broede, P., Kampmann, B., Jendritzky, G. (2012) UTCI-Fiala multi-node model of human heat transfer and temperature regulation. *Int. J. Biometeorol.* **56**(3): 429–441.

Fiala, D., Lomas, K. J., Stohrer, M. (1999) A computer model of human thermoregulation for a wide range of environmental conditions: The passive system. *J. Appl. Physiol.* **87**(5): 1957–1972.

Fiala, D., Psikuta, A., Jendritzky, G., Frihns, A. (2010) Physiological modeling for technical, clinical and research applications. *Front. Biosci.* **2**(3): 939–968.

Flynn, D. R., Healy, W. M., Zarr, R. R. (2005) High-temperature guarded hot plate apparatus control of edge heat loss. In: McElroy, D. L. (ed.), Proceedings of the 28th International Thermal Conductivity Conference, Brunswick, Canada.

Flynn, D., Zarr, R. R., Healy, Filliben J. (2002) Design concepts for a new guarded hot plate apparatus for use over an extended temperature range. In: Desjarlais, A. O. and Zarr, R. R. (Eds.), *Insulation Materials: Testing and Application*: 4th Volume, ASTM STP 1426, ASTM International, West Conshohocken, PA.

Foda, E., Siren, K. (2011) A new approach using the Pierce two-node model for different body parts. *Int. J. Biometeorol.* **55**(4): 519–532.

Foda, E., Sirén, K. (2012) A thermal manikin with human thermoregulatory control: Implementation and validation. *Int. J. Biometeorol.* **56**(5): 959–971.

Fonseca, G. (1974) Heat transfer properties of military protective headgear, US Army Natick laboratory Report 74-29-CE, pp. 1–32.

Fonseca, G. F. (1975) Sectional dry heat transfer properties of clothing in wind. *Text. Res. J.* **45**(1): 30–34.

Fourt, L., Harris, M. (1947) Diffusion of water vapor through textiles. *Text. Res. J.* **17**(5): 256–263.

Fukazawa, T., Lee, G., Matsuoka, T., Kano, K., Yutaka, T. (2004) Heat and water vapour transfer of protective clothing systems in a cold environment, measured with a newly developed sweating thermal manikin. *Eur. J. Appl. Physiol.* **92**(6): 645–648.

Gagge, A. P., Burton, A. C., Bazett, H. C. (1941) A practical system of units for the description of the heat exchange of man with his environment. *Science* **94**(2445): 428–430.

Gao, C., Kuklane, K., Holmér, I. (2005) Using 3D whole body scanning to determine clothing area factor. In: Holmér, I., Kuklane, K., and Gao, C. (eds.) *Proceedings of 11th International Conference on Environmental Ergonomics*, Ystad, Sweden, pp. 452–456.

Gao, C., Kuklane, K., Holmér, I. (2010) Thermoregulatory manikins are desirable for evaluations of intelligent clothing and smart textiles. In: The Eighth International Meeting for Manikins and Modeling (8I3M), Victoria, BC, Canada.

Hanada, K. (1979) Studies on the regional thermal resistance of clothing system, Part I. *J. Jap. Res. Assoc. Text. End-Uses.* **20**: 273–304.

Havenith, G., Heus, R., Lotens, W. A. (1990) Resultant clothing insulation: A function of body movement, posture, wind, clothing fit and ensemble thickness. *Ergonomics* **33**(1): 67–84.

Havenith, G., Richards, M. G., Wang, X., Bröde, P., Candas, V., Hartog, E., Holmér, I., Kuklane, K., Meinander, H., Nocker, W. (2008a) Apparent latent heat of vaporization from clothing: Attenuation and "heat pipe" effects. *J. Appl. Physiol.* **104**(1): 142–149.

Havenith, G., Richards, M. G., Wang, X., Bröde, P., Candas, V., Hartog, E., Holmér, I., Kuklane, K., Meinander, H., Nocker, W. (2008b) Use of clothing vapor resistance values derived from manikin mass losses or isothermal heat losses may cause severe under and over estimation of heat stress. In: Proceedings of the 7th International Thermal Manikin and Modeling Meeting (7I3M). Coimbra, Portugal: University of Coimbra, pp. 1–2.

Holmér, I. (2000) Thermal manikins in research and standards. In: Proceedings of the Third International Meeting on Thermal Manikin Testing, 3IMM, Arbete och Hälsa, pp. 1–7.

Holmér, I. (2004) Thermal manikin history and applications. *Eur. J. Appl. Physiol.* **92**(6): 614–618.

Holmér, I., Nilsson, H. (1995) Heated manikins as a tool for evaluating clothing. *Ann. Occup. Hyg.* **39**(6): 809–818.

Huang, J. (2006a) Sweating guarded hot plate test method. *Polym. Test.* **25**(5): 709–716.

Huang, J. (2006b) Thermal parameters for assessing thermal properties of clothing. *J. Therm. Biol.* **31**: 461–466.

Huang, J. (2008) Calculation of thermal insulation of clothing from mannequin test. *Meas. Tech.* **51**(4): 428–435.

Huang, J. (2012) Theoretical analysis of three methods for calculating thermal insulation of clothing from thermal manikin. *Ann. Occup. Hyg.* **56**(6): 728–735.

Huang, J., Xu, W. (2006) A new practical unit for the assessment of the heat exchange of human body with the environment. *J. Therm. Biol.* **31**(4): 318–322.

ISO 9920 (2007) Ergonomics of the thermal environment—Estimation of thermal insulation and water vapour resistance of a clothing ensemble. International Organization for Standardization, Geneva, Switzerland.

ISO 11902 (2014) Textiles—Physiological effects—Measurement of thermal and water-vapour resistance under steady-state conditions (sweating guarded-hotplate test). International Organization for Standardization, Geneva, Switzerland.

ISO 15831 (2004) Clothing—Physiological effects—Measurement of thermal insulation by means of a thermal manikin. International Organization for Standardization, Geneva, Switzerland.

Kar, F., Fan, J., Yu, W. (2007) Comparison of different test methods for the measurement of fabric or garment moisture transfer properties. *Meas. Sci. Technol.* **18**(7): 2033–2038.

Kilinc-Balci, F. S. (2011) Testing, analyzing and predicting the comfort properties of textiles. In: Song, G. (ed.) *Improving Comfort in Clothing.* Cambridge: Woodhead Publishing, pp. 138–162.

Kim, C. S., McCullough, E. A. (2000) Static and dynamic insulation values for cold weather protective clothing. In: Nelson, C. N. and Henry, N. W. (eds.), *Performance of Protective Clothing: Issues and Priorities for the 21st Century*: Seventh volume, ASTM STP1386, American Society for Testing and Materials, West Conshohocken, PA.

Kim, J., Powell, J., Roberge, R., Shepherd, A., Coca, A. (2014) Evaluation of protective ensemble thermal characteristics through sweating hot plate, sweating thermal manikin, and human tests. *J. Occup. Environ. Hyg.* **11**(4): 259–267.

Koelblen, B., Psikuta, A., Bogdan, A., Annaheim, S., René, M. (2017) Comparison of fabric skins for the simulation of sweating on thermal manikins. *Int. J. Biometeorol.* **61**(9): 1519–1529.

Konarska, M., Soltyński, K., Sudol-Szopińska, I., Chojnacka, A. (2007) Comparative evaluation of clothing thermal insulation measured on a thermal manikin and on volunteers. *Fibres Text. East. Eur.* **15**(2): 73–79.

Konarska, M., Soltyński, K., Sudol-Szopińska, I., Dariusz, M., Chojnacka, A. (2006) Aspects of standardisation in measuring thermal clothing insulation on a thermal manikin. *Fibres Text. East. Eur.* **14**(4): 58–63.

Kuklane, K., Gao, C., Holmér, I. (2007) Calculation of clothing insulation by serial and parallel methods: Effects on clothing choice by IREQ and thermal responses in the cold. *Int. J. Occup. Saf. Ergon.* **13**(2): 103–116.

Kuklane, K., Gao, C., Wang, F., Holmér, I. (2012) Parallel and serial methods of calculating thermal insulation in European manikin standards. *Int. J. Occup. Saf. Ergon.* **18**(2): 171–179.

Kuklane, K., Sandsund, M., Reinertsen, R. E., Tochihara, Y., Fukazawa, T., Holmér, I. (2004) Comparison of thermal manikins of different body shapes and size. *Eur. J. Appl. Physiol.* **92**(6): 683–688.

Lai, D., Wei, F., Lu, Y., Wang, F. (2017) Evaluation of a hybrid personal cooling system using a manikin operated in constant temperature mode and thermoregulatory model control mode in warm conditions. *Text. Res. J.* **87**(1): 46–56.

Lee, J. Y., Ko, E. S., Lee, H. H., Kim, J. Y., Choi, J. W. (2011) Validation of clothing insulation estimated by global and serial methods. *Int. J. Cloth. Sci. Technol.* **23**(2/3): 184–198.

Lees, C. (1982) On the thermal conductivities of crystals and other bad conductors. *Philos. Trans. R. Soc. Lond.* **183**: 481–509.

Malina, R. M. (1973) Thermal comfort: Analysis and applications in environmental engineering by P. O. Fanger. *Hum. Biol.* **45**(1): 116–117.

Martínez, N., Psikuta, A., Rossi, R. M., Corberán, J. M., Annaheim, S. (2016) Global and local heat transfer analysis for bicycle helmets using thermal head manikins. *Int. J. Ind. Ergonom.* **53**: 157–166.

Mattle, N. (1999) Use sweating articulated manikin SAM for thermophysiological assessment of complete garments. In: Proceedings of the Third International Meeting on Thermal Manikin Testing (3IMM), Stockholm, Sweden, pp. 100–101.

McCullough, E. A. (2005) Evaluation of protective clothing systems using manikins. In: Scott, R. A. (ed.), *Textiles for Protection*. Cambridge: Woodhead Publishing, pp. 217–232.

McCullough, E. A., Hong, S. (1992) A database for determining the effect of walking on clothing insulation. In: Lotens, W. A. and Havenith, G. (eds.), Proceedings of the Fifth International Conference on Environmental Ergonomics, Maastricht, The Netherlands. pp. 68–69.

McCullough, E. A., Hong, S. (1994) A database for determining the decrease in clothing insulation due to body motion. *ASHRAE Trans.* **100**(1): 765–775.

McCullough, E. A., Huang, J., Deaton, S. (2005) Methods for measuring the clothing area factor. In: Holmér, I., Kuklane, K., and Gao, C. (eds.), Proceedings of the 11th International Conference on Environmental Ergonomics, Ystad, Sweden, pp. 433–436.

McCullough, E. A., Huang, J., Kim, C. (2004) An explanation and discussion of sweating hot plate standards. *J. ASTM Int.* **1**(7): 1–13.

McCullough, E. A., Jones, B. W., Huck, J. (1985) A comprehensive data base for estimating clothing insulation. *ASHRAE Trans.* **91**(2): 29–47.

McCullough, E. A., Jones, B. W., Tamura, T. (1989) A data base for determining the evaporative resistance of clothing. *ASHRAE Trans.* **95**: 316–328.

Mecheels, J., Umbach, K. H. (1977) Thermo-physiological properties of clothing system. *Melliand Textilber.* **57**: 1029–1032.

Melikov, A. (2004) Breathing thermal manikins for indoor environment assessment: Important characteristics and requirements. *Eur. J. Appl. Physiol.* **92**(6): 710–713.

Meinander, H. (1992a) Determination of clothing comfort properties with the sweating thermal manikin. In: Proceedings of The Fifth International Conference on Environmental Ergonomics. pp. 40–41.

Meinander, H. (1992b) Coppelius—A sweating thermal manikin for the assessment of functional clothing. Proceedings, NOKOBETEF IV, Kittila, Finland. 157–161.

Meinander, H. (1999) Extraction of data from sweating manikin tests. In: The Third International Meeting on Thermal Manikin Testing (3IMM), Stockholm, Sweden, 12–13 October 1999, Arbete och Hälsa, Sweden, pp. 95–99.

Meinander, H., Anttoene, H., Bartels, V., Holmér, I., Reinertsen, R. E., Soltynski, K., Varieras, S. (2004) Manikin measurements versus wear trials of cold protective clothing (Subzero project). *Eur. J. Appl. Physiol.* **92**(6): 619–621.

Meinander, H., Hellsten, M. (2004) The influence of sweating on the heat transmission properties of cold protective clothing studied with a sweating thermal manikin. *Int. J. Occup. Saf. Ergon.* **10**(3): 263–269.

Mert, E., Psikuta, A., Bueno, M., Rossi, R. (2015) Effect of heterogeneous and homogenous air gaps on dry heat loss through the garment. *Int. J. Biometeorol.* **59**: 1701–1710.

Nilsson, H., Anttonen, H., Holmér, I. (2000) New algorithms for prediction of wind effects on cold protective clothing. In: Kuklane, K., and Holmér, I. (Eds.), Proceedings of Nokobetef 6 and the 1st European Conference on Protective Clothing (ECPC), National Institute for Working Life, Stockholm, Sweden, pp. 17–20.

O'Brien, C., Blanchard, L., Cadarette, B., Endrusick, T., Xu, X., Berglund, L., Sawka, M., Hoyt, R. W. (2011) Methods of evaluating protective clothing relative to heat and cold stress: Thermal manikin, biomedical modeling, and human testing. *J. Occup. Environ. Hyg.* **8**(10): 588–599.

Olesen, B. W. (1982) Effect of body posture and activity on the insulation of clothing: Measurement by a movable thermal manikin. *ASHRAE Trans.* **88**(2): 791–805.

Oliveira, A. V. M., Branco, V. J., Gaspar, A. R., Quintela, D. A. (2008b) Measuring thermal insulation of clothing with different manikin control methods. Comparative analysis of the calculation methods. In: Proceedings of The 7th International Thermal Manikin and Modelling Meeting (7I3M), University of Coimbra, Portugal.

Oliveira, A. V. M., Gaspar, A. R., Quintela, D. A. (2008a) Measurements of clothing insulation with a thermal manikin operating under the thermal comfort regulation mode: Comparative analysis of the calculation methods. *Eur. J. Appl. Physiol.* **104**(4): 679–688.

Oliveira, A., Gaspar, A., Quintela, D. (2011) Dynamic clothing insulation measurements with a thermal manikin operating under the thermal comfort regulation mode. *Appl. Ergon.* **42**(6): 890–899.

Ouahrani, D., Itani, M., Ghaddar, N., Ghali, K., Khater, B. (2017) Experimental study on using PCMs of different melting temperatures in one cooling vest to reduce its weight and improve comfort. *Energ. Build.* **155**: 533–545.

Pamuk, O. (2008) Thermal manikins: an overview. *e-J. New World Sci. Acad. Nat. Appl. Sci.* **13**(1): 124–132.

Parsons, K. C. (2014) *Human thermal environments: The effects of hot, moderate, and cold environments on human health, comfort, and performance.* Third Ed., London: CRC Press.

Psikuta, A. (2009) Development of an 'artificial human' for clothing research. Ph.D. Thesis. De Montfort University, Leicester, England.

Psikuta, A., Allegrini, J. B., Koelblen, B., Bogdanc, A., Annaheima, S., Martíneza, N., Deromeb, D., Carmelietb, J., Rossi, R. M. (2017b) Thermal manikins controlled by human thermoregulation models for energy efficiency and thermal comfort research—A review. *Renew. Sust. Energ. Rev.* **78**: 1315–1330.

Psikuta, A., Koelblen, B., Mert, E., Fontana, P., Annaheim, S. (2017a) An integrated approach to develop, validate and operate thermo-physiological human simulator for the development of protective clothing. *Ind. Health* **55**(6): 500–512.

Psikuta, A., Richards, M., Fiala, D. (2008) Single-sector thermophysiological human simulator. *Physiol. Meas.* **29**(2): 181–192.

Psikuta, A., Wang, L., Rossi, R. (2013) Prediction of the physiological response of humans wearing protective clothing using a thermophysiological human simulator. *J. Occup. Environ. Hyg.* **10**(4): 222–232.

Richards, M. G. M. (2003) Modelling fire-fighter responses to exercise and asymmetric IR-radiation using a dynamic multi-mode model of human physiology and results from the sweating agile thermal manikin (SAM). In: Candas, V. (ed.), The 5th International Meeting on Manikins and Modelling (5I3M), Strasbourg, France.

Richards, M., Fiala, D. (2004) Modelling fire-fighter responses to exercise and asymmetric infrared radiation using a dynamic multi-mode model of human physiology and results from the sweating agile thermal manikin. *Eur. J. Appl. Physiol.* **92**: 649–653.

Richards, M. G. M., Mattle, N. G. (2001) A sweating agile thermal manikin (SAM) developed to test complete clothing systems under normal and extreme conditions. In: RTO HFM Symposium on Blowing Hot and Cold: Protecting Against Climatic Extremes, Dresden, Germany, RTO-MP-076.

Richards, M. G. M., Mattle, N. G., Becker, C. (2003) Assessment of the protection and comfort of fire fighters' clothing using a sweating manikin. In: Makinen, H., and Rossi, R. (eds.), The 2nd European Conference on Protective Clothing (ECPC) and NOKOBETEF 7, Montreaux, Switzerland.

Richards, M. G. M., McCullough, E. A. (2005) Revised interlaboratory study of sweating thermal manikins including results from the sweating agile thermal manikin. *J. ASTM Int.* **2**(4): 1–13.

Richards, M. G. M., Psikuta, A., Fiala, D. (2006) Current development of thermal sweating manikins at Empa. In: Fan, J. (ed.), Proceedings of the 6th International Thermal Manikin and Modelling Meeting (6I3M), Hong Kong, pp. 173–179.

Ross, K. A. (2005) Evaluation of an instrumented sweating manikin for predicting heat stress in firefighters' turnout ensembles. M.Sc. Thesis, North Carolina State University, Raleigh, NC.

Rossi, R. M., Psikuta, A. (2010) Assessment of the coupled heat and mass transfer through protective garments using manikins and other advanced measurement devices. In: NATO Advanced Study Institute on Defence-related Intelligent Textiles and Clothing for Ballistic and NBC Protection, Split, Croatia.

Rugh, J. P., Bharathan, D. (2005) Predicting human thermal comfort in automobiles. In: Proceedings of the Vehicle Thermal Management Systems Conference. Toronto, Canada.

Rugh, J. P., Farrington, R. B., Vlahinos, D. B., Burke, R., Huizenga, C., Zhang, H. (2004) Predicting human thermal comfort in a transient nonuniform thermal environment. *Eur. J. Appl. Physiol.* **92**(6): 721–727.

Rugh, J., King, C., Paul, H., Trevino, L., Bue, G. (2006) Phase II Testing of liquid cooling garments using a sweating manikin, controlled by a human physiological model. In: The 2006 International Conference on Environmental Systems, Norfolk, Virginia, pp. 1–9.

Rugh, J. P., Lustbader, J. (2006) Application of a sweating manikin controlled by a human physiological model and lessons learned. In: Fan, J. (ed.), Proceedings of the 6th International Thermal Manikin and Modeling Meeting (6I3M). Hong Kong, pp. 303–312.

Salmon, D. (2001) Thermal conductivity of insulations using guarded hot plates, including recent developments and sources of reference materials. *Meas. Sci. Technol.* **12**(12): R89–R98.

Silva, M., Coelho, J. (2002) Convection coefficients for the human body parts—Determined with a thermal mannequin. In: The 8th International Conference on Air Distribution in Rooms, Copenhagen, Denmark, pp. 277–280.

Stolwijk, J. A. J. (1970) *Mathematical model of thermoregulation, physiological and behavioral temperature regulation.* Springfield: Charles C. Thomas Publication, pp. 703–721 (Chapter 48).

Stolwijk, J. A. J. (1971) A mathematical model of physiological temperature regulation in man. Technical Report, NASA-CR-1855, NASA, WA.

Sun, C., Fan, J. (2017) Comparison of clothing thermal comfort properties measured on female and male sweating manikins. *Text. Res. J.* **87**(18): 2214–2223.

Tanabe, S., Kobayashi, K., Nakano, J., Ozeki, Y., Konishi, M. (2002) Evaluation of thermal comfort using combined multi-node thermoregulation (65MN) and radiation models and computational fluid dynamics (CFD). *Energ. Build.* **34**(6): 637–646.

Ueno, S., Sawada, S. (2012) Correction of the evaporative resistance of clothing by the temperature of skin fabric on a sweating and walking thermal manikin. *Text. Res J.* **82**(11): 1143–1156.

Wang, F. (2017a) Measurement of clothing evaporative resistance using a sweating thermal manikin: An overview. *Ind. Health* **55**(6): 473–484.

Wang, F. (2017b) Effect of body movement on the thermophysiological responses of an adaptive manikin and human subjects. *Measurement* **116**: 251–256.

Wang, F., Ferraro, S., Lin, L. Y., Mayor, T., Molinaro, V., Ribeiro, M., Gao, C., Kuklane, K., Holmér, I. (2012a) Localized boundary air layer and clothing evaporative resistances for individual body segments. *Ergonomics* **55**(7): 799–812.

Wang, F., Ferraro, S., Molinaro, V., Morrissey, M., Rossi, R. (2014) Assessment of body mapping sportswear using a manikin operated in constant temperature mode and thermoregulatory model control mode. *Int. J. Biometeorol.* **58**(7): 1673–1682.

Wang, F., Gao, C., Kuklane, K., Holmér, I. (2011) Determination of clothing evaporative resistance on a sweating thermal manikin in an isothermal condition: Heat loss method or mass loss method? *Ann. Occup. Hyg.* **55**(7): 775–783.

Wang, F., Lai, D., Shi, W., Fu, M. (2017) Effects of fabric thickness and material on apparent "wet" conductive thermal resistance of knitted fabric "skin" on sweating manikins. *J. Therm. Biol.* **70**: 69–76.

Wang, F., Kuklane, K., Gao, C., Holmér, I. (2010a) Development and validity of a universal empirical equation to predict skin surface temperature on thermal manikins. *J. Therm. Biol.* **35**(4): 197–203.

Wang, F., Kuklane, K., Gao, C., Holmér, I. (2012b) Effect of temperature difference between manikin and wet fabric skin surfaces on clothing evaporative resistance: How much error is there? *Int. J. Biometeorol.* **56**(1): 177–182.

Wang, F., Kuklane, K., Gao, C., Holmér, I., Havenith, G. (2010b) Development and validation of an empirical equation to predict wet fabric skin surface temperature of thermal manikins. *J. Fiber Bioeng. Inform.* **3**(1): 9–15.

Wang, F., Peng, H., Shi, W. (2016) The relationship between air layers and evaporative resistance of male Chinese ethnic clothing. *Appl. Ergon.* **56**: 194–202.

Wang, F., Song, W. (2017) An investigation of thermophysiological responses of human while using four personal cooling strategies during heatwaves. *J. Therm. Biol.* **70**: 37–44.

Wang, F., Zhang, C., Lu, Y. (2015) Correction of the heat loss method for calculating clothing real evaporative resistance. *J. Therm. Biol.* **52**: 45–51.

Wu, Y. S. (2010) Development of a sweating fabric manikin with sedentary and supine postures. Ph.D. Thesis, The Hong Kong Polytechnic University, Hong Kong.

Wyon, D. P. (1989) Use of thermal manikins in environmental ergonomics. *Scand. J. Work Environ. Health.* **15**(1): 84–94.

Xu, X., Endrusick, T., Gonzalez, J., Santee, W., Hoyt, R. (2008) Comparison of parallel and serial methods for determining clothing insulation. *J ASTM Int.* **5**(9): 1–6.

Xu, X., Werner, J. (1997) A dynamic model of the human/clothing/environment-system. *Appl. Human Sci.* **16**(2): 61–75.

Yang, J., Weng, W., Fu, M. (2014) Coupling of a thermal sweating manikin and a thermal model for simulation human thermal response. *Procedia Eng.* **84**: 893–897.

Yasuhiko, D., Yoshio, A., Toshitada, S., Takenishi, S. (1991) A model of sweating thermal manikin. *J. Tex. Mach. Soc.* **37**(4): 101–112.

Yi, W., Zhao, Y., Chan, A. (2017) Evaluation of the ventilation unit for personal cooling system (PCS). *Int. J. Ind. Ergonom.* **58**: 62–68.

Zarr, R. (1997) "History of the Guarded-Hot-Plate Apparatus at NIST," available at <http://www.bfrl.nist.gov/863/hotplate>, Gaithersburg MD: National Institute of Standards and Technology.

Zarr, R. (2001) A history of testing heat insulators at the national institute of standards and technology. *ASHRAE Trans.* **107**(2): 661–671.

Zhou, X., Zheng, C., Qiang, Y., Holmér, I., Gao, C., Kuklane, K. (2010) The thermal insulation difference of clothing ensembles on the dry and perspiration manikins. *Meas. Sci. Technol.* **21**: 085203.

Zwolińska, M., Bogdan, A., Fejdyś, M. (2014) Influence of different types of the internal system of the ballistic helmet shell on the thermal insulation measured by a manikin headform. *Int. J. Ind. Ergonom.* **44**(3): 421–427.

8

Human Wear Trials for Assessing Comfort Performance of Firefighter Protective Clothing

Faming Wang and Udayraj

8.1 Background

Firefighters work under extremely hot environmental conditions involving flame and radiant heat exposures of different intensities (Udayraj et al., 2016; Barker, 2005). Use of thermal protective clothing is a must under such extreme conditions for survival of firefighters. Thermal protective clothing usually consists of more than one clothing layer as shown in Figure 8.1. The typical structure of a firefighting clothing system consists of three layers: outer shell (outer layer), moisture barrier (mid-layer), and thermal liner (inner layer). Protective performance of the clothing depends significantly on thickness and density of various clothing layers. Higher fabric thickness and density result in higher protective performance (Shalev and Barker, 1983; Udayraj et al., 2016; Udayraj et al., 2017). So, in the process of developing clothing of higher thermal protective performance, usually, the firefighter protective clothing becomes bulky. To perform their duties such as smoke-diving and rescuing operations effectively, firefighters also carry personal protective equipment such as gloves, goggles, helmet, boots, hand tools, backpack pump, and self-contained breathing apparatus (SCBA), weighing as high as 20–30 kg (Duncan et al., 1979; Sköldström, 1987; Smith et al., 1997; Smith and Petruzzello, 1998; Dreger et al., 2006). These loads,

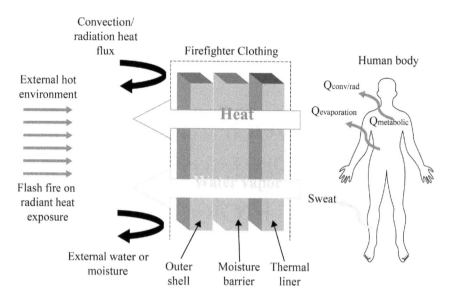

Figure 8.1

Firefighter clothing: Basic mechanism of heat and mass transfer through clothing.

along with the weight of protective clothing, contribute significantly to the physiological strain in firefighters (Sköldström, 1987; Faff and Tutak, 1989). Another significant contributor to the physiological strain and heat stress of firefighters are the physically demanding activities required as a part of firefighting and rescue operations under extreme environmental conditions.

The above-mentioned factors elicit profuse sweating, which further contributes to the discomfort of the firefighters. The main function of firefighter clothing is to avoid transmission of external heat and moisture to the human body as shown in Figure 8.1. A good and comfortable firefighter clothing system should be able to transmit body heat and water vapor (present due to sweating) from the body to external environment as shown in Figure 8.1. Although protective clothing is essential to block external heat to transfer from the environment to human body, it also restricts body heat and sweat from moving toward the external environment (White and Hodous, 1988). Thus, the use of thermal protective clothing presents an intensive challenge to the human thermoregulatory system, and this results in great physiological strain by minimizing the heat and moisture transfer from the body to the environment and adding extra loading to the body (Wen et al., 2015). Avoiding physiological strain and providing comfort to the wearers are thus important criteria (apart from thermal protective performance) for evaluating the performance of firefighting protective clothing.

At present, there are many bench-scale tests available to analyze the protective performance of fabrics used in firefighter clothing based on various international standards for flame and radiant heat exposures (Udayraj et al., 2016). Studies have been conducted to analyze heat and moisture transfer through firefighter clothing under heat exposures of different types and intensities (Udayraj et al., 2016; Baitinger, 1979; Udayraj et al., 2014; Udayraj et al., 2017; Shalev and Barker, 1984; Lee and Barker, 1986). To determine the protective performance of the entire firefighter ensemble, full-scale mannequin tests were developed

(Camenzind et al., 2007; Dale et al., 1992). Both flame mannequins and radiant heat mannequins were employed to analyze the protective performance of firefighter protective clothing.

It is an important requirement for firefighter clothing that it should be able to provide good thermal protection for at least a short duration during actual firefighting and rescue operations in fire scenarios. Under such scenarios, protection from heat and burn injuries is a top concern. It is to be noted that firefighters actually perform such operations under extreme conditions only for a small fraction of time in their duty. Most of the time they have to perform usual works such as walking, running, attending false alarm, and other accidents under relatively normal environmental conditions. During these times, comfort is a major concern rather than thermal protection. Thus, firefighters not only expect their turnout gear to be thermally protective but also comfortable (Barker et al., 2010). However, almost all existing tests and studies are just limited to determining the thermal protective performance of fabrics or firefighter clothing and not really bothered about the thermophysiological discomfort, cardiovascular strain, psychological responses, and heat stress faced by firefighters. Even the heat and mass transfer models developed so far for firefighter clothing (Torvi, 1997; Mell and Lawson, 2000; Chitrphiromsri and Kuznetsov, 2005; Udayraj et al., 2016a; Udayraj et al., 2017a) do not integrate human thermoregulation model to analyze the effect of different firefighter clothing on physiological responses. Human wear trials are therefore the only option available at present to analyze the effect of various firefighter clothings on human comfort and to access comfort-related performance of firefighter protective clothing.

In this chapter, fundamental knowledge on test protocols and experimental design of human wear trials for assessing comfort performance of protective clothing is presented. Various parameters used to characterize human thermophysiological responses (e.g., skin temperature, core temperature, mean body temperature, heat storage, heart rate, oxygen consumption, metabolic heat, sweating) are addressed and their measurement methodologies are explicitly discussed. Perpetual responses (e.g., thermal sensation, comfort sensation, skin wetness sensation, perceived exertion) and associated rating scales, as well as heat strain indices (e.g., physiological strain index, psychological strain index), are also presented. Finally, various human trial case studies conducted to analyze comfort performance of protective clothing under different environmental conditions are summarized.

8.2 Comfort Performance Evaluation of Firefighting Clothing

Performance denotes how well an apparel product performs in actual use conditions. For firefighting protective clothing, the two most important performance elements are thermal protective performance and comfort performance. To examine the thermal protective performance of protective clothing against various hazards, it is unethical to use human subjects to evaluate the actual performance. Comfort performance can conveniently be assessed by human subjects (Wang and Gao, 2014). Various environmental conditions and activity levels may be used in human trial studies. Human trial studies will be mainly used to examine the human physiological responses and psychological perceptions. Generally, such vital physiological parameters are widely used to characterize the thermophysiological response: heart rate, oxygen uptake, skin and body core temperature, and sweat production. Psychological responses are usually assessed using subjective rating scales, for instance, thermal sensation scale, comfort sensations scale, and perceived exertion scale. Human trials can also be used to evaluate the ergonomic comfort performance of firefighting protective clothing by performing the range of motion (ROM) tests and collecting perceived subjective sensations while performing certain task-related activities (Huck, 1988; Adams and Keyserling, 1995; Havenith and Heus, 2004; Coca et al., 2008; Wang and Gao, 2014).

8.2.1 Experimental Design

A good experimental design for human trial studies is necessary to accurately access comfort-related performance of firefighter clothing. Proper selection of subjects for the study and development of appropriate test protocol are two major criteria for a good experimental design.

8.2.1.1 Selection of Subjects and Ethical Issues

The selection of subjects (test persons) is one of the most important aspects of human trials, which has been highly unrated over the years. This is perhaps due to unavailability of general guidelines for selection of subjects. The following are some of the major points that should be kept in mind while selecting subjects for human wear trials (Goldman, 2013):

i. The number of subjects: the number of human subjects should be determined based on the objectives of human trials. For physiological testing, it is generally recommended that a minimum of eight subjects (ASTM F2668, 2016) shall be used. It is always good to have a large number of subjects to eliminate individual differences. It has been observed that 8 to 42 subjects have been used over the years for analyzing comfort performance of firefighting protective clothing using human trials as shown in Table 8.1. For subjective assessment of the comfort performance of firefighting protective clothing, a minimal number of 20–30 subjects shall be used because different subjects may rate quite different subjective perceptions even for the same clothing and thermal environments.

ii. Gender selection: select either all male or all female subjects. Do not mix male and female subject for any particular study. Even if both male and female subjects were selected for the trials due to unavailability of enough participants or due to other circumstances (McLellan and Selkirk, 2004; Horn et al., 2013), they should be analyzed separately.

iii. Experimentalists (investigators) should ensure that all the selected participants are of the same somatotype and of almost the same age.

iv. If required, familiarization or acclimatization sessions should be conducted before beginning the study.

v. Special care needs to be taken while selecting subjects for the human trial study related to firefighter clothing. Selected subjects should be active, professional, and physically and mentally healthy firefighters or well-trained persons.

vi. If trials are to be conducted under harsh environmental conditions (e.g., extremely hot or cold environments), the safety of subjects should be ensured and an experienced physician shall be readily available.

vii. All the participants (subjects) should be checked for any contraindications to the test conditions required as part of the study, and they should be given detailed information about the study before tests.

Reference	Clothing	Environmental Conditions	Physical Activity	Subjects	Measured Parameters	Major Objective
Leyenda et al., FP 2017	4 firefighter clothing with 20 kg weight (backpack pump)	30°C and 30% RH (warm condition)	Submaximal walking test (moderate exercise intensity)	8 healthy firefighters	Heart rate, respiratory gas exchange, gastrointestinal temperature, blood lactate concentration, rating of perceived exertion (RPE), humidity and temp under clothing	To analyze effect of 4 different firefighter clothing on comfort performance
Holmér et al., 2006	4 different firefighting turnout gear used in Nordic countries	55°C and 30% RH	Cycling at 50 W for 20 min and treadmill walking at 5 km/h for 30 min	5 student firefighters in a rescue training school	Skin temperature, rectal temperature, heart rate, oxygen uptake, sweat production	To investigate the thermal stress of different turnout gear during moderate work in an extremely hot and humid condition
Sköldström, 1987	Firefighter clothing with breathing apparatus	15°C and 45°C, 15% RH	Treadmill walking for 60 min at speed of 3.5 km/h	8 firefighters	Heart rate (HR), oxygen uptake, deep body temperature, skin temperature	Effect of extra weight (firefighter equipment) and environmental temperature on comfort
Smith et al., 1997	Standard firefighter turnout gear including multilayer clothing, helmet, hood, boots, gloves, SCBA	13.7°C (neutral condition) and 89.6°C (hot condition with live fire)	Ceiling overhauling activity for 16 min followed by a 10 min recovery period	16 male firefighters	HR, tympanic temperature, lactate and blood glucose levels, RPE, perceptions of respiration, thermal sensations and state anxiety	Effect of different thermal environment on thermal comfort of firefighters
Duncan et al., 1979	Normal lightweight uniform and firefighter uniforms along with boots and breathing apparatus	17.8°C, 16.3°C, and 41.8°C	Treadmill walking for 15 min at a speed of 4.0 km/h	11 male firefighters	HR, mean skin temperature, rectal temperature, oxygen uptake	Effect of additional weight and insulation offered by firefighter clothing on heat stress

(Continued)

Table 8.1 (Continued) Details of Selected Documented Studies Dealing with Human Trials for Accessing the Performance of Firefighter Clothing

Reference	Clothing	Environmental Conditions	Physical Activity	Subjects	Measured Parameters	Major Objective
Dreger et al., 2006	Firefighter clothing with SCBA and normal clothing	Normal environment (21°C to 24°C)	Treadmill walking at constant speed	12 healthy male subjects	Respiratory gas exchange	Effect of firefighter clothing and SCBA on oxygen uptake
Smith and Petruzzello, 1998	Two different firefighting gears (NFPA 1500 gear and hip-boot configuration gear) with SCBA	A training structure containing live fire with temperature up to 78.7°C	Various physical activities involved during firefighting	10 healthy career firefighters	Tympanic membrane temperature, HR, RPE, Thermal sensation	Effect of two different firefighter clothing on physiological and psychological responses
Kim and Lee, 2016	Firefighter clothing without SCBA (7.75 kg weight)	Sinusoidal variation of ambient air temperature between 29.5°C and 35.5°C	20 min rest, then treadmill walking for 60 min at a speed of 4.5 km/h, followed by 10 min recovery	8 healthy professional male firefighters	HR, VO_2, rectal temperature, skin temperature	Effect of changing ambient temperature on physiological parameters with firefighter clothing
Walker et al., 2015	Firefighter clothing with SCBA (approximately 22 kg weight)	A hot chamber maintained at 100°C	Two work bouts of 20 min each, separated by 10 min rest	42 male firefighters	Intestinal temperature, HR, RPE, thermal sensation, platelet count, leukocyte count	Effect of repeated work bout of firefighters on physiological, psychological, inflammation, and immune responses under hot conditions

(Continued)

Table 8.1 (Continued) Details of Selected Documented Studies Dealing with Human Trials for Accessing the Performance of Firefighter Clothing

Reference	Clothing	Environmental Conditions	Physical Activity	Subjects	Measured Parameters	Major Objective
White et al., 1991	Normal clothing with SCBA (15.6 kg) and chemical protective clothing with SCBA (16.7 kg)	Cool (10.6°C, 60% RH), neutral (22.6°C, 54% RH), and hot (34°C, 55% RH) environmental conditions	Treadmill walking for 120 min at a walking speed of 4 km/h	9 male firefighters/emergency service personnel	HR, rectal temperature, skin temperature, RPE	Effect of chemical protective clothing at different environmental conditions
Baker et al., 2000	Sports ensemble and firefighter ensemble without SCBA	21°C and 55% RH	Treadmill exercises: 1. 5 km/h for 6 min followed by 7 km/h for next 6 min 2. 6 km/h for 60 min	18 male firefighters	HR, rectal temperature, VO₂, RPE	Effect of exercise intensity on physiological and psychological responses
McLellan and Selkirk, 2004	Firefighter gear with SCBA (approximate 22 kg weight)	35°C and 50% RH	Treadmill exercises: 1. Heavy—At 4.8 km/h, 5% elevation, 2. Moderate—4.5 km/h, 2.5% elevation, 3. Light—4.5 km/h, 0% elevation, 4. Very light—2.5 km/h	22 male and 2 female firefighters	HR, rectal temperature, mean skin temperature, VO₂, RPE, rating of thermal comfort	Effect of long and short pant was analyzed on heat stress of firefighters at three different exercise intensities
Wang et al., 2011, 2013	RB90 Firefighting clothing (2.58 clo)	30°C, 47% RH and 40°C, 30% RH	Treadmill walking at 4.5 km/h	8–10 college students	Heart rate, rectal and skin temperatures, oxygen uptake, sweat production, subjective sensation	Effect of environmental conditions on human thermophysiological responses and prediction performance of models

Experimental design and test protocol should be formulated well in advance in accordance with the declaration of Helsinki and submitted to an independent ethical review committee for consideration and approval. The study should be approved by the research and ethics committee. During the trials, experimentalists must strictly adhere to relevant regulations for the protection of human subjects and comply with recommendations provided by their institutional board of ethics committee. It is ethically unacceptable to expose subjects to hazardous test environments (for instance, smoke, harmful particles) that firefighters actually face while on duty. Besides, complete anonymity and confidentiality shall be given to each participant. Prior to starting the trials, detailed information regarding the tests and potential risks shall be explained to the subjects, and a written consent from the participants shall be signed. Subjects should participate in the human trial study voluntarily. Moreover, they should not be forced to continue with the trials if at any point during the trial they want to quit. Experimentalists should not bother even if any subject terminates the trial at any time.

8.2.1.2 Test Protocol

The test protocol and test conditions must be carefully considered to provide wearers with accurate information on the actual clothing performance. Subjects should wear firefighter clothing, performance of which needs to be examined. To simulate actual situations in the experimental design, additional weight or equipment may be assigned to the subjects. Several studies are available in the literature where apart from firefighter clothing, boots, helmet, SCBA, and pump have been used during trials. Clothing and equipment used in various studies over the years are mentioned in Table 8.1. Two crucial factors in the experimental design for analyzing firefighter clothing are the physical activity to be performed by subjects and environmental conditions under which tests are to be performed. Comfort characteristics of firefighter clothing need to be tested under realistic wear conditions. Firefighters are usually involved in rescuing and firefighting operations. These operations involve heavy physical activities under hot environmental conditions. To simulate heavy physical activities of firefighters under hot environmental conditions, three approaches have been adopted over the years.

The first approach involves conducting human trials under controlled environmental conditions in climate chambers. Temperature and relative humidity are set at a predefined level. Environmental temperatures ranging from cool to hot conditions such as 15°C, 30°C, and 55°C have been used. To simulate the heavy physical activity of firefighters, subjects were asked to perform running exercise at a particular speed on a treadmill inside the climate chamber. Duration and speed of running exercise can vary depending on the desired intensity of physical activity. In the second approach, subjects are exposed to radiant heat. Subjects wearing firefighter clothing were asked to perform the exercise while they were exposed to radiant heat flux (e.g., 1.0 kW/m²) provided by infrared heating panels (Levels et al., 2012). The third approach, which is more realistic, involves drills under live fire conditions. The drill may simulate ceiling overhaul activity (Smith et al., 1997) or other firefighting activities (Smith and Petruzzello, 1998; Smith et al., 2001; Bruce-Low et al., 2007; Horn et al., 2013). Tests were performed in a concrete building room with wooden fire at few places to simulate real operation scenarios. Human trials have been performed in environmental conditions involving temperature as high as 60.5°C (Smith et al., 2001), 78.7°C (Smith and Petruzzello, 1998), or 89.6°C (Smith et al., 1997). Typical environmental conditions used and physical activity performed by the subjects in various documented studies related to firefighter protective clothing are shown in Table 8.1.

It should be ensured that the test protocol shall not involve too high activity level for subjects continuously for a long time during trials. Gaps between physical activities should be provided. Water should be provided to the participants during rest period to avoid dehydration. If trials involve high-intensity physical activity or exercise, subjects should be asked to refrain from alcohol, drugs, nicotine, heavy physical work (running, cycling, swimming, weightlifting, etc.) 24 h before trials. Subjects should not be left alone and should be accompanied and monitored by instructors throughout the entire trial period. To judge the performance of firefighter clothing during human trials, certain physiological and psychological parameters are measured. Physiological and psychological responses of subjects should be recorded continuously or after a certain interval during the entire trial duration. Efforts should be made and it should be ensured that the subjects are blind to physiological response measurements and unaware of the expected outcome of the study. Testing or trials should be stopped at any point in time during trials if conditions become harmful to the subject to continue or the subject wants to stop continuing the trial. Trials should also be terminated immediately if any of the following conditions is observed (ASTM F2668, 2016).

- Heart rate (HR) exceeds a certain percentage of the maximum heart rate, that is, 90% or 95% of the maximum HR (Robergs and Landwehr 2002);
- Core temperature reaches or exceeds 38.5–39.5°C;
- Subjects have intolerant pain on any body site or have almost physically exhausted or developed any other acute health symptoms.

8.2.2 Physiological Parameters

Physiological parameters are important parameters as they represent the real-time state of human health. Measurements of those physiological parameters are also important due to the fact that any sudden abnormal change in any of those physiological parameters can be diagnosed by continuously monitoring those parameters during any physical activity. In this way, any accident or undesirable medical situations on subjects can be avoided during the testing. This is even more important in case of firefighters who wear heavy and relatively impermeable protective clothing, work under strenuous circumstances, and are prone to excessive heat stress. Some of the major physiological parameters used in the firefighter clothing–related research will be discussed in detail and the effect of various parameters on these physiological parameters will also be discussed.

8.2.2.1 Skin Temperature

The skin is the largest organ of the human body, accounting for around 15% of the total body weight in adults (Kanitakis, 2002). Skin protects the body tissue against injuries and it contains thermal receptors that participate in thermoregulatory control, which affects human thermal and comfort sensations. Skin serves as the interface between the body and the clothing/environment, and hence, its temperature is largely influenced by the body, clothing, and environmental conditions. Skin temperature at different body parts may vary significantly due to

Table 8.2 Relation of Skin Temperature with termal and Pain Sensations

Skin Temperature, T_{sk}	Sensation State
29°C–31°C	Uncomfortable cold while sedentary
30°C–32°C	Thermal neutrality at 3–6 METs exercise
32°C–33°C	Thermal neutrality at 2–4 METs exercise
33°C–34°C	Comfortable (thermal neutrality at rest)
35°C–37°C	Initial warm sensation
35°C–39°C	Hot sensation
39°C–41°C	Threshold of transient pain
41°C–43°C	Threshold of burning pain
>45°C	Rapid tissue damage

Source: Gagge, A. P., Nishi, Y., Heat exchange between human skin surface and thermal environment, in: *Handbook of Physiology-Comprehensive Physiology*, American Physiological Society, Bethesda, MD, pp. 69–92, 1977.

the difference in local body heat balance (Houdas and Ring, 1982). Skin temperature is directly correlated with thermal (hot and cold) sensation and under extreme environmental conditions, pain sensation. Table 8.2 shows the relation between sensation states and skin temperature.

Skin temperature can be measured with the help of either contact devices or noncontact devices. In both cases, the accuracy of skin temperature measurement should be within ±0.1°C (ISO 9886, 2004). Contact devices used for measuring skin temperature include thermocouples and thermistors. Traditional thermocouples or thermistors are placed over the skin surface in close contact with the skin and are connected to data acquisition systems with the help of wires. Thermocouples and thermistors have been used significantly over the years to measure skin temperature in human trials involving firefighter clothing (Duncan et al., 1979; Sköldström, 1987). Although measurement of the skin temperature by the contact devices is cheap, reliable, and accurate (James et al., 2014), this measurement technique has some major drawbacks. To be able to measure skin temperature properly, there should be proper contact between the skin and the sensing point. The type of sensors, attachment methods, measurement environments, and clothing all affect the results of temperature measurements, and therefore it is suggested to standardize these influence factors to enhance the precision and accuracy of skin temperature measurements (Cheung and Sweeney, 2001; Tyler, 2011; Psikuta et al., 2014). In firefighter drills and during exercise, it is sometimes difficult to maintain proper contact between the temperature sensors and skin. Perhaps a bigger hurdle in using contact devices for measuring skin temperature during physical work is the presence of wires in it. It is almost impossible to avoid wire entanglement without compromising with exercise or firefighter drills, which are to be performed by subjects. Due to above complexity of skin temperature measurements with wired contact devices, several studies have applied wireless contact devices called telemetry/wireless thermistors (see Figure 8.2) to measure human skin temperature to avoid wire entanglement problem (van Marken Lichtenbelt et al., 2006; Smith et al., 2010; Stewart et al., 2014; James et al., 2014). It was found that the wireless iButton thermistors have a mean accuracy of (−0.09)–(−0.4)°C with a precision of 0.05–0.09°C for the iButton DS1291H type (van Marken Lichtenbelt et al., 2006). Smith et al. (2010) compared the measurement accuracy of iButton (type: DS1922L) and wired thermistors (measurement accuracy: 0.01°C) and

Figure 8.2

Wireless skin temperature thermistors iButton. (Courtesy of Maxim Integrated Products Inc., San Jose, CA.)

discovered that the mean bias was +0.121°C and +0.045°C for iButton and wired thermistors, respectively. They concluded that wireless iButton provides a relatively good alternative for human skin temperature measurements in both laboratory and field investigations.

With the advancement of modern technology, contactless infrared devices are increasingly used to measure human skin temperature. On the other hand, inherent limitations of contact devices also led to the advancement and application of noncontact devices for measuring skin temperature (Bernard et al., 2013; James et al., 2014; Ludwig et al., 2014). The noncontact devices are basically optoelectronic devices that involve skin temperature measurement using infrared thermal imaging. According to Prevost's theory, each object (the human skin in this case) that is at a temperature above absolute zero (i.e., –273.15°C) emits radiation. The intensity of emitted infrared radiation depends on the temperature of the object. These noncontact devices incorporate infrared camera that converts infrared radiation emitted by objects into electric signals. These electrical signals are then processed to display thermal images where temperature distribution of the object can be obtained. With this technique, skin temperature can be obtained without any actual contact with the human body and without any hardwire connection. Thus, skin temperature can be monitored remotely, which provides freedom of movement to subjects during trials. This measurement technique requires an infrared camera that is costly as compared to a thermocouple or thermistor. Another disadvantage of this approach is that the value of measured skin temperature depends significantly on the emissivity of the skin (Bernard et al., 2013). The emissivity value of the skin needs to be provided accurately to obtain accurate skin temperature. The emissivity of clean (uncontaminated) skin is well known and found to be close to 0.98 (Bernard et al., 2013; Ludwig et al., 2014). However, most of the time, skin is contaminated with some dust or other particles, which results in significant change in skin surface emissivity and thus measured temperature may deviate significantly from the actual temperature. Also, moisture, which is generally found over the skin surface due to sweating, also results in changes in the skin surface emissivity (Herman, 2016). These issues make an accurate measurement of skin temperature using infrared camera challenging. It should be ensured while measuring skin temperature using an infrared camera that the skin is clean. James et al. (2014) analyzed reliability of the three approaches discussed above (i.e., wired thermistor, telemetry thermistor, and infrared camera) by measuring the skin temperature of subjects exercising in hot and humid environments using the three approaches simultaneously. Temperatures measured using the three approaches were compared and found that both the telemetry/wireless thermistor and the thermo-infrared camera are quite reliable with acceptable deviation from skin temperature measured by wired thermistors. Conversely, some researchers reported different findings. Bach et al. (2015) performed a comprehensive study on the comparison of conductive and infrared devices for measuring the mean skin temperature of 30 healthy subjects while resting, exercising, and recovering in different environments (24°C and 38°C). It was found that the infrared devices showed poor agreement with conductive temperature sensors and, hence, infrared devices may not be suitable for monitoring skin temperature of human subjects. Nevertheless, the selection of skin temperature measuring devices should be made with caution. The selection of devices should be determined based on the test protocol and application environments. For example, it may not be suitable and accurate to measure skin temperature using infrared devices for situations involving body movement of clothed subjects.

Human skin temperature at different body sites may vary greatly due to local heat balance difference. It would be meaningful to use a single value to represent mean skin temperature, and this will be helpful to define the body heat balance more accurately. Unfortunately, the ideal concept to measure mean skin temperature can never be realized. Measuring devices can only measure local skin temperature. Thus, calculated mean skin temperature (from a series of local skin temperature measurements) is used to represent the conceptual measured mean skin temperature. Local skin temperature can be measured at various body sites, and the challenge is how to obtain the most precise mean skin temperature to represent the unknown true value. In practice, the less number of local skin points used, the better. However, a sufficient number of points are still required to obtain precise mean skin temperature. To calculate mean skin temperature, a weighting system should be assigned to each local temperature measurement. The general formula used to calculate mean skin temperature may be written as (Mitchell and Wyndham, 1969)

$$\overline{T_{sk}} = a_1 \times T_{sk,1} + a_2 \times T_{sk,2} + \cdots + a_n \times T_{sk,n} \tag{8.1}$$

where $a_1, a_2, \ldots, a_n$ are the weighting factors assigned to $T_{sk,1}, T_{sk,2}, \ldots$ and $T_{sk,n}$; n is the total number of local skin temperature measurement points.

Depending on the number of locations selected for averaging local skin temperatures, mean skin temperature ($\overline{T}_{sk}$) values differ. Over the years, various formulas ranging from 3 to 15 points were proposed and used for computing the mean skin temperature. Table 8.3 lists some widely used mean skin temperature calculation formulas.

Various formulas developed for calculating mean skin temperature were compared under a wide range of environmental conditions by Mitchell and Wyndham (1969). They recommended the use of 4-point mean correlation proposed by

Table 8.3 Mean Skin Temperature Calculation Formulas

3-point formula (Burton, 1935)	$\overline{T}_{sk} = 0.5T_{chest} + 0.36T_{leg} + 0.14T_{lower\ arm}$
4-point formula (Ramanathan, 1964)	$\overline{T}_{sk} = 0.3T_{chest} + 0.3T_{arm} + 0.2T_{thigh} + 0.2T_{leg}$
4-point formulas (ISO 9886, 2004)	$\overline{T}_{sk} = 0.28(T_{neck} + T_{scapula} + T_{shin}) + 0.16T_{hand}$
7-point formula (Hardy and DuBois, 1938)	$\overline{T}_{sk} = 0.35T_{trunk} + 0.19T_{thigh} + 0.14T_{arm} + 0.13T_{leg} + 0.07T_{head} + 0.07T_{foot} + 0.05T_{hand}$
8-point formula (ISO 9886, 2004)	$\overline{T}_{sk} = 0.07T_{forehead} + 0.175T_{scapula} + 0.175T_{upper\ chest} + 0.07T_{upper\ arm} + 0.07T_{lower\ arm}$ $+ 0.05T_{hand} + 0.19T_{anterior\ thigh} + 0.2T_{calf}$
15-point formula (Winslow, 1936)	$\overline{T}_{sk} = \frac{1}{15}(T_{forehead} + T_{thorax} + T_{posterior\ arm} + T_{lateral\ arm} + T_{head} + T_{abdomen} + T_{anterior\ thigh}$ $+ T_{medial\ thigh} + T_{posterior\ thigh} + T_{anterior\ leg} + T_{posterior\ leg} + T_{upper\ foot} + T_{neck}$ $+ T_{middle\ back} + T_{lumbar\ back})$

Ramanathan (1964) for calculating mean skin temperature considering its simplicity and accuracy. Nevertheless, it is highly recommended that at least eight body points including hands/fingers and feet/toes should be measured to obtain mean skin temperature for experiments performed in cold environments.

For a given environmental condition, the mean skin temperature increases while wearing firefighter turnout gear as compared to using normal clothing (see Figure 8.3 [Sköldström, 1987]). Duncan et al. (1979) also found that the mean skin temperature increases due to the additional weight of firefighter clothing and equipment. They also observed that the mean skin temperature increases

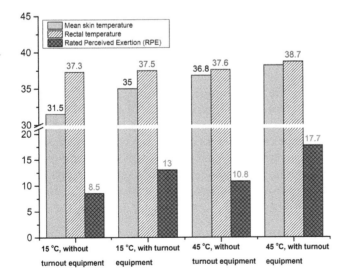

Figure 8.3

Effect of firefighter clothing and environmental temperature on the mean skin temperature, rectal temperature, and rated perceived exertion (RPE). (From Sköldström, B., *Ergonomics*, **30**(11): 1589–1597, 1987.)

as the environmental temperature increases. In fact, the effect of environmental temperature is more significant on the mean skin temperature as compared to the effect of additional weight due to firefighter clothing on the mean skin temperature, which can also be seen from Figure 8.3. It has been found that the impacts of additional weight and insulation offered by firefighter ensemble on the mean skin temperature are more significant at low-intensity physical activities compared to heavy physical exercises (McLellan and Selkirk, 2004).

8.2.2.2 Core (Deep) Temperature

The term "core" is generally referred to as all tissues at a sufficient depth and that are not easily affected by a temperature difference through the body surface tissue (e.g., skin). Core temperature is one of the most important physiological parameters that depicts the thermal state of inner body and organs. Core temperature is a commonly used vital indicator of human health as well as endurance performance. For example, core temperature has been consistently used to serve as a key indicator to terminate the human trials in both laboratories and fields. WHO recommends the threshold core temperature to terminate work is 38.0°C (ISO 7933, 2004). For human physiological testing, most researchers adopt 38.5°C as the threshold temperature to terminate human trials (Havenith, 1999; McLellan, 2001; Wang et al., 2010; Wang et al., 2013). For some human trial studies involved with physically active and well-trained subjects (e.g., firefighters, soldiers), a threshold temperature of 39.0°C or 39.5°C may be used as the criterion to conclude the trial (Beachy and Repasky, 2013; Xu et al., 2016; ISO 9886, 2004).

Obviously, there can be no one deep core temperature, and several sites may be used to represent deep core temperature. It is believed that hypothalamic temperature induces human thermoregulatory effector responses, and thus, the hypothalamus has been considered as the ideal area for deep body measurement. Unfortunately, the hypothalamic temperature is rather difficult to measure. Thus, researchers have sought more accessible deep body sites to obtain core temperatures. Choice of core temperature measurement may depend on the age, accessibility, and comfort status of subjects. Core temperatures vary slightly in a narrow temperature range depending on the location of measurement (Pušnik and Miklavec, 2009) and the age of subjects apart from periodic temperature change from morning to evening (i.e., circadian variation) and changes linked to female menstrual cycles (Houdas and Ring, 1982). The normal "core" temperature under a normal condition at various deep body sites varies slightly (see Table 8.4). However, under heavy physical activities or extremely hot and hostile environmental conditions as found during firefighting, the core temperature may exceed well above the normal core temperature range. Commonly used body sites for core temperature measurements are oral cavity (mouth), ear (auditory) canal, tympanum, esophagus, gastrointestinal tract (i.e., stomach, intestines), rectum, and urinary meatus. Deep core temperatures have been named according to sites used for measuring temperature.

8.2.2.3 Rectal Temperature

The rectum is the most commonly used site for core temperature measurement. Rectal temperature has been widely used in many documented studies dealing with the analysis of firefighter clothing by human trials (Duncan et al., 1979; Sköldström, 1987; White et al., 1991; Levels et al., 2012). To obtain a true rectal temperature, the temperature probe must be inserted into the rectum at least 8 cm in adults (the depth is determined from the anal sphincter edge to the

Table 8.4 Deep Core Temperature at Various Sites in a Resting Subject Under Normal Condition (Compared with Rectal Temperature: 36.85°C, Unit: °C)

Site	Low	High
Mouth	−0.45	−0.30
Esophagus	−0.30	−0.20
Stomach	−0.20	−0.10
Liver	−0.25	−0.05
Vagina	−0.05	+0.05
Urine (measured at the urinary meatus)	−0.15	−0.10
Hypothalamus	−0.25	+0.05
Nasopharynx	−0.45	−0.40
Tympanic membrane		
T_{air} = +10.0°C	−0.40	−0.40
T_{air} = +40.0°C	+0.40	+0.40
External auditory meatus	−0.50	−0.10
Pulmonary artery	−0.25	−0.15
Jugular vein		
Higher part	−0.05	0.00
Lower part	−0.25	−0.20
Vena cava	−0.30	−0.25
Renal vein	−0.25	−0.15
Coronary sinus	−0.05	+0.05
Right atrium	−0.25	−0.15
Right ventricle	−0.20	−0.15
Left ventricle	−0.25	−0.15

Source: Hondas, Y., Ring, E. F., *Human body temperature: its measurement and regulation*, New York: Plenum Press, 1982.

temperature sensing point; Houdas and Ring, 1982). If the insertion depth was shallower than 8 cm, the rectal temperature will be more responsive to the change of body activities as well as the change of external environments. On the other hand, there will be no significant difference in the measured rectal temperature for any insertion depth deeper than 8 cm (Lee et al., 2010). For children from birth up to 15 years, an insertion depth of about 5 cm should be used to ensure the measurement accuracy (Karlberg, 1949). It should be ensured while taking the rectal temperature measurement that a new thermometer probe is used each time or the probe is sterilized before use to avoid medical risks. Due to cultural and ethnic differences, people from Eastern countries do not normally accept the use of rectal sensors. Alternative core temperature measurements (e.g., intestinal temperature or ear canal temperature) shall be used if the subjects would not be interested in taking the rectal temperature measurement.

Firefighter clothing and environmental temperature affect rectal temperature as shown in Figure 8.3. It can be observed that the rectal temperature is higher for the cases with firefighter clothing as compared to normal clothing. This is mainly because firefighter clothing, compared to normal clothing, has a much greater thermal insulation, which greatly restricts body heat being dissipated to external environments; heat accumulates inside the body and this results in the rising rectal temperature. In addition, firefighting clothing is normally much heavier, and this can increase the metabolic production, which also contributes to a greater rectal temperature than those in normal clothing. Similarly, Duncan et al. (1979) observed a significant impact of the additional weight of firefighter clothing and equipment on rectal temperature. Rectal temperature increased 0.23°C during 15 min exercise in case of firefighter clothing, whereas it increased just 0.06°C in case of normal clothing.

It is also clear from Figure 8.3 that higher environmental temperatures result in higher rectal temperature. In fact, Duncan et al. (1979) noted that the environment temperature significantly affects the rectal temperature. They found that the rectal temperature increased by only 0.23°C in case of the 16.3°C environmental temperature, whereas it increased by 0.56°C in case of the 41.8°C environmental temperature after a 15 min treadmill running at 40 km/h with an inclination grade of 10%. Similar effect of protective clothing on rectal temperature was observed by White et al. (1991). Moreover, Sköldström (1987) found that the effect of the weight of firefighter gear is even larger at the higher environmental temperature (45°C) as compared to the lower environmental temperature (15°C), which can be noticed from Figure 8.3.

It should be also noted that rectal temperature is normally the highest among all core temperatures measured at different deep body sites if the conditions are correct and the subject is resting. Rectal temperature is mainly influenced by the venous blood from the limbs, and it is one of the slowest responsive core temperatures. The rectal temperature will not change if the rectum is sterilized by antibiotics.

8.2.2.4 Gastrointestinal Temperature

Gastrointestinal temperature involves measurements of deep body (core) temperature at the stomach or the intestines. It is usually measured using ingestible telemetric "pills" or "capsules" (Byrne and Lim, 2007; Levels et al., 2012; Bongers et al., 2015; Walker et al., 2015; Leyenda et al., 2017), and hence it is sometimes referred to as pill temperature. These pills are temperature sensitive and transmit data to an external receiver. The telemetric pills are commercially available (see Figure 8.4), and presently such telemetric core temperature measurement serves as one of the best ways to monitor core temperature non-invasively. This test method has gained wide applications in sports and occupational

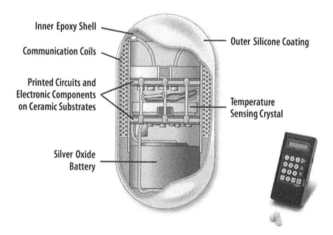

Figure 8.4

The ingestible CorTemp® telemetric pill and monitor. (Courtesy of HQ Inc., Palmetto, FL.)

settings such as deep-sea diving and soldier training (Byrne and Lim, 2007; Gant et al., 2006; Ruddock et al., 2014). Before swallowing each pill, it is necessary to validate the temperature measured by each pill to check its accuracy. For this, water bath immersion study at different temperatures should be conducted beforehand (Hunt and Stewart, 2008; Travers et al., 2016). Subjects are normally requested to swallow the pill with tepid water about 3–8 h before the experiments to facilitate the capsule transition through the stomach into the intestine tract (Domitrovich et al., 2010; Kolka et al., 1997) as well as to reach a relative stable core temperature by avoiding the interference of food and/or drinks (Wilkinson et al., 2008). This method has demonstrated close agreement with rectal and esophageal temperatures (Byrne and Lim, 2007). Also, the intestinal temperature is sometimes preferred over rectal temperature as a measure of core temperature because it responds more rapidly to any change in core temperature as compared to rectal temperature (Byrne and Lim, 2007; Teunissen et al., 2012).

Another advantage of this technique is that the core temperature can be measured continuously while performing exercise or any other outdoor activity without interruptions. Nevertheless, such telemetric pills used for measuring the intestinal temperature are costly and are not retrievable. Hence, it can be quite expensive if a large number of human trials need to be performed. Apart from this, if pills are to be used during trials, consumption of hot or cold water may result in significant error in core temperature readings. Further, such telemetry pill systems can easily be disturbed by electromagnetic interference (e.g., between the temperature monitoring system and the treadmill or other electronic devices), which causes it to lose temperature data. To avoid these aforementioned problems associated with core temperature measurement using pills, alternative approaches were used where core temperature can be determined from the heart rate with reasonable accuracy (Buller et al., 2015).

8.2.2.5 Esophageal Temperature

The esophageal temperature probe is used to measure the temperature in the esophagus. In this technique, subjects swallow a temperature probe through the nostril into the esophagus. The positioning of the temperature probe is critical to the measurement accuracy (Mekjavic and Rempel, 1990). The esophageal temperature probe should be positioned in roughly the lower third of the esophagus, which is in contact over a length of 50–70 mm with the front of the left auricle and with the rear surface of the descending aorta (see Figure 8.5; Domitrovich et al., 2010; ISO 9886, 2004). Thus, the esophageal temperature should be close to the cardiac (i.e., arterial blood) temperature. This is to say, the esophageal temperature represents the most accurate reflection of temperature variations in the blood leaving the heart as well as the temperature of the blood irrigating the thermoregulatory centers in the hypothalamus. To facilitate a smooth insertion, the temperature probe should be softened by placing it in warm water for a few minutes. After the insertion, the probe should be secured by taping it to the nose. It has been found that both the rectal and intestinal temperatures react slower to a rapidly changing body temperature than the esophageal temperature (Byrne and Lim, 2007). Therefore, the use of disposable core temperature capsule and rectal thermometer has certain limitations. Nevertheless, these two methods are still widely used for measuring the

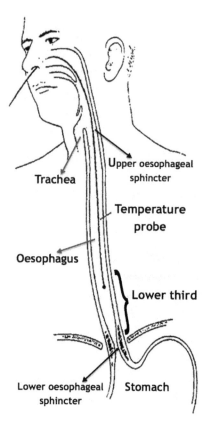

Figure 8.5

The approximate positioning of the esophageal temperature probe.

core temperature because the core temperature measurement using esophageal temperature probes involves health risks, and it may also result in discomfort and inconvenience for subjects.

8.2.2.6 Tympanic Temperature

The tympanic membrane is one of the popular sites for measuring deep core temperature (Childs et al., 1999). Tympanic temperature reflects the variations in the arterial blood temperature (which affects the human thermoregulatory center) because the eardrum has a rather low thermal inertia and its mass is low but vascularity is high. The tympanic temperature can be measured using a thermistor or a thermocouple. The contact between the sensor and the tympanum is identified by the sensation felt by the subject. Attention should be paid to the properties of temperature probes (e.g., shape, stiffness of the probe) to avoid injuries to the tympanic membrane. Also, it has been found that ambient temperature, particularly cold air, largely affects readings of the measured tympanic temperature. Thus, the thermal probes should be well insulated using an insulation cover for those trials performed in cool and cold environments (i.e., T_{air}<15.0°C) to ensure good measurement precision (Brinnel and Cabanac, 1989). ISO 9886 (2004) recommends that the infrared thermometers may only be used at the range of air temperatures between 18°C and 30°C.

With the modern infrared technology, the tympanic temperature can easily and rapidly be measured using an infrared tympanic thermometer (Smith et al., 1997; Smith and Petruzzello, 1998). More importantly, this method measures the tympanic temperature in a non-invasive way. These infrared thermometers measure radiant heat emitted by the tympanic membrane of the ear. Hence, it is able to measure tympanic membrane temperature without actually touching it. It is also a safer approach to access core temperature as it does not involve insertion of a thermometer deep in the eardrum and thus avoids pain and any risk of perforation of eardrum due to the pressure of thermocouple. Care should be taken while measuring tympanic membrane temperature that the reading should be taken either on the left or right ear throughout the experiment. Studies have been found that tympanic temperatures measured using the infrared thermometer are in good agreement with the axillary temperature measured using a mercury glass thermometer (Gasim et al., 2013). Nevertheless, applying the infrared method often encounters significant problems. For instance, investigators should carefully examine whether the infrared thermometer measures the tympanic membrane but not the walls of the auditory meatus. Also, it seems difficult to apply the infrared thermometer to continuously monitor and record time-course tympanic temperatures. Besides, most infrared thermometers have a rather wide optical angle, and there is no control over the sensor focus, which makes it rather too difficult to obtain accurate tympanic temperature using infrared thermometers. It is well recognized that readings obtained from commercially available ear thermometers are lacking in repeatability and accuracy (Barnett et al., 2011; Yeoh et al., 2017). Other factors affecting measurement precision include the structure of the auditory meatus (narrow and bent), hair in the ear canal, and the presence of cerumen.

Tympanic temperatures have been used to reflect the deep core temperature of firefighters while performing simulation tasks. Smith et al. (1997) examined the impact of thermal environments on physiological responses of 16 male firefighters in a training drill. The rise of tympanic temperature is more pronounced at a much hotter environmental condition (i.e., 89.6°C) compared to the normal condition (i.e., 13.7°C). During the 16 min firefighting activity (i.e., simulated ceiling overhaul task), an increase of 3.15°C in the mean tympanic membrane temperature was observed in the extreme hot condition involving fire, whereas the mean tympanic temperature increased by only 0.31°C under the normal condition. The development of tympanic temperature is also influenced by the type of firefighter clothing. Smith and Petruzzello (1998) found that the tympanic temperature increased by 1.5°C during live-fire trials in case of NFPA 1500 firefighter clothing, whereas it increased only by 0.9°C with hip-boot firefighter gear.

8.2.2.7 Mean Body Temperature

Mean body temperature (MBT) is commonly defined as the mass-weighted average of tissue temperatures throughout the body. MBT is a fundamental characterization of an individual's thermal status. It also serves as a required parameter for calculating body heat balance. Ideally, mean body temperature is measured by averaging all temperatures measured in both the deep body tissues as well as skin surface temperatures. In practice, this could not be directly measured. Therefore, MBT is normally calculated from the mean skin temperature and one core temperature such as rectal temperature. Coefficients should be given for the mean skin temperature and core temperature to demonstrate the relative importance of body heat distribution (core and peripheral tissues). Burton (1935) first proposed that MBT can be computed using the below formula:

$$MBT = a \times T_{core} + (1-a) \times \overline{T_{skin}} \tag{8.2}$$

where a is the weighting factor that varies as a function of ambient temperature (strictly speaking, it varies with the body heat load).

The main concept of Burton's formula is based on the logic that core tissues are pretty homogeneous, whereas the temperature of peripheral tissues decreases parabolically from the body core to the skin (Lenhardt and Sessler, 2006). In Burton's work, the coefficient was estimated by simultaneously measuring the changes in the body heat content, core temperature, as well as mean skin temperature. Burton's equation (i.e., Equation 8.3) has demonstrated good accuracy in computing MBT for trials performed in cold conditions and/or when the mean skin temperature of the subject drops below 32.0°C. At higher air temperatures, the contribution of core temperature increases due to peripheral vasodilation. The formula used to calculate MBT for warm/neutral environments is presented in Equation 8.4. For human trials involving firefighters, Equation 8.4 shall be used to calculate MBT because firefighters have relatively warm skin due to the high metabolic heat production while performing duties as well as the limited body heat dissipation capability caused by high insulating and impermeable firefighting turnout gear. Nevertheless, this core-shell approach to estimate MBT has been widely criticized for underestimating the actual MBT (Jay and Kenny, 2007).

$$MBT = 0.64 \times T_{core} + 0.36 \times \overline{T_{skin}} \tag{8.3}$$

$$MBT = 0.80 \times T_{core} + 0.20 \times \overline{T_{skin}} \tag{8.4}$$

8.2.3 Body Heat Storage

Body heat storage reflects the body heat balance between heat gains and heat losses. Also, the heat storage is closely correlated with the tolerance time in hot environments and hence, body heat storage has been used as a physiological parameter for maximal heat tolerance. In hot environments, the heat balance can be easily disturbed by the limited body heat dissipation capability (mainly due to the limited environmental evaporation capacity). If the heat gains overweigh the heat loss, body heat storage inside the body changes and subsequently a change of body heat content occurs. Theoretically, the change of body heat content should be determined by measuring both the total heat production (determined by indirect calorimetry, i.e., estimation of body heat production based on the measurement of gaseous exchange) and the total body heat loss (determined using direct calorimetry, e.g., the whole body calorimeter [Jéquier, 1986]). Unfortunately, the direct calorimeters are expensive and not easily accessible. Often the body heat content is determined using the thermometry method. The body heat storage (body heat content) may be given by

$$\Delta S = C_p \times W_b \times \overline{\Delta T_b} \qquad (8.5)$$

where $\overline{\Delta T_b}$ is the change of mean body temperature (MBT) during an interval of time, which may be obtained using Equation 8.3 or Equation 8.4; W_b is the total body mass; and C_p is the average specific heat capacity of body tissues, which is normally taken as 3.47 kJ/(kg·K). In fact, the specific heat capacity of body varies with the percentage of body fat. For example, the specific heat capacity of the body of a skinny man containing 12% fat is about 3.20 kJ/(kg·K), whereas for an obese body containing 50% fat, it is 2.73 kJ(kg·K).

To calculate the body heat content, the change of MBT should be determined first. As mentioned earlier, estimation of MBT using the two components "core-shell" concept significantly underestimated the MBT and thereby the body heat storage will also be underestimated (Vallerand et al., 1992). Therefore, application of the estimated body heat storage using the thermometry method to determine the maximal heat tolerance time of subjects should be performed with caution. If possible, the partitional calorimetry method (i.e., calculating the body heat storage and heat gains, heat losses via the dry and evaporative heat transfer pathways) is recommended to determine the body heat storage because it generates a rather accurate prediction on the actual body heat storage.

8.2.3.1 Heart Rate

Heart rate (HR) is used to describe the frequency of cardiac cycle, and it has been widely used to predict health and endurance performance. HR is also a key component to calculate/predict physiological indices such as the Physiological Strain Index (Moran et al., 1998) and VO_{2max} (maximal oxygen uptake). HR can be considered as the sum of several independent components, which may be expressed as (ISO 9886, 2004)

$$HR = HR_0 + \Delta HR_M + \Delta HR_s + \Delta HR_T + \Delta HR_N + \Delta HR_\varepsilon \qquad (8.6)$$

where HR_0 is the mean heart rate of the subject at rest while sitting in neutral conditions; ΔHR_M is the increase in HR linked to work metabolism; ΔHR_s is the increase in HR linked with static exertion; ΔHR_T is the increase in HR due to thermal strain experienced by the subject; ΔHR_N is the increase in HR due to psychological factors; and ΔHR_ε is the increase in HR connected with breathing rhythm, circadian rhythm, etc.

It can be easily deduced from Equation 8.6 that HR is influenced by many factors (e.g., exercise, environmental and psychological stressors); hence, HR is less sensitive to be used to judge individual endurance performance. During the human trials, subjects should not be too excited (e.g., listening to rock music may cause emotional change) to eliminate the effect of psychological factors on measured HR. Investigators should also avoid having too much unnecessary discussion/conversations with the subject. Besides, ventilation inside the climatic chamber should always be turned on to ensure the oxygen concentration inside the chamber is maintained at a normal range. Abnormal oxygen concentration may affect breathing rate of the subject and brings errors to measured HR.

Heart rate is normally calculated as the number of heart beats in 1 min, and the unit of HR is "beats per minute" (bpm). Presently various techniques have been used to measure HR during human trials, which involves

- Physiological monitoring system (Levels et al., 2012),
- Wireless heart rate monitor (Smith et al., 1997),
- ECG electrodes (Respironics) and a cardiometer (Excentric) (Sköldström, 1987), and
- Optical heart rate monitoring devices (Smith and Petruzzello, 1998; Parak et al., 2015; Horn et al., 2013).

A number of different metrics have been used to describe heart rate. The most widely used metrics are basal heart rate, resting heart rate, maximal heart rate, and target heart rate. Basal heart rate (HR_{basal}) is the heart rate at rest, which is normally determined when the subject is sleeping. HR_{basal} is usually taken as a major indicator to evaluate parasympathetic functions. Resting heart rate (HR_{rest}) is a vital metric to determine fitness level and cardiovascular health, which is defined as the heart rate when the subject is aware in a thermoneutral condition, and the subject has not been subject to any exertion or stimulation. HR_{rest} varies from individual person to person (factors affecting resting HR includes age, gender, heart size, body size, activity level, fitness level, environmental condition, body position, emotion, and medication status), and the average resting heart rate is usually between 60 and 80 bpm. For endurance athletes or people who often perform cardiovascular activities, HR_{rest} can be as low as around 30–50 bpm. For obese (overweight) people, HR_{rest} can be significantly higher than the normal range. The decrease in HR_{rest} on trained subjects is referred to as training bradycardia. Maximal heart rate (HR_{max}) is the rate at which a person subjectively feels that he/she is at maximum effort (i.e., all efforts to exhaustion). It is an important physiological variable to evaluate maximal exertion of a subject during an exercise performance test. In fact, HR_{max} denotes the temporary state of the cardiovascular and nervous systems of an individual person. HR_{max} does not change with training but it decreases with age (Londeree and Moeschberger, 1982). HR_{max} can be determined by a cardiac stress test accurately. HR_{max} can also conveniently be predicted using the well-known Haskell and Fox formula (error range: ±10–13 bpm), which may be read as

$$HR_{max} = 220 - age \qquad (8.7)$$

HR serves as a tool for individual training. One way to monitor individual activity intensity is to determine whether a person's HR is within the target HR zone during the physical activity. Target heart rate (THR) is defined as the desired HR range during aerobic exercise, which enables the heart and lungs of an individual to receive the most benefit from a workout. As per American College of Sports Medicine (ACSM) and American Heart Association (AHA), an individual shall maintain his/her HR within a safe range to improve cardiovascular fitness. THR should be 50–70% of his/her maximum heart rate for a moderate intensity physical activity. Two steps shall be used to determine the THR: (i) determination of HR_{max}; (ii) calculate THR by multiplying the percentage of physical activity intensity level. For vigorous intensity physical activity, a person's THR shall be 70–85% of his/her maximum heart rate.

As mentioned earlier, the formula $HR_{max} = 220$-age has been criticized because it does not always yield accurate HR_{max}. Hence, the two-step calculation of THR, calculated based on this formula, will be inaccurate. Karvonen (Karvonen and Vuorimaa, 1988) proposed the most effective method to calculate THR based on the resting heart rate and HR_{max}, which may be written as

$$THR = HR_{rest} + \left(HR_{max} - HR_{rest} \right) \times \%intensity \qquad (8.8)$$

where %*intensity* is the percentage of training/exercise intensity.

The Karvonen formula is the most widely used and reliable method to determine the THR for appropriate activity intensity of a person (King and Senn, 1996). It is highly recommended to apply Karvonen formula to compute THR for medium to high levels of aerobic training as well as human physiological testing performed in laboratories and fields (Goldberg et al., 1988).

HR metrics have been widely applied to examine the health and performance of firefighters. During continuous heavy physical exercise, HR may consistently rise and eventually approach HR_{max} of the individual subject (Barnard and Duncan, 1975; Walker et al., 2015). This phenomenon is even more pronounced among firefighters on duty. Higher HR was noticed in firefighter clothing with equipment as compared to normal clothing by Duncan et al. (1979). Also, the significant impact of environmental temperature is observed on HR. As the environmental temperature increased from 16.3°C to 41.8°C, HR increased from 136.4 to 172.7 bpm after only 15 min of exercises. Effect of protective clothing and associated equipment on HR was also analyzed by Sköldström (1987). It is found that the added insulation due to clothing and added weight due to equipment result in an increase in HR as shown in Figure 8.6. The effect of turnout gear on HR is even more significant at higher environmental temperature conditions as evident from Figure 8.6. HR increases as environmental

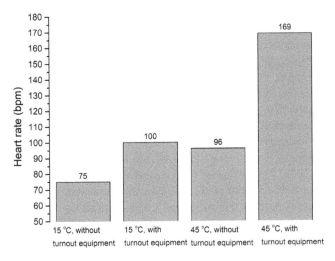

Figure 8.6

Effects of clothing and environmental temperature on HR. (From Sköldström, B., *Ergonomics*, **30**(11): 1589–1597, 1987.)

temperature increases (White et al., 1991). Similar results are also discovered by Smith et al. (1997), who investigated the effect of firefighter physical activity on HR under different environmental conditions. HR increased by 47 bpm while performing a 16 min task in case of the hot environmental condition with live fire, whereas it increased by only 10 bpm under the lower temperature condition. McLellan and Selkirk (2004) analyzed the effect of wearing shorts and long pants along with the firefighter clothing ensemble (with SCBA) on the development of HR. Higher HRs were reported in case of long pants. The effect is especially significant at lower intensity exercises in hot environments.

8.2.3.2 Oxygen Consumption (VO$_2$)

Oxygen is inhaled in the form of air into the lungs. The oxygen molecules are then transported to the cells by red blood cells (RBCs) through the cardiovascular network. Some part of this oxygen is then utilized to convert food (or glucose) into adenosine triphosphate (ATP) through the process of cellular respiration. ATPs then react with water molecules to produce work or muscle movement required during physical activities. Higher oxygen consumption means more ATP generation and more ATP generation denotes higher energy production. Hence, higher oxygen consumption ultimately means the higher ability of body muscles to keep working and to continue physical activities for long durations. In other words, in case of heavy physical activity or long duration exercise, more energy (more number of ATPs) is required. To generate such high energy (or to produce a larger number of ATPs), more oxygen (which means higher VO$_2$) is required. Therefore, oxygen consumption (VO$_2$) represents the efficiency of a subject's body to produce work. It is usually expressed as volume of oxygen consumed per unit time (e.g., ml/min). Maximal oxygen consumption (VO$_{2max}$) is the maximum possible value of the oxygen consumption (VO$_2$) by a subject. VO$_{2max}$ is a good indicator of physical fitness, endurance capacity, and health of subjects. Higher VO$_{2max}$ of a subject represents higher cardiorespiratory fitness level of the subject and it is a good standard for judging the same. It actually means that subjects with higher VO$_{2max}$ can more efficiently take oxygen to their cells, which utilize oxygen to produce ATP (energy). This implies that individuals with higher VO$_{2max}$ can produce a higher amount of work and can perform physical activities for longer periods. Value of VO$_{2max}$ of subjects depends on age, sex, fitness, training, and body composition (e.g., % of body fat). VO$_{2max}$ generally decreases with age (Pollock et al., 1987). It is found approximately 15% higher for men as compared to women (Sparling, 1980). Similarly, VO$_{2max}$ of well-trained athletes is higher than that of a normal person (Gollnick et al., 1972).

VO$_{2max}$ is measured by performing incremental exercise (i.e., graded exercise test) performed on a treadmill (Zwiren et al., 1991; Baker et al., 2000; Vehrs et al., 2007), cycle ergometer (Astrand and Rodahl, 1970; Zwiren et al., 1991), or field tests (Zwiren et al., 1991). In this approach, the exercise intensity of subjects is increased continuously, which results in an increase in VO$_2$ value. As the exercise intensity increases continuously, a state of maximal exertion is subsequently reached where it is not possible for subjects to be able to perform exercise at a higher exercise intensity. VO$_2$ obtained at this state of maximal exertion is defined as the maximal oxygen consumption (VO$_{2max}$). Oxygen consumption does not increase beyond this even if exercise intensity is somehow increased further. The average untrained healthy males and females have a relative VO_{2max} of approximately 35–40 mL/(kg·min) and 27–31 mL/(kg·min), respectively. The VO_{2max} of world elite male and female athletes (e.g., cross-country skiers) can be as high as 85 and 77 mL/(kg·min), respectively.

VO_2 and VO_{2max} can be obtained easily with the help of the Fick equation (Fick, 1870). According to the Fick equation, oxygen exchange can be obtained as follows:

$$VO_2 = Q\left(C_{O_2,a} - C_{O_2,v}\right)$$ (8.9)

$$VO_2 = \left[SV \times HR\right] \times \left[C_{O_2,a} - C_{O_2,v}\right]$$ (8.10)

where Q is the cardiac output (i.e., the amount of blood pumped per min); $C_{O_2,a}$ and $C_{O_2,v}$ are the oxygen contents in the arterial and venous blood vessels, respectively; $\left(C_{O_2,a} - C_{O_2,v}\right)$ is the oxygen present in the blood, which is used by cells to make ATPs (i.e., the oxygen extraction or the arterio-venous difference); SV is the stroke volume, that is, the blood being pumped per heart beat; and HR is the heart rate.

The three parameters (i.e., Q, $C_{O2,a}$, and $C_{O2,v}$) should be measured during exercise. Depending on whether the above three parameters are obtained during normal condition or at the stage of maximal exertion, VO_2 or VO_{2max} can be obtained from Equations 8.9 and 8.10.

Application of incremental exercise to determine VO_{2max} may expose the subject to face potential risks (e.g., exhaustion). Over the years, various methods have been proposed. For instance, Bradshaw et al. (2005) proposed a regression equation to estimate VO_{2max} based on non-exercise data. Similarly, to avoid risks faced by subjects during maximal exertion state involving extremes of cardiorespiratory system in the determination of VO_{2max}, submaximal test protocols were also developed (Zwiren et al., 1991; Vehrs et al., 2007).

The determination of oxygen consumption has always been performed on firefighters or laboratory testing with subjects wearing firefighter protective clothing. Significant effects of additional insulation and weight due to firefighter clothing and equipment were observed on VO_2 consumption and VO_{2max}. While VO_2 increases (Sköldström, 1987; Baker et al., 2000; Dreger et al., 2006; Bruce-Low et al., 2007), VO_{2max} decreases (Dreger et al., 2006) due to the use of protective clothing and equipment. Sköldström (1987) found a 0.4 L/min increase in VO_2 consumption due to firefighter gear as compared to normal clothing as shown in Figure 8.7. However, no effect of an increase in environmental temperature on VO_2 was noticed. Similar results were obtained in other studies as well (Duncan et al., 1979). Duncan et al. (1979) also found a significant impact of clothing weight and firefighter clothing on VO_2, whereas VO_2 increases only a little with the increasing environmental temperature. VO_2 does not depend much on different types of firefighter clothing (Leyenda et al., 2017). Bruce-Low et al. (2007) later confirmed that additional weight of firefighter protective clothing is the main factor that results in higher VO_2 in case of firefighter clothing rather than the impermeable nature of firefighter clothing. Dreger et al. (2006) noticed significantly lower VO_{2max} in case of protective clothing with SCBA as compared to normal clothing, whereas the opposite effect of protective clothing on VO_2 was noticed.

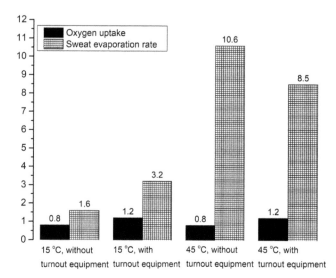

Figure 8.7

Effect of firefighter clothing and environmental temperature on VO$_2$ and sweat evaporation. (From Sköldström, B., *Ergonomics*, **30**(11): 1589–1597, 1987.)

8.2.3.3 Metabolic Heat Production

Thermal comfort of humans is directly related to the heat balance and ability of the human body to maintain thermal equilibrium. Heat balance of the human body is basically a balance of the metabolic heat produced by the body, convective and radiative heat exchange of body with the surroundings, and latent heat of vaporization due to perspiration. When there is net heat storage in the body, core body temperature and skin temperature rise. Protective clothing worn by firefighters acts as a barrier in the heat exchange between body and environment as shown in Figure 8.1. Metabolic heat plays an important role in overall heat balance. The rate of metabolic heat production depends on the activity level of the individual (ISO 8996, 2004). The higher rate of metabolic heat production due to heavy activities of firefighters and the less heat exchange with the surroundings due to higher thermal insulation contribute significantly to the heat stress of firefighters. The additional weight of firefighter clothing and equipment further results in an increase in the rate of metabolic heat production. The hot atmospheric temperature may also cause heat transfer toward the body. These factors escalate chances of heat stress. It is therefore important to measure and monitor the rate of metabolic heat production of firefighters during operation.

The rate of metabolic heat production (M) is correlated well with the respiratory exchange ratio (RQ) and rate of oxygen consumption (VO_2). It can be calculated as (Gagge and Nishi, 1977; Gagge and Gonzalez, 1996; Leyenda et al., 2017)

$$M = 5.873 \cdot VO_2 \cdot (60/A_D)\{0.23RQ + 0.77\} \quad (W/m^2) \qquad (8.11)$$

where RQ is the respiratory quotient value, which varies from 0.83 (at rest) to 1.0 (during heavy exercise) and A_D is the surface area of the nude body (m²), which can be obtained using the Dubois height-weight formula (Du Bois and Du Bois, 1989):

$$A_D = 0.202(W)^{0.425}(H)^{0.725} \qquad (m^2) \qquad (8.12)$$

where W is the body weight in kg and H is the height of the subject in m.

The rate of metabolic heat production depends on loading carried by the subjects including the weight of protective clothing and equipment (Givoni and Goldman, 1971; Dorman and Havenith, 2009; Wang et al., 2010). It also depends on the activity level. Performance of protective clothing is usually analyzed with the help of treadmill exercise in a controlled environment (e.g., climate chamber). Under such cases, the rate of metabolic heat production can be obtained from the load carriage equation (Pandolf et al., 1977; Stewart et al., 2014)

$$M = 1.5W + 2(W+L)(L/W)^2 + n(W+L)(1.5V^2 + 0.35VG) \qquad (8.13)$$

where L is the load carried by the subject including clothing and equipment weight in kg, V is the speed of treadmill in m/s, G is the gradient in %, and n is a terrain coefficient, which is equal to 1.0 in case of the treadmill. Apart from these, various other methods (e.g., screening, observation; see Table 8.5 for example) can be used to estimate the rate of metabolic heat production (ISO 8996, 2004).

Table 8.5 Metabolic Rate for Various Kinds of Activities

Class	Average Metabolic Rate		Examples
	W/m²	W	
0	65	115	Resting, sitting at ease
Resting	(55–70)	(100–125)	
1	100	180	Light manual work (writing, typing, drawing, sewing, book-keeping); hand and
Low metabolic rate	(70–130)	(125–235)	arm work (small bench tools, inspection, assembly or sorting of light materials); arm and leg work (driving vehicle in normal conditions, operating foot switch or pedal); standing drilling (small parts); milling machine (small parts); coil winding; small armature winding; machining with low power tools; casual walking (speed up to 2.5 km/h)
2	165	295	Sustained hand and arm work (hammering in nails, filing); arm and leg work
Moderate metabolic rate	(130–200)	(235–360)	(off-road operation of lorries, tractors, or construction equipment); arm and trunk work (work with pneumatic hammer, tractor assembly, plastering, intermittent handling of moderately heavy material, weeding, hoeing, picking fruits or vegetables, pushing or pulling lightweight carts or wheelbarrows, walking at 2.5–5.5 km/h, forging)
3	230	415	Intense arm and trunk work; carrying heavy material; shovelling;
High metabolic rate	(200–260)	(360–465)	sledgehammer work; sawing; planning or chiselling hardwood; hand mowing; digging; walking at 5.5–7.0 km/h; pushing or pulling heavily loaded handcarts or wheelbarrows; chipping castings; concrete block laying
4	290	520	Very intense activity at fast to maximum pace; working with an axe; intense
Very high metabolic rate	(>260)	(>465)	shovelling or digging; climbing stairs, ramp or ladder; walking quickly with small steps; running; walking at a speed >7.0 km/h

Source: ISO 8996, Ergonomics of the thermal environment—Determination of metabolic rate, International Organization for Standardization, Geneva, Switzerland, 2004.

It is clear from Equation 8.13 that the rate of metabolic heat production is a function of both clothing weight and physical activity level. In fact, Dorman and Havenith (2009) found that more than half of the increase in the rate of metabolic heat production was due to the protective clothing weight. The rate of metabolic heat production varies significantly for different protective clothing for various physical activities, and for a given clothing, the rate of metabolic heat production varies with physical activity level as well. Similarly, Givoni and Goldman (1971) observed an increase in the rate of metabolic heat production with the increase in weight.

8.2.3.4 Sweat Production and Evaporation

Sweat is a liquid secreted by sweat glands. Sweat evaporation plays an important role in human thermoregulation. When the body temperature increases, sweat is secreted over the body to provide evaporative cooling. Generally, moisture may be lost from the skin by either insensible sweating and sensible sweating (or active sweating; Ohhashi et al., 1998). Insensible sweat loss includes water loss from the respiratory passages, the skin, and gaseous exchanges in the lungs. Sweating becomes sensible when the sweat production rate exceeds about 100 g/h and it is possible for a healthy body to produce a sweat rate of 1–2 kg/h under extreme conditions. The amount of excreted sweat is affected by many factors such as body location, body temperature, environmental conditions (e.g., air temperature, RH, air velocity, radiation), and psychological stressors (e.g., emotion).

Sweating becomes the only effective approach to dissipate excess body heat in hot environments (i.e., $T_{air} > T_{sk}$) (Candas et al., 1979). Basically, sweat evaporation has two phases (Tam et al., 1976). When the sweat production is low, complete evaporation occurs, and thus, the skin is relatively dry. This phase is called the total evaporation phase. In this phase, the evaporation rate is mainly determined by the sweat rate. As the sweat production increases, the sweat evaporation rate cannot keep up with the production rate, partial evaporation occurs, and there will be sweat accumulation on the skin. This phase is called the partial evaporation phase. The evaporation rate in this phase (Tam et al., 1976) is mainly determined by the maximum evaporative capacity. In practice, only some of the produced sweat can be evaporated, due to the limited evaporative capacity of the environments. The remaining sweat will either drip off the body or be absorbed by the material of any overlying clothing. The dripped sweat does not contribute to cooling the body, whereas absorbed sweat by clothing may still contribute to evaporative cooling. Sweating efficiency (or efficiency of sweating/sweat evaporative efficiency, η_{sw}) is an indicator for the proportion of sweat evaporated, which is defined as the ratio of the amount of sweat evaporated to the total sweat production. Sweating efficiency may be expressed as

$$\eta_{sw} = \frac{EVPA}{SWp} \times 100\%$$

(8.14)

The sweating efficiency decreases with increasing sweat production as well as the skin wittedness (fraction of the body covered by liquid sweat at the skin temperature, skin wittedness is expressed as a decimal fraction with 1 representing fully wet skin while 0.06 representing the minimal value due to the insensible sweating on the skin). For example, the sweating efficiency drops to 0.67 for fully wet skin, whereas it is maintained at 0.74–1.0 for partially wet skin (Candas et al., 1979). The sweating efficiency is equal to the unity before any liquid accumulation on the skin (i.e., all produced sweat was evaporated).

At a given environment, the maximum capacity of evaporation E_{max} is computed by

$$E_{max} = h_e \times (p_{sk} - p_a) = a \times v^b \times (p_{sk} - p_a) \qquad (8.15)$$

where h_e is the evaporative heat transfer coefficient; p_{sk} and p_a are the water vapor pressure on the skin surface and in the air, respectively; v is the air speed; a is a constant, which is related to body shape; and b is also a constant, which is the power of the air speed.

The evaporative heat transfer coefficient may be computed by (Clifford et al., 1959)

$$h_e = 109.37 \cdot v^{0.63} \qquad (8.16)$$

It can be deduced from Equation 8.15 that the maximum evaporative capacity is determined by the body shape, air velocity, and water vapor pressure gradient between the skin surface and the ambient air. The water vapor pressure is computed by the temperature and the RH (relative humidity).

Sweat evaporation provides a powerful physiological cooling mechanism, taking up 0.58 kcal (kilocalorie) of heat for each gram of liquid sweat (Nunneley, 1989). This cooling power can only be applied to the amount of sweat being evaporated on the skin surface. If the secreted sweat were wicked away by clothing worn and evaporated inside clothing or on the outer surface of the clothing, the actual cooling power to the skin decreases (Kerslake, 1972; Wang et al., 2014). The evaporative cooling efficiency is then used to describe the actual cooling performance of sweat evaporation, which is defined as the ratio of actual evaporative cooling to the latent heat of total evaporated sweat.

For human trials performed in laboratories, the measurement of total sweat production (SWp) can be easily determined by the body weight change before and after the trials. Clothing weight change before and after the trial is called sweat residue (SWr), and the difference between SWp and SWr is defined as the total sweat evaporation (EVAP). Sweat evaporation rate ($\dot{m}_e$) is calculated from the total sweat evaporated during the trial (EVAP in g) by dividing the total duration of the trial (Δt in min).

$$\dot{m}_e = \frac{EVAP}{\Delta t} = \frac{SWp - SWr}{\Delta t} \qquad (\text{g/min}) \qquad (8.17)$$

In the above calculation of sweat evaporation rate, respiratory losses were ignored. The calculation should be modified appropriately if fluid intake during trials was allowed (Young et al., 1987). Sweat evaporation rate can be used further to determine the evaporative heat loss, E (Bröde et al., 2008; Holmér, 2009; Gagge and Gonzalez, 2011; Kofler et al., 2015)

$$E = \frac{\dot{m}_e \times \lambda}{D \times A_D} \qquad (\text{W/m}^2) \qquad (8.18)$$

---

where λ is enthalpy of evaporation, which is equal to 2425 J/g at 32.0°C; and A_D is the sweating body surface area.

Protective clothing causes a significant reduction in sweat evaporation (McLellan and Selkirk, 2004) due to its impermeable nature and therefore causing lower evaporative heat loss. Lower evaporative heat loss results in more body heat storage and increases in core temperature, which increases the risk of heat stress and reduces work capacity of wearers (Kofler et al., 2015). The weight of protective clothing/equipment and environmental temperature are other major factors that govern sweat production and evaporation. Faff and Tutak (1989) noted significantly higher sweat rate in case of protective equipment due to the additional weight of equipment. It was found to be 811 g/m²/h for the case involving firefighting equipment as compared to 564 g/m²/h without equipment. The significant impact of clothing weight and environmental temperature was also observed on weight loss (SWp) by Duncan et al. (1979). Weight loss increases as clothing weight and environmental temperature increases. However, Stewart et al. (2014) found no significant effect of environmental temperatures on sweat rate. In fact, sweat evaporation rate is also not affected by environmental temperature as shown in Figure 8.7. Sweat evaporation rate increases as the weight carried by firefighters increases. Sköldström (1987) discovered higher sweat evaporation in case of firefighters carrying equipment as compared to the case without equipment as shown in Figure 8.7. Protective clothing composition also affects the sweat evaporation rate (Leyenda et al., 2017).

8.3 Psychological Perceptions and Assessment Scales

Physiological responses are related to physiology and can be easily characterized with the help of physiological parameters like core temperature, skin temperature, heart rate, etc. described above. However, the physiological parameters cannot provide a good indication of individual subjective perception. Psychological responses are therefore necessary as they provide real perception or judgement of the end users. For human trials performed with an aim to assess clothing performance, some widely used psychological perceptions are thermal sensations, comfort sensations, skin humidity sensations, perceived exertion, and wearer acceptability. Psychological sensations of subjects are usually assessed in the form of rating scales (ISO 10551, 2001). Different rating scales were designed and used for different sensations. Subjects are provided with a scale where they can rate their different perceptions according to their judgment. In this way, psychological responses of subjects are collected in the form of rating throughout the trial. Rating scales are provided to subjects to allow immediate reference to the scales during trials. Rating scales should be fully explained to subjects prior to collecting data. Subjects should be instructed that the sensation rating should be based on whole body sensation and it should not be based on local body sensation; local body sensation may differ significantly depending on the body location (Havenith and Heus, 2004). Nevertheless, sometimes local body sensations are also collected. Various sensations and their associated rating scales are discussed below.

8.3.1 Thermal Sensation

Thermal sensation refers to a conscious thermal feeling commonly graded into seven categories: cold, cool, slightly cool, neutral, slightly warm, warm, and hot. Hence, the thermal sensation is a sensory experience, that is, a psychological response to the state of thermoreceptors in the human body. Human thermal sensation requires subjective evaluation. Information on thermal sensation provides perceptions of the feeling of warmth and coldness of participants, and it is related to the skin and ambient temperatures (Gagge et al., 1969; Young et al., 1987). Various scales were developed for assessing subjects' thermal sensations (Gagge et al., 1967; Young et al., 1987). It is usually assessed with the help of either 7-point or 9-point rating scales. The 7-point scale was proposed by Gagge et al. (1967) with perceptions ranging from cold (rating = 1) to hot (rating = 7). The numerical values in the Gagge's thermal sensation scale were deducted by four so that the modified 7-point scale ranges from −3 to +3 instead of 1 to 7, which is shown in Figure 8.8a. The modified scale is easier to remember because it is symmetrical around the zero point (corresponds to "Neutral" sensation, i.e., the subject prefers neither warmer nor cooler surroundings and he/she is not able to respond to whether a change of thermal environment is needed; Fanger, 1970), and the positive values denote warm sensations while negative values mean cold sensations. This psychophysical 7-point voting scale has been adopted by ASHRAE (The American Society of Heating, Refrigerating and Air Conditioning Engineers) for assessing indoor thermal environmental conditions. It should be noted that the 7-point ASHRAE does not provide any information on comfort. Thus, the ASHRAE scale is often combined with an acceptability scale or question on thermal preference (e.g., preferably having a warmer condition, no change, or prefer having a cooler condition). Another common thermal sensation rating scale is the 9-point scale, which may first be proposed by Young et al. (1987) with perceptions ranging from "Very cold" (revised rating = −4) to "Very hot" (revised rating = +4) as displayed in Figure 8.8b. The 9-point thermal sensation scale has been widely used to assess thermal sensations of subjects while wearing various types of clothing in different thermal environments (Smith et al., 1997; Smith and Petruzzello, 1998; Walker et al., 2015; Wang et al., 2010; Song and Wang, 2016).

Rating of perceived thermal sensations increases during demanding physical activity (Walker et al., 2015; Song and Wang, 2016). Similar effects of physical activity on thermal sensation were also observed by other researchers (Smith et al., 1997; Smith et al., 2001; Smith and Petruzzello, 1998; Wang et al., 2010).

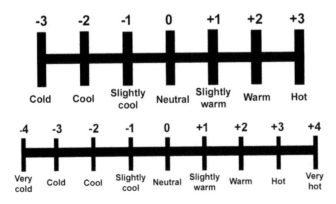

Figure 8.8

(a) The modified 7-point thermal sensation scale. (From Gagge, A. P. et al., *Environ. Res.*, **1**(1): 1–20, 1967.) (b) Modified 9-point rating scale used for assessing human thermal sensations. (From Young, A. J. et al., *J. Appl. Physiol.*, **63**(3): 1218–1223, 1987.)

Moreover, it was found that the increase in thermal sensation with time during physical activity was even more significant in case of hot environmental condition as compared to neutral condition (Smith et al., 1997; von Heimburg et al., 2017). In fact, it has been found that the thermal sensation correlates well with the environmental temperature (Gagge et al., 1967; Gagge et al., 1969). Smith and Petruzzello (1998) analyzed the effect of two different firefighting clothing ensembles on perceived thermal sensations. It was found that the NFPA 1500 standard ensemble that had a higher thermal insulation resulted in significantly higher thermal sensations as compared to a hip-boot configuration ensemble. Hence, it may be concluded that the type of protective clothing plays an important role in perceived thermal sensations by the subjects.

8.3.2 Comfort Sensation

According to ASHRAE standard 55 (2013), thermal comfort for a person is defined as "that condition of mind which expresses satisfaction with the thermal environment." An individual person may be comfortable over a range of temperatures, but it is rather complicated to define a comfort range because of the biological variance of the individual person (Fanger, 1970). A person may feel comfortable in one condition but another person may feel uncomfortable in the exact same condition. Hence, comfort sensation may be assessed by thermal discomfort (too warm or too cool) from the thermal viewpoint.

Similar to the assessment of thermal sensation, comfort sensation should also be subjectively evaluated using psychophysical scales. Perhaps Winslow et al. (1937) are the first to use verbal scales to assess comfort sensation of subjects (i.e., the five-point scale: Very Pleasant, Pleasant, Indifferent, Unpleasant, Very Unpleasant). It was concluded that the subjective sensations of pleasantness (comfort) are closely correlated with skin temperature in hot environments and are even more closely correlated with sweat secretion. Gagge et al. (1967) developed a four-point voting scale to evaluate human comfort sensations ranging from "Comfortable" (+0) to "Very comfortable" (+3) as shown on the left side of Figure 8.9. Similar to the findings of Winslow et al. (1937), it was found that the ambient temperature plays an important role in affecting comfort sensations of subjects. Discomfort increases as ambient temperature increases or decreases from normal temperature (Gagge et al., 1967). In addition to Gagge et al.'s four-point voting scale, ISO 10551 (2001) standard also proposes a five-point comfort sensation voting scale, which ranges from "Comfortable" (+0) to "Extremely uncomfortable" (+4).

Perceptions of overall comfort on protective clothing are much more complicated than perceptions of body comfort based merely on physical stimuli such as heat and moisture (which is usually investigated on nude or seminude subjects). The sensory comfort of subjects while wearing protective clothing involves sensory channels of senses, visual, auditory, smell, taste, and touch (interaction between skin and clothing materials). In addition to the above-mentioned comfort sensation ratings, sensory comfort assessments of protective clothing should also use other descriptors such as *snug, loose, heavy, lightweight, limp, stiff, sticky, non-absorbent, clammy, damp, hot, clingy, sultry, prickly, painful, rough, scratchy,* and *itchy* to comprehensively evaluate the overall comfort sensation while using the protective clothing. Though subjective evaluation of clothing comfort properties may be carried out using synthetic test equipment, the perception of clothing comfort performance must be investigated by human subjects in actual wear

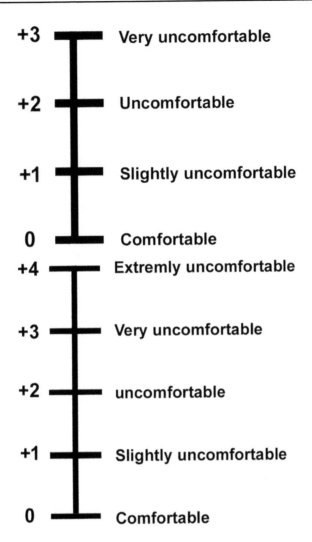

+3 — Very uncomfortable

+2 — Uncomfortable

+1 — Slightly uncomfortable

0 — Comfortable

+4 — Extremly uncomfortable

+3 — Very uncomfortable

+2 — uncomfortable

+1 — Slightly uncomfortable

0 — Comfortable

Figure 8.9

Comfort sensation voting scales; Left: the four-point scale proposed by Gagge et al. (From Gagge, A. P. et al., *Environ. Res.*, **1**(1): 1–20, 1967.); Right: a five-point scale suggested by ISO 10551. (From ISO 10551, Ergonomics of the thermal environment–Assessment of the Influence of the Thermal Environment Using Subjective Judgement scales, International Organization for Standardization, Geneva, Switzerland, 2001.)

conditions. Consequently, human wear trial is the most important technique to evaluate clothing comfort performance.

8.3.3 Rated Perceived Exertion

Rated perceived exertion (RPE) is used to quantify the subjective response of an individual's sensation who is subjected to stress or fatigue due to either physically demanding activities or harsh environmental conditions. Apart from the estimation mode as in human trial studies with protective clothing, RPE can also be used in the prescription mode to prescribe the exercise intensity level to patients during medication. Responses of subjects are recorded with the help of certain rating scales. Various scales and their modifications have been proposed over the years for rating perceived exertion (Borg, 1970, 1980). The two most widely used rating scales for perceived exertion are the 15-point Borg scale and the category-ratio scale (CR10), which are listed in Table 8.6.

The 15-point Borg's RPE scale was proposed by Borg in 1970 and is a voting scale with numerical values ranging from "No exertion" (6) to "Maximal exertion" (20). Later, Borg (1980) proposed another RPE voting scale (rating range 0–10), which is a category scale. This scale is generally referred as the CR10 scale with the lowest rating 0 representing "Nothing at all" and the highest rating 10 representing "Very, very strong" exertion. In this CR10 scale, subjects are allowed to use decimal rating and rating can go beyond 10. These two scales were later compared (Borg, 1982) and it was recommended that the 15-point Borg scale should be used as it is a better indicator of perceived exertion. Since then, the 15-point Borg scale has been used in most of the studies of interest here to estimate RPE (Sköldström, 1987; Smith et al., 1997; Smith and Petruzzello, 1998; McLellan and Selkirk, 2004; Barker et al., 2010; Wang et al., 2010; Levels et al., 2012; Wang et al., 2013; Kofler et al., 2015; Walker et al., 2015; Leyenda et al., 2017).

RPE increases during trials while subjects are wearing protective clothing and performing physical activities (Smith et al., 1997; Smith and Petruzzello, 1998; Smith et al., 2001; von Heimburg et al., 2017; Levels et al., 2012; Walker et al., 2015). The increase in RPE during physical activity is more significant at hot environmental conditions (White et al., 1991; Smith et al., 1997; von Heimburg et al., 2017). The extent of increase in RPE during physical activity also depends on the type of protective clothing used (Smith and Petruzzello, 1998; Wen et al., 2015). Protective clothing with higher insulation results in greater RPE ratings (Smith and Petruzzello, 1998). Bruce-Low et al. (2007) analyzed the effect of exercise intensity on RPE for three different types of clothing. It was observed that the RPE increases as the intensity of physical activity increases for all types of clothing. However, the increase in RPE was found to be more significant at higher levels of physical activity. Significant effects of clothing insulation offered by protective clothing and additional weight due to SCBA were found at higher

Table 8.6 The 15-Point and CR10 Borg RPE Rating Scales

15-point Borg Scale		Category-Ratio (CR10) Scale	
Rating	Perceived Exertion	Rating	Perceived Exertion
6	No exertion	0	Nothing at all
7	Extremely light	0.5	Very, very weak
8			(just noticeable)
9	Very light	1	Very weak
10		2	Weak (light)
11	Light		
12		3	Moderate
13	Somewhat hard		
14		4	Somewhat strong
15	Hard (heavy)	5	Strong (heavy)
16		6	
17	Very hard	7	Very strong
18		8	
19	Extremely hard	9	Very, very strong (almost max)
20	Maximal exertion	10	Maximal

exercise intensity. Environmental temperature also affects RPE in case of protective clothing, especially when the ambient RH is high. Sköldström (1987) noted that the RPE increases as environmental temperature increases. Effect of additional weight due to equipment in case of protective ensemble and environmental temperature on RPE is also visible in Figure 8.3. It is clear from Figure 8.3 that the RPE increases as the weight of ensemble and environmental temperature increases. Effect of additional weight due to equipment on RPE is more pronounced at higher environmental temperature. Study shows that the RPE is correlated well with HR (Sköldström, 1987) as well as with restriction to arm and leg movements (Wen et al., 2015).

Psychological responses and physiological parameters are in a way related. However, the relation between the two is not straightforward, and psychological responses depend on many factors. Still, efforts have been made to correlate the RPE and the physiological strain index (PSI or PhSI) and perceptual strain index (PeSI) in case of firefighter turnout gear (Baker et al., 2000). PSI is a parameter proposed by Moran et al. (1998) and it has been widely used to evaluate heat strain of subjects. PSI is calculated based on physiological parameters, HR and rectal temperature (T_{re}), and it may be written as (Moran et al., 1998; Baker et al., 2000; Tikuisis et al., 2002; Stewart et al., 2014)

$$PSI = 5 \times \frac{(T_{re,t} - T_{re,initial})}{(39.5 - T_{re,initial})} + 5 \times \frac{(HR_t - HR_{initial})}{(180 - HR_{initial})} \tag{8.19}$$

where the subscript initial and t denote time instances at the beginning of the trial and at any time t, respectively. The application limits of T_{re} and HR in the above equation are $36.5 \leq T_{cr} \leq 39.5$ and $60 \leq HR \leq 180$ bpm, respectively. Normally, a true $HR_{initial} = HRrest$ is not obtained because the subjects experience elevated HR at the beginning of the trials due to the encumbrance of protective clothing. Therefore, an arbitrary HR value of 60 bpm can be assigned to $HR_{initial}$ in Equation 8.19.

In Equation 8.19, maximum T_{re} and HR are assumed to be 39.5°C and 180 bpm, respectively. Hence, PSI is basically calculated by normalizing increase in HR and T_c with equal weightage to both HR and T_c. PSI is thus obtained as a 10-point scale (ratings range from 1 to 10) with the lowest rating 1 representing "No strain/Little strain" and the highest rating 10 representing "Very high heat strain" as shown in Table 8.7. It has been found that the RPE and PSI correlate very well in case of firefighters' protective clothing ensembles during prolonged exercise (Baker et al., 2000). However, PSI overestimated the RPE for short duration exercises. Correlating psychological responses with physiological parameters is justified as it may enable researchers to determine the level of exertion by just measuring few physiological parameters that might simplify the analysis.

While dealing with RPE, proper care should be taken and the rating scale should be explained to subjects before actually beginning the trial. The experimentalist (investigator) should also ensure that the participants are providing RPE judgement in the scale based on their overall (whole body) exertion and not based on any local body sensation. When carrying heavy equipment or respiratory support systems on the shoulders, the subjects (e.g., firefighters) may report RPE based on perceived exertion of local muscles such as shoulders and the back rather than general (i.e., whole body) fatigue, which could be misleading.

Table 8.7 The PSI (Physiological Strain Index) Scale

Ratings	Strain
1	No strain/Little strain
2	
3	Low strain
4	
5	Moderate strain
6	
7	High strain
8	
9	Very high strain
10	

Source: Moran, D. S. et al., *Am. J. Physiol.,* **275**(1 Pt. 2): R129–R134, 1998.

One of the major limitations for the PSI index is that it requires a measurement of core temperature. To solve this issue, the perceptual strain index (PeSI) has been proposed as a universal applicable measure of body heat strain during physical activities/exercises in both laboratory and fields based only on subjective data on thermal sensation and perceived exertion (Tikuisis et al., 2002). PeSI may be calculated by

$$PeSI = \left[5 \times \frac{(TS-1)}{6} \right] + \left[5 \times \frac{RPE}{10} \right] \tag{8.20}$$

where *TS* is a modified version of Gagge et al.'s rating of thermal sensations (TS) from comfortable to intolerably hot (on a scale from 7 to 13, respectively); *RPE* is a modified version of the Borg rating of physical exertion on a scale from 0 to 10.

The PeSI is accurate for untrained subjects, but the index tends to underestimate perceived heat strain on endurance-trained athletes. Thus, it is suggested to use both Psi and PeSI indices to evaluate the heat strain in well-trained athletes. PeSI has been found to correlate well with PSI, RPE, WBGT (wet bulb globe temperature), air temperature, and relative heart rate, and it could be used as a simple and robust tool to assess heat strain of occupational workers while wearing various protective clothing (Sartang and Dehghan, 2015; Chan and Yang, 2016; Borg et al., 2017). Compared to the WBGT index, the PeSI index has greater ability to assess the thermal strain. Petruzzello et al. (2009) examined the physiological and perceptual strain associated with working in personal protective equipment and performing firefighting activities in a hot environment using PSI and PeSI indices. It was found that both PSI and PeSI increase significantly over the time and both two indices give moderate to high levels of heat strain for relatively brief bouts of exercise on subjects while wearing heavy impermeable clothing or while performing simulated firefighting activities in hot environments. In summary, the PeSI may be used as an effective tool to evaluate heat strain and performance of occupational workers such as firefighters while performing demanding activities in hot environments if other assessment methods (e.g., measurement of core temperature) were not accessible.

Table 8.8 The Five-Point and Nine-Point Wet Sensation Scales

Rating	Wet Sensation	Rating	Wet Sensation
0	Dry	1	Dry
1	Slightly wet	2	Slightly moist
2	Wet	3	
3	Very wet	4	Moist
		5	
		6	Wet
		7	
		8	Soaked
		9	Totally soaked

Sources: ISO 10551, Ergonomics of the thermal environment–Assessment of the Influence of the Thermal Environment Using Subjective judgement scales, International Organization for Standardization, Geneva, Switzerland, 2001; Havenith, G., Heus, R., *Appl. Ergon.*, 35 (1): 3–20, 2004; Leyenda, B. C. et al., *Front. Physiol.*, **8**: 618, 2017.

8.3.4 Skin Humidity (Wetness) Sensation

There is no humidity (skin wetness) receptor in the human skin and the sense of skin humidity (wetness) is indirect through perceptions from thermoreceptors (skin cooling due to sweat evaporation causes skin temperature to change) as well as mechanoreceptors (mechanical pressure, friction between skin and clothing materials due to movement). Human beings are believed to "learn" to perceive the wetness experience when the skin is in contact with the wet surface or when sweat is secreted on the skin surface (Fillingeri and Havenith, 2015). Skin humidity (wetness) greatly affects the comfort sensation of an individual person. This is even more pronounced on firefighters who are performing heavy duties while wearing relatively impermeable protective clothing. Humid (wet) skin may result in unease or discomfort in wearers of protective clothing. This unease or discomfort due to sweat can be estimated with the help of subjective responses based on subjective rating scales. Table 8.8 presents two mostly used voting scales for skin humidity sensations.

During heavy physical activity, sweating is a common phenomenon. If protective clothing is not permeable enough, wetness in clothing and skin wet sensation increases during physical activity (Wen et al., 2015; de Rome et al., 2016). Hence, permeable protective clothing should be preferred. However, at present, engineering permeable but thermally protective clothing is still a great challenge to material scientists and engineers.

8.3.4.1 Wearer Acceptability

Wearer acceptability test is important to determine how subjects feel about the protective clothing and also how they perceive the fit and comfort of clothing. This test is especially important in case of firefighting protective clothing because if a firefighting protective clothing system lacks acceptability of firefighters, they may refuse to wear the protective clothing, which can lead to serious accident. Protective clothing should also ensure wearer's movability and freedom of movement, which can help greatly while rescuing or moving out quickly from hostile or hazardous environmental conditions. In the wearer acceptability test, subjects are asked to complete a nine-point wearer acceptability questionnaire after completing a range of body movements (Table 8.9) (Huck et al., 1997). Data collected after trials are then analyzed to assess performance and acceptability of clothing. Clothing with higher ratings should always be preferred.

Table 8.9 The Wearer Acceptability Voting Scale

Please place a check (tick) between each pair of adjectives at the locations that best describes how you feel about the clothing:

1	Comfortable	9	8	7	6	5	4	3	2	1	Uncomfortable
2	Acceptable	9	8	7	6	5	4	3	2	1	Unacceptable
3	Rested	9	8	7	6	5	4	3	2	1	Tired

Please place a check (tick) between each pair of adjectives at the locations that best describes the clothing you are wearing:

4	Flexible	9	8	7	6	5	4	3	2	1	Stiff
5	Easy to put on	9	8	7	6	5	4	3	2	1	Hard to put on
6	Freedom of movement of arms	9	8	7	6	5	4	3	2	1	Restricted movement of arms
7	Easy to move in	9	8	7	6	5	4	3	2	1	Hard to move in
8	Satisfactory fit	9	8	7	6	5	4	3	2	1	Unsatisfactory fit
9	Freedom of movement of legs	9	8	7	6	5	4	3	2	1	Restricted movement of legs
10	Freedom of movement of torso	9	8	7	6	5	4	3	2	1	Restricted movement of torso
11	Like	9	8	7	6	5	4	3	2	1	Dislike
12	Loose	9	8	7	6	5	4	3	2	1	Tight
13	Crotch of overall right distance from body	9	8	7	6	5	4	3	2	1	Crotch of overall too close or too far from body

Source: Huck, J., Maganga, O., Kim, Y. (1997) Protective overalls: Evaluation of clothing design and fit. *Int. J. Cloth. Sci. Technol.* **9**(1): 45–61.

8.4 Human Trial Case Studies on Firefighting Protective Clothing

Firefighters face serious challenges not only of protection against extreme fire and flame scenarios during operations but also of physiological strain during normal duties. Some major contributing factors to the pronounced physiological strain (e.g., high body temperatures, relative high heart rate, heavy sweating) of firefighters are protective clothing, associated equipment of firefighters, high-intensity work, and hostile environmental conditions (Montain et al., 1994). A large number of studies have been conducted over the years for analyzing comfort-related performance of firefighter clothing and to analyze the effect of different firefighter clothing, equipment, environmental conditions, and activity levels on physiological and psychological responses of wearers. Studies have been performed under controlled conditions in climate chambers as well as under live fire situations. Some of the major studies dealing with comfort analysis of firefighter clothing along with various details will be discussed in this section.

8.4.1 Effect of Protective Clothing and Equipment

Firefighter clothing offers much better thermal protection to the wearer than normal clothing. Unique features that make firefighter clothing so special as compared to normal clothing are that firefighting clothing has higher thermal insulation, higher weight, and lower air/water vapor permeability. Protective equipment and SCBA further add to the weight of the firefighter ensemble, which can be as high as 20–30 kg, as mentioned earlier. These features of fire-fighter ensemble result in higher physiological strain and undesirable psychological responses relative to normal clothing (Louhevaara et al., 1985; White and Hodous, 1987).

Various studies have been conducted over the years to analyze the effect of firefighter clothing and equipment on physiological parameters and psychological responses of firefighters. White et al. (1991) analyzed the effect of protective clothing over normal clothing on physiological and psychological responses at three different environmental conditions. Subjective responses of participants indicated that the subjects preferred normal clothing over protective clothing. Effects of additional weight and insulation offered by firefighter clothing on heat stress were analyzed (Duncan et al., 1979). It was found that under hot conditions, additional weight and insulation offered by firefighter clothing and equipment (boot and breathing apparatus) result in increased heat stress as compared to normal clothing with significantly negative effect on physiological parameters. A recent study conducted on chemical protective clothing also indicates that physical burden increased as the weight of clothing increases (Wen et al., 2015). Figure 8.10 shows the effect of protective clothing as compared to normal clothing on work tolerance. It can be observed that although there is almost no effect of protective clothing at neutral and cold conditions, the total work duration is affected significantly by protective clothing at hot environmental conditions.

Apart from clothing weight and insulation, the effect of the additional weight of firefighting protective equipment was also analyzed on work tolerance and physiological parameters (Faff and Tutak, 1989). It was found that the additional weight due to equipment results in a significant reduction in the work capacity of firefighters. Firefighter turnout gear consists of many components such as clothing, helmet, gloves, boot, SCBA, etc. All of these components contribute to overall discomfort. In recent years, efforts were also made to analyze the effect of individual components of firefighter protective ensemble on heat stress

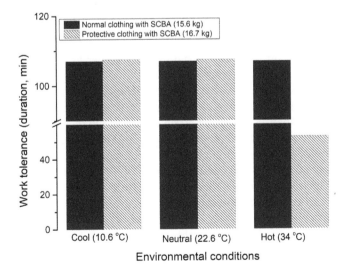

Figure 8.10

Effect of protective clothing on work tolerance (i.e., tolerance time) for different environmental conditions. (From White, M. K. et al., *Ergonomics*, **34**(4): 445–457, 1991.)

by monitoring human physiological and psychological responses (Taylor et al., 2012; Kim et al., 2014; Lee et al., 2014).

8.4.2 Effect of Different Types of Protective Clothing

Thermal comfort or physiological strain of firefighters depends on the heat and moisture transfer ability of firefighter clothing apart from physical activity, environmental conditions, and weight of firefighter ensemble. Heat and moisture transfer through clothing depends on various properties of clothing such as fabric material, fabric thermal conductivity, evaporative resistance (or water vapor resistance), fabric thickness, fabric density, porosity, etc. It also depends on clothing size and whether single or multilayer protective clothing is used. Havenith et al. (2011) found that air permeability plays an important role in heat strain with higher air permeability leading to lower heart rate and lower skin and core temperatures. Air permeability is a function of fabric weave pattern, areal density, and porosity (Udayraj et al., 2017). Various studies were conducted in the past where the effect of different protective clothing was analyzed on the thermophysiological comfort of wearers (Smith and Petruzzello, 1998; Havenith and Heus, 2004).

Smith and Petruzzello (1998) analyzed the effect of two different firefighter clothing ensembles on physiological and psychological responses of firefighters under live-fire conditions in a building. Performance of three different single and multilayer chemical protective coveralls made up of different materials was analyzed (Wen et al., 2015). Good correlation of fabric evaporative resistance and fabric thickness was found with physiological and perpetual responses. Effect of two different barrier liners was analyzed by White and Hodous (1988) on physiological responses of firefighters under two different activity levels in a hot environment. Significant effect on skin temperature was observed due to liner but no significant effect of the same on work tolerance was obtained. Insulative properties of clothing affect physiological and psychological responses of the firefighters negatively as noticed by Smith et al. (1995), who compared the performance of two different firefighter ensembles: hip boot configuration (used before 1987, prior to the proposal of NFPA 1500 standard) and NFPA 1500 standard firefighter ensembles. They found that the NFPA 1500 standard ensemble results in significantly higher heart rate, rectal temperature, VO_2 (oxygen consumption), and RPE (rating of perceived exertion) as compared to the hip boot configuration ensemble. The inferior performance of NFPA 1500 standard ensemble in terms of physiological parameters and psychological responses is mainly attributed to the higher insulation offered by NFPA 1500 standard ensemble. In contrast, Holmér et al. (2006) examined four types of Swedish firefighting turnout gear using human subjects in an extremely hot and humid condition (i.e., 55°C, 30% RH). It was discovered that small differences between the firefighting turnout gears in terms of design, thickness, and insulation have almost no effect on the physiological strain in the studied condition.

For acceptance of protective clothing by end users (e.g., firefighters), it is essential that protective clothing provides better comfort while not compromising with the thermal protective performance (Holmér 2006). It can be achieved by properly modifying the design of conventional firefighter clothing and by incorporating permeable membranes and lightweight thermal liner (Barker et al., 2000). Hence, by proper selection and optimization of various parameters associated with the fabric of firefighter clothing, the comfort of firefighters can be improved while simultaneously meeting the thermal protection requirements.

8.4.3 Effect of Environmental Conditions

Firefighters work under different environmental conditions ranging from thermoneutral conditions to extremely hot conditions while performing duties. Protective clothing may behave differently under different environmental conditions (White et al., 1991). It can also be observed from Figure 8.10 that although no effect of protective clothing was observed in lower environmental temperatures, protective clothing affects work tolerance significantly at higher environmental temperatures. To analyze the effect of different environmental conditions on comfort performance of protective clothing, various studies were conducted. Environmental conditions involving a fixed temperature in a climate chamber (Sköldström, 1987; White et al., 1991; McLellan and Selkirk, 2004; Dreger et al., 2006; Kofler et al., 2015), radiant heat exposure (Havenith and Heus, 2004; Levels et al., 2012), and live firefighting drills (Smith et al., 1997; Smith and Petruzzello, 1998; Smith et al., 2001; Bruce-Low et al., 2007; Horn et al., 2013) were analyzed.

Increase in environmental temperature of firefighters wearing protective ensemble results in an increase in heat stress and, thereby, a reduction in the work tolerance. Increased environmental temperature affects the physiological parameters negatively (Duncan et al., 1979; Levels et al., 2012). Stewart et al. (2014) analyzed the effect of three different environmental conditions (i.e., 24°C and 50% RH, 32°C and 60% RH, 48°C and 20% RH) on physiological parameters and work tolerance for different walking speeds. Decreasing trend of work tolerance with environmental temperature was observed at various treadmill walking speeds. External environment faced by firefighters during operations does not involve constant and uniform temperature conditions. In actual situations faced by firefighters, environmental temperature changes throughout the operation. Apart from the studies considering constant temperature ambient conditions where the environmental temperature was maintained constant throughout the trial, studies were conducted considering the effect of varying environmental conditions on firefighter protective clothing performance. Kim and Lee (2016) analyzed the effect of environmental temperature fluctuation (fluctuating between 29.5°C and 35.5°C) on core and skin temperatures of subjects wearing firefighter protective clothing. Apart from those, studies were also carried out to analyze the effect of actual fire conditions on overall comfort of firefighters (Sothman et al., 1992; Smith et al., 1997).

8.4.4 Effect of Physical Activity

As mentioned in the "Background" section, firefighters usually have to perform duties under different circumstances involving a range of physical activities with different levels of work intensities and work durations. Thus, it is important to analyze the effect of physical activity on physiological and psychological responses of firefighters. A large number of studies were devoted to analyzing the effect of different physical activities on overall comfort of firefighters (Baker et al., 2000; McLellan and Selkirk, 2004; Stewart et al., 2014). It has been found that heat stress of firefighters wearing protective clothing increases as the level of physical activity or work duration increases. Baker et al. (2000) conducted a study to analyze the effect of exercise duration on physiological parameters and perceived exertion in firefighters wearing protective clothing. One exercise was for a short duration (i.e., 12 min) and another one for a longer duration (i.e., 60 min). It was found that prolonged physical exercise results in higher physiological strain.

Performance of firefighter clothing may vary depending upon the activity level. A protective clothing suitable for one particular activity may not be suitable for another activity. Effect of two different clothing configurations on human thermal comfort was analyzed by McLellan and Selkirk (2004) for four different intensities of physical activity. Two clothing configurations involved shorts and long pant. Four different levels of exercises considered were heavy (i.e., 4.8 km/h, 5% grade), moderate (i.e., 4.5 km/h, 2.5% grade), light (i.e., 4.5 km/h), and very light (i.e., 2.5 km/h). Although no significant effect of replacing long pant with shorts was found on perceived exertion and thermal comfort at high-intensity physical exercises, it affected physiological parameters significantly at low-intensity physical exercises. Similar results were also found by Stewart et al. (2014). They found that the work tolerance in the treadmill walking reduces significantly at higher walking speeds.

Considering the fact that the protective clothing actually results in higher heat stress and the wearer has to do heavy physical work for a longer duration, it is recommended that the entire physical activity should be split into parts and workers should be allowed to take rest in between. Effect of repeated work bout of firefighters was analyzed on physiological, psychological, inflammation, and immune responses under hot conditions (Walker et al., 2015). Subjects performed two work bouts of 20 min period each in a hot chamber maintained at 100°C with the help of gas burners. The work bouts were separated by a rest period of 10 min at an environmental temperature of 13.5°C. Horn et al. (2013) also analyzed the effect of repeated bouts of firefighting activities on firefighters under live fire conditions. Heart rate and intestinal temperature were recorded during entire 3 h trials. It was found that prolonged firefighting activity increases the risk of exertional heat stress even if rest is provided in between activities.

It can be observed from the above discussion that weight, impermeability, and insulation of protective clothing along with additional weight due to equipment such as SCBA contribute significantly to the heat stress of the wearers under various physical activities. While the protection of firefighters cannot be compromised either, steps should be taken to counter the negative effect caused by the weight of firefighter clothing and equipment on comfort and performance of firefighters. It is, therefore, the need of the hour to provide lightweight firefighter clothing and equipment without compromising their protective performance. Innovative clothing design and necessary equipment design modifications should be made to realize lightweight protective firefighter clothing. It has been found from a recent survey conducted on firefighters that heat stress is still a challenge to the existing protective clothing during firefighter operations (Barker et al., 2010). Efforts were made to incorporate suggestions of end-users of protective clothing (firefighters) to improve its design (Barker et al., 2010). This way a lightweight, better thermally protective, and comfortable firefighter clothing can be developed, which can also provide better ease of body motion.

8.5 Conclusions

Although the concept of human trial is not new, human wear trials serve as the most important and widely used method to evaluate clothing performance.

Human trials provide valuable information on actual comfort performance of various types of protective clothing. First, they provide precise information on physiological strain resulted from thermal stressors due to the protective clothing system, activities, and/or environmental conditions. Information on individual psychological perceptions may also be obtained from wear trials. Second, human trials add significant information about individual variations. Due to the presence of individual variations, a large number of subjects may be needed to observe significant statistical differences between independent variables such as the tested protective clothing systems and test conditions. Third, wear trials can accurately assess the ergonomic performance of protective clothing systems, which seems rather too difficult to assess using synthetic test equipment. Particularly, with the great progress of modern technology in wearable electronics and clothing, measurement of human physiological responses will become more convenient. Further, human wear trials offer precise data for mathematical models dealing with human comfort and thermophysiological strain. Traditional fabric/clothing material tests are two-dimensional and differences found from such tests may always be inaccurate to predict the actual performance of three-dimensional clothing systems. Presently, it seems that existing advanced models still fail to precisely predict the actual comfort performance of clothing systems. Therefore, human wear trials will continue serving as the primary method to assess the actual comfort performance of protective clothing systems.

References

Adams, P. S., and Keyserling, W. M. (1995) The effect of size and fabric weight of protective coveralls on range of gross body motions. *Am. Ind. Hyg. Assoc. J.* **56**(4): 333–340.

ANSI/ASHRAE 55 (2013) Thermal environmental conditions for human occupancy. American Society of Heating, Refrigerating, and Air-Conditioning Engineers, Atlanta, GA.

ASTM F2668 (2016) Standard practice for determining the physiological responses of the wearer to protective clothing ensembles. American Society for Testing and Materials, Western Conshohocken, PA.

Åstrand, P. O., and Rodhal, K. (1970) *Textbook of work physiology*. New York: McGraw-Hill Book Company.

Bach, A. J. E., Stewart, I. B., Disher, A. E., Costello, J. T. (2015) A comparison between conductive and infrared devices for measuring mean skin temperature at rest, during exercise in the heat, and recovery. *PLoS ONE* **10**(2): e0117907.

Baitinger, W. F. (1979) Product engineering of safety apparel fabrics: Insulation characteristics of fire-retardant cottons. *Text. Res. J.* **49**(4): 221–225.

Baker, J., Grice, J., Roby, L., Matthews, C. (2000) Cardiorespiratory and thermoregulatory response of working in fire-fighter protective clothing in a temperate environment. *Ergonomics* **43**(9): 1350–1358.

Barker, R. L. (2005) A review of gaps and limitations in test methods for first responder protective clothing and equipment. Final report presented to national personal protection technology laboratory. National Institute for Occupational Safety and Health (NIOSH).

Barker, R., Deaton, S., Liston, G., Thompson, D. (2010) A CB protective firefighter turnout suit. *Int. J. Occup. Saf. Ergon.* **16**(2): 135–152.

Barnard, R. J., Duncan, H. W. (1975) Heart rate and ECG responses of fire fighters. *J. Occup. Med.* **17**(4): 247–250.

Barnett, B. J., Nunberg, S., Tai, J., Lesser, M. L., Fridman, V., Nichols, P., Powell, R., Silverman, R. (2011) Oral and tympanic membrane temperatures are inaccurate to identify fever in emergency department adults. *West J. Emerg. Med.* **12**(4): 505–511.

Beachy, S. H., Repasky, E. A. (2013) Toward establishment of temperature thresholds for immunological impact of heat exposure in humans. *Int. J. Hyperthermia* **27**(4): 344–352.

Bernard, V., Staffa, E., Mornstein, V., Bourek, A. (2013) Infrared camera assessment of skin surface temperature—Effect of emissivity. *Phys. Med.* **29**(6): 583–591.

Bongers, C. C., Hopman, M. T., Eijsvogels, T. M. (2015) Using an ingestible telemetric temperature pill to assess gastrointestinal temperature during exercise. *J. Vis. Exp.* (104): e53258, doi:10.3791/53258

Borg, D. N., Costello, J. T., Bach, A. J., Stewart, I. B. (2017) Perceived exertion is as effective as the perceived strain index in predicting physiological strain when wearing personal protective clothing. *Physiol. Behav.* **169**: 216–223.

Borg, G. A. V. (1970) Perceived exertion as an indicator of somatic stress. *Scand. J. Rehabil. Med.* **2**(2): 92–98.

Borg, G. A. V. (1980) A category scale with ratio properties for intermodal and interindividual comparisons. In: Proceedings of the 22nd International Congress of Psychology, VEB Deutscher Verlag, Leipzig.

Borg, G. A. V. (1982) Psychophysical bases of perceived exertion. *Med. Sci. Sports Exerc.* **14**(5): 377–381.

Bradshaw, D. I., George, J. D., Hyde, A., LaMonte, M. J., Vehrs, P. R., Hager, R. L., Yanowitz, F. G. (2005) An accurate VO2max nonexercise regression model for 18–65-year-old adults. *Res. Q. Exerc. Sport.* **76**(4): 426–432.

Brinnel, H., Cabanac, M. (1989) Tympanic temperature is a core temperature in humans. *J. Therm. Biol.* **14**(1): 47–53.

Bröde, P., Havenith, G., Wang, X., Candas, V., den Hartog, E. A., Griefahn, B., Holmér, I., Kuklane, K., Meinander, H., Nocker, W., Richards, M. (2008) Non-evaporative effects of a wet mid layer on heat transfer through protective clothing *Eur. J. Appl. Physiol.* **104**(2): 341–349.

Bruce-Low, S. S., Cotterrell, D., Jones, G. E. (2007) Effect of wearing personal protective clothing and self-contained breathing apparatus on heart rate, temperature and oxygen consumption during stepping exercise and live fire training exercises. *Ergonomics* **50**(1): 80–98.

Buller, M. J., Tharion, W. J., Duhamel, C. M., Yokota, M. (2015) Real-time core body temperature estimation from heart rate for first responders wearing different levels of personal protective equipment. *Ergonomics* **58**(11): 1830–1841.

Burton, A. C. (1935) Human calorimetry II. The average temperature of the tissues of the body. *J. Nutr.* **9**(3): 261–280.

Byrne, C., Lim, C. L. (2007) The ingestible telemetric body core temperature sensor: A review of validity and exercise applications. *Br. J. Sports Med.* **41**(3): 126–133.

Camenzind, M. A., Dale, D. J., Rossi, R. M. (2007) Manikin test for flame engulfment evaluation of protective clothing: Historical review and development of a new ISO standard. *Fire Mater.* **31**(5): 285–295.

Candas, V., Libert, J. P., Vogt, J. J. (1979) Human skin wettedness and evaporative efficiency of sweating. *J. Appl. Physiol.* **46**(3): 522–528.

Chan, A. P. C., Yang, Y. (2016) Practical on-site measurement of heat strain with the use of a perceptual strain index. *Int. Arch. Occup. Environ. Health* **89**(2): 299–306.

Cheung, S. S., Sweeney, D. H. (2001) The influence of attachment method and clothing on skin temperature sensor accuracy. *Med. Sci. Sports Exerc.* **33**(5): 161.

Childs, C., Harrison, R., Hodkinson, C. (1999) Tympanic membrane temperature as a measure of core temperature. *Arch. Dis. Child.* **80**(3): 262–266.

Chitrphiromsri, P., Kuznetsov, A. V. (2005) Modeling heat and moisture transport in firefighter protective clothing during flash fire exposure. *Heat Mass Transf.* **41**(3): 206–215.

Clifford, J., McK Kerslake, D., Waddell, J. L. (1959) The effect of wind speed on maximum evaporative capacity in man. *J. Physiol.* **147**(2): 253–259.

Coca, A., Roberge, R., Shepherd, A., Powell, J. B., Stull, J. O., Williams, W. J. (2008) Ergonomic comparison of a chem/bio prototype firefighter ensemble and a standard ensemble. *Eur. J. Appl. Physiol.* **104**(2): 351–359.

Dale, J. D., Crown, E. M., Ackerman, M. Y., Leung, E., Rigakis, K. B. (1992) Instrumented Mannequin Evaluation of Thermal Protective Clothing. In: McBriarty, J. P., and Henry, N. W. (eds.), *Performance of Protective Clothing, the Fourth Volume*, ASTM STP1133, American Society for Testing and Materials, Western Conshohocken, PA, pp. 717–733.

de Rome, L., Taylor, E. A., Croft, R. J., Brown, J., Fitzharris, M., Taylor, N. A. S. (2016) Thermal and cardiovascular strain imposed by motorcycle protective clothing under Australian summer conditions. *Ergonomics* **59**(4): 504–513.

Delgado-Gonzalo, R., Parak, J., Renevey, P., Bertschi, M., Korhonen, I. (2015) Evaluaton of accuracy and reliability of PulseOn optical heart rate monitoring device. In: Proceedings of the 37th Annual International Conference of the IEEE, Milan, Italy, pp. 430–433, doi:10.1109/EMBC.2015.7320273

Domitrovich, J. W., Cuddy, J. S., Ruby, B. C. (2010) Core-temperature sensor ingestion timing and measurement variability. *J. Athl. Train.* **45**(6): 594–600.

Dorman, L. E., Havenith, G. (2009) The effects of protective clothing on energy consumption during different activities. *Eur. J. Appl. Physiol.* **105**(3): 463–470.

Dreger, R. W., Jones, R. L., Petersen, S. R. (2006) Effects of the self-contained breathing apparatus and fire protective clothing on maximal oxygen uptake. *Ergonomics* **49**(10): 911–920.

Du Bois, D., Du Bois, E. F. (1989) A formula to estimate the approximate surface area if height and weight be known. *Nutrition* **5**(5): 303–311.

Duncan, H. W., Gardner, G. W., Barnard, R. J. (1979) Physiological responses of men working in fire fighting equipment in the heat. *Ergonomics* **22**(5): 521–527.

Faff, J., Tutak, T. (1989) Physiological responses to working with fire fighting equipment in the heat in relation to subjective fatigue. *Ergonomics* **32**(6): 629–638.

Fanger, P. O. (1970) *Thermal comfort: Analysis and applications in environmental engineering*. New York: McGraw-Hill Book Company.

Fick, A. (1870) Uber die messung des blutquantums in den hertzvent rikeln. (On the measurement of blood mass in the heart ventricles.) *Sitzber Physik-Med Ges Wurzburg* **36**: 16–28. (in German)

Filingeri, D., Havenith, G. (2015) Human skin wetness perception: Psychophysical and neurophysiological bases. *Temperature* **2**(1): 86–104.

Gagge, A. P., Gonzalez, R. R. (1996) Mechanisms of heat exchange: Biophysics and physiology. In: *Handbook of Physiology-Comprehensive Physiology*, American Physiological Society, Bethesda, MD, pp. 45–84.

Gagge, A. P., Nishi, Y. (1977) Heat exchange between human skin surface and thermal environment. In: *Handbook of Physiology-Comprehensive Physiology*, American Physiological Society, Bethesda, MD, pp. 69–92.

Gagge, A. P., Stolwijk, J. A., Hardy, J. D. (1967) Comfort and thermal sensations and associated physiological responses at various ambient temperatures. *Environ. Res.* **1**(1): 1–20.

Gagge, A. P., Stolwijk, J. A. J., Saltin, B. (1969) Comfort and thermal sensations and associated physiological responses during exercise at various ambient temperatures. *Environ. Res.* **2**(3): 209–229.

Gant, N., Atkinson, G., Williams, C. (2006) The validity and reliability of intestinal temperature during intermittent running. *Med. Sci. Sports Exerc.* **38**(11): 1926–1931.

Gasim, G. I., Musa, I. R., Abdien, M. T., Adam, I. (2013) Accuracy of tympanic temperature measurement using an infrared tympanic membrane thermometer. *BMC Res. Notes* **6**(1): 194.

Givoni, B., Goldman, R. F. (1971) Predicting metabolic energy cost. *J. Appl. Physiol.* **30**(3): 429–433.

Goldberg, L., Elliot, D. L., Kuehl, K. S. (1988) Assessment of exercise intensity formulas by use of ventilator threshold. *Chest* **94**(1): 95–98.

Goldman, R. F. (2013) A guide to the conduct of severe human studies on human subjects. In: Cotter, J. D., Lucas, S. J. E., Mündel, T. (eds.), Proceedings of The 15th International Conference on Environmental Ergonomics (ICEE), Queenstown, New Zealand, pp. 107–110.

Gollnick, P. D., Armstrong, R. B., Saubert, IV, C. W., Piehl, K., Stalin, B. (1972) Enzyme activity and fiber composition in skeletal muscle of untrained and trained men. *J. Appl. Physiol.* **33**(3): 312–319.

Hardy, J. D., Dubois, E. F. (1938) The technic of measuring radiation and convection. *J. Nutr.* **15**(5): 461–475.

Havenith, G. (1999) Heat balance when wearing protective clothing. *Ann. Occup. Hyg.* **43**(5): 289–296.

Havenith, G., Hartog, E. D., Martin, S. (2011) Heat stress in chemical protective clothing: Porosity and vapour resistance. *Ergonomics* **54**(5): 497–507.

Havenith, G., Heus, R. (2004) A test battery related to ergonomics of protective clothing. *Appl. Ergon.* **35**(1): 3–20.

Herman, I. P. (2016) *Physics of the human body.* Zug: Springer International Publishing.

Holmér, I. (2006) Protective clothing in hot environments. *Ind. Health* **44**(3): 404–413.

Holmér, I. (2009) Human wear trials for cold weather protective clothing systems. In: Williams, J. T. (ed.) *Textiles for Cold Weather Apparel*, Cambridge: Woodhead Publishing, pp. 256–273.

Holmér, I., Kuklane, K., Gao, C. (2006) test of firefighter's turnout gear in hot and humid air exposure. *Int. J. Occup. Saf. Ergon.* **12**(3): 297–305.

Hondas, Y., Ring, E. F. (1982) *Human body temperature: its measurement and regulation.* New York: Plenum Press.

Horn, G. P., Blevins, S., Fernhall, B., Smith, D. L. (2013) Core temperature and heart rate response to repeated bouts of firefighting activities. *Ergonomics* **56**(9): 1465–1473.

Huck, J. (1988) Protective clothing systems—A technique for evaluating restriction of wearer mobility. *Appl. Ergon.* **19**(3): 185–190.

Huck, J., Maganga, O., Kim, Y. (1997) Protective overalls: Evaluation of clothing design and fit. *Int. J. Cloth. Sci. Technol.* **9**(1): 45–61.

Hunt, A. P., Stewart, I. B. (2008) Calibration of an ingestible temperature sensor. *Physiol. Meas.* **29**(11): 71–78.

ISO 7933 (2004) Ergonomics of the thermal environment-analytical determination and interpretation of heat stress using calculation of the predicted heat strain. International Organization for Standardization, Geneva, Switzerland.

ISO 8996 (2004) Ergonomics of the thermal environment—Determination of metabolic rate. International Organization for Standardization, Geneva, Switzerland.

ISO 9886 (2004) Ergonomics—Evaluation of thermal strain by physiological measurements. International Organization for Standardization, Geneva, Switzerland.

ISO 10551 (2001) Ergonomics of the thermal environment—Assessment of the Influence of the Thermal Environment Using Subjective judgement scales. International Organization for Standardization, Geneva, Switzerland.

James, C. A., Richardson, A. J., Watt, P. W., Maxwell, N. S. (2014) Reliability and validity of skin temperature measurement by telemetry thermistors and a thermal camera during exercise in the heat. *J. Therm. Biol.* **45**: 141–149.

Jay, O., Kenny, G. (2007) The determination of changes in body heat content during exercise using calorimetry and thermometry. *J. Hum. Environ. Syst.* **10**(1): 19–29.

Jéquier, E. (1986) Human whole body direct calorimetry. *IEEE Eng. Med. Biol. Mag.* **5**(2): 12–14.

Kanitakis, J. (2002) Anatomy, histology and immunohistochemistry of normal human skin. *Eur. J. Dermatol.* **12**(4): 390–399.

Karlberg, P. (1949) The significance of depth of insertion of the thermometer for recording rectal temperatures. *Acta Paediatr.* **38**(1): 359–366.

Karvonen, J., Vuorimaa, T. (1988) Heart rate and exercise intensity during sports activities: Practical application. *Sports Med.* **5**(5): 303–311.

Kerslake, D. McK. (1972) *The stress of hot environments*. Cambridge: Cambridge University Press.

Kim, S., Jang, Y. J., Baek, Y. J., Lee, J. Y. (2014) Influences of partial components in firefighters' personal protective equipment on subjective perception. *Fashion Text.* **1**(3): 1–14.

Kim, S., Lee, J. Y. (2016) Skin sites to predict deep-body temperature while wearing firefighters' personal protective equipment during periodical changes in air temperature. *Ergonomics* **59**(4): 496–503.

King, C. N., Senn, M. D. (1996) Exercise testing and prescription. *Sports Med.* **21**(5): 326–336.

Kofler, P., Burtscher, M., Heinrich, D., Bottoni, G., Caven, B., Bechtold, T., Herten, A. T., Hasler, M., Faulhaber, M., Nachbauer, W. (2015) Performance limitation and the role of core temperature when wearing light-weight workwear under moderate thermal conditions. *J. Therm. Biol.* **47**: 83–90.

Kolka, M. A., Levine, L., Stephenson, L. A. (1997) Use of an ingestible telemetry sensor to measure core temperature under chemical protective clothing. *J. Therm. Biol.* **22**(4/5): 343–349.

Lee, Y. M., Barker, R. L. (1986) Effect of moisture on the thermal protective performance of heat-resistant fabrics. *J. Fire Sci.* **4**(5): 315–331.

Lee, J. Y., Kim, S., Jang, Y. J., Baek, Y. J., Park, J. (2014) Component contribution of personal protective equipment to the alleviation of physiological strain in firefighters during work and recovery. *Ergonomics* **57**(7): 1068–1077.

Lee, J. Y., Wakabayashi, H., Wijayanto, T., Tochihara, Y. (2010) Differences in rectal temperatures measured at depths of 4–19 cm from the anal sphincter during exercise and rest. *Eur. J. Appl. Physiol.* **109**(1): 73–80.

Lenhardt, R., Sessler, D. I. (2006) Estimation of mean-body temperature from mean-skin and core temperature. *Anesthesiology* **105**(6): 1117–1121.

Levels, K., Koning, J. J., Foster, C., Daanen, H. A. M. (2012) The effect of skin temperature on performance during a 7.5-km cycling time trial. *Eur. J. Appl. Physiol.* **112**(9): 3387–3395.

Leyenda, B. C., Villa, J. G., Satué, J. L., Marroyo, J. A. R. (2017) Impact of different personal protective clothing on wildland firefighters' physiological strain. *Front. Physiol.* **8**: 618.

Londeree, B. R., Moeschberger, M. L. (1982) Effect of age and other factors on maximal heart rate. *Res. Q. Exerc. Sport* **53**(4): 297–304.

Louhevaara, V., Smolander, L., Tuomi, T., Korhonen, O., Jaakkola, J. (1985) Effects of an SCBA on breathing pattern, gas exchange, and heart rate during exercise. *J. Occup. Med.* **27**(3): 213–216.

Ludwig, N., Formenti, D., Gargano, M., Alberti, G. (2014) Skin temperature evaluation by infrared thermography: Comparison of image analysis methods. *Infrared Phys. Technol.* **62**: 1–6.

McLellan, T. M. (2001) The importance of aerobic fitness in determining tolerance to uncompensable heat stress. *Comp. Biochem. Physiol. Part A Mol. Integr. Physiol.* **128**(4): 691–700.

McLellan, T. M., Selkirk, G. A. (2004) Heat stress while wearing long pants or shorts under firefighting protective clothing. *Ergonomics* **47**(1): 75–90.

Mekjavic, I. B., Rempel, M. E. (1990) Determination of esophageal probe insertion length based on standing and sitting height. *J. Appl. Physiol.* **69**(1): 376–379.

Mell, W. E., Lawson, J. R. (2000) A heat transfer model for fire fighter's protective clothing. *Fire Technol.* **36**(1): 39–68.

Mitchell, D., Wyndham, C. H. (1969) Comparison of weighting formulas for calculating mean skin temperature. *J. Appl. Physiol.* **26**(5): 616–622.

Montain, S. L., Sawka, M. N., Cadarette, B. S., Quigley, M. D., Mckay, J. M. (1994) Physiological tolerance to uncompensable heat stress: Effects of exercise intensity, protective clothing, and climate. *J. Appl. Physiol.* **77**(1): 216–222.

Moran, D. S., Shitzer, A., Pandolf, K. B. (1998) A physiological strain index to evaluate heat stress. *Am. J. Physiol.* **275**(1 Pt. 2): R129–R134.

Nunneley, S. A. (1989) Heat stress in protective clothing. Interactions among physical and physiological factors. *Scand. J. Work Environ. Health* **15**(Suppl. 1): 52–57.

Ohhashi, T., Sakaguchi, M., Tsuda, T. (1998) Human perspiration measurement. *Physiol. Meas.* **19**(4): 449–461.

Pandolf, K. B., Givoni, B., Goldman, R. F. (1977) Predicting energy expenditure with loads while standing or walking very slowly. *J. Appl. Physiol.* **43**(4): 577–581.

Parak, J., Tarniceriu, A., Renevey, P., Bertschi, M., Delgado-Gonzalo, R., Korhonen, I. (2015) Evaluaton of the heat-to-beat detection accuracy of PulseOn wearable optical heart rate monitor. In: Proceedings of the 37th Annual International Conference of the IEEE, Milan, Italy, pp. 8099–8102, doi:10.1109/EMBC.2015.7320273

Petruzzello, S. J., Gapin, J. I., Snook, E., Smith, D. L. (2009) Perceptual and physiological heat strain: Examination in firefighters in laboratory- and field-based studies. *Ergonomics* **52**(6): 747–754.

Pollock, M. L., Foster, C., Knapp, D., Rod, J. L., Schmidt, D. H. (1987) Effect of age and training on aerobic capacity and body composition of master athletes. *J. Appl. Physiol.* **62**(2): 725–731.

Psikuta, A., Niedermann, R., Rossi, R. M. (2014) Effect of ambient temperature and attachment method on surface temperature measurements. *Int. J. Biometeorol.* **58**(8): 877–885.

Pušnik, I., Miklavec (2009) Dilemmas in measurement of human body temperature. *Instrum. Sci. Technol.* **37**(5): 516–530.

Ramanathan, N. L. (1964) A new weighting system for mean surface temperature of the human body. *J. Appl. Physiol.* **19**(3): 531–533.

Robergs, R. A., Landwehr, R. (2002) The supervising history of the "HRmax=220-age" equation. *J. Exerc. Physiol.* **5**(2): 1–10.

Ruddock, A., Tew, G. A., Purvis, A. (2014) Reliability of intestinal temperature using an ingestible telemetry pill system during exercise in a hot environment. *J. Streng. Condition. Res.* **28**(3): 861–869.

Sartang, A. G., Dehghan, H. (2015) Investigating relationship between perceptual strain index with indices heat strain score index, wet bulb globe temperature in experimental hot condition. *Int. J. Health Eng.* **4**: 37.

Shalev, I., Barker, R. L. (1983) Analysis of heat transfer characteristics of fabrics in an open flame exposure. *Text. Res. J.* **53**(8): 475–482.

Shalev, I., Barker, R. L. (1984) Protective fabrics: A comparison of laboratory methods for evaluating thermal protective performance in convective/radiant exposures. *Text. Res. J.* **54**(10): 648–654.

Sköldström, B. (1987) Physiological responses of fire fighters to workload and thermal stress. *Ergonomics* **30**(11): 1589–1597.

Smith, A. D. H., Crabtree, D. R., Bilzon, J. L. J., Walsh, N. P. (2010) The validity of wireless iButton and thermistors for human skin temperature measurement. *Physiol. Meas.* **31**(1): 95–114.

Smith, D. L., Manning, T. S., Petruzzello, S. J. (2001) Effect of strenuous live-fire drills on cardiovascular and psychological responses of recruit firefighters. *Ergonomics* **44**(3): 244–254.

Smith, D. L., Petruzzello, S. J. (1998) Selected physiological and psychological responses to live-fire drills in different configurations of firefighting gear. *Ergonomics* **41**(8): 1141–1154.

Smith, D. L., Petruzzello, S. J., Kramer, J. M., Misner, J. E. (1997) The effects of different thermal environments on the physiological and psychological responses of firefighters to a training drill. *Ergonomics* **40**(4): 500–510.

Smith, D. L., Petruzzello, S. J., Kramer, J. M., Warner, S. E., Bone, B. G., Misner, J. E. (1995) Selected physiological, psychophysical, and psychological responses to physical activity in different configurations of firefighting gear. *Ergonomics* **38**(10): 2065–2077.

Song, W., Wang, F. (2016) The hybrid personal cooling system (PCS) could effectively reduce the heat strain while exercising in a hot and moderate humid environment. *Ergonomics* **59**(8): 1009–1018.

Sothmann, M. S., Saupe, K., Jasenof, D., Blaney, J. (1992) Heart rate responses of firefighters to actual emergencies: Implications for cardiorespiratory fitness. *J. Occup. Med.* **34**(8): 797–800.

Sparling, P. B. (1980) A meta-analysis of studies comparing maximal oxygen uptake in men and women. *Res. Q. Exerc. Sport.* **51**(3): 542–552.

Stewart, I. B., Stewart, K. L., Worringham, C. J., Costello, J. T. (2014) Physiological tolerance times while wearing explosive ordnance disposal protective clothing in simulated environmental extremes. *PLoS ONE* **9**(2): e83740.

Tam, H. S., Darling, R. C., Downey, J. A., Chek, H. Y. (1976) Relationship between evaporation rate of sweat and mean sweating rate. *J. Appl. Physiol.* **41**(5 Pt. 1): 777–780.

Taylor, N. A. S., Lewis, M. C., Notley, S. R., Peoples, G. E. (2012) A fractionation of the physiological burden of the protective equipment worn by firefighters. *Eur. J. Appl. Physiol.* **112**(8): 2913–2921.

Teunissen, L. P. J., de Haan, A., de Koning, J. J., Daanen, H. A. M. (2012) Telemetry pill versus rectal and esophageal temperature during extreme rates of exercise-induced core temperature change. *Physiol. Meas.* **33**(6): 915–924.

Tikuisis, P., McLellan, T. M., Selkirk, G. (2002) Perpetual versus physiological heat strain during exercise-heat stress. *Med. Sci. Sports Exerc.* **34**(9): 1454–1461.

Torvi, D. A. (1997) Heat transfer in thin fibrous materials under high heat flux conditions (Ph.D. thesis). University of Alberta, Edmonton, Alberta, Canada.

Travers, G. J. S., Nichols, D. S., Farooq, A., Racinais, S., Périard, J. D. (2016) Validation of an ingestible temperature data logging and telemetry system during exercise in the heat. *Temperature* **3**(2): 208–219.

Tyler, C. J. (2011) The effect of skin thermistor fixation method on weighted mean skin temperature. *Physiol. Meas.* **32**(10): 1541–1547.

Udayraj, Talukdar, P., Alagirusamy, R., Das, A. (2014) Heat transfer analysis and second degree burn prediction in human skin exposed to flame and radiant heat using dual phase lag phenomenon. *Int. J. Heat Mass Transf.* **78**: 1068–1079.

Udayraj, Talukdar, P., Das, A., Alagirusamy, R. (2016) Heat and mass transfer through thermal protective clothing – A review. *Int. J. Therm. Sci.* **106**: 32–56.

Udayraj, Talukdar, P., Das, A., Alagirusamy, R. (2016a) Simultaneous estimation of thermal conductivity and specific heat of thermal protective fabrics using experimental data of high heat flux exposure. *Appl. Therm. Eng.* **107**: 785–796.

Udayraj, Talukdar, P., Das, A., Alagirusamy, R. (2017) Effect of structural parameters on thermal protective performance and comfort characteristic of fabrics. *J. Text. Inst.* **108**(8): 1430–1441.

Udayraj, Talukdar, P., Das, A., Alagirusamy, R. (2017a) Numerical modeling of heat transfer and fluid motion in air gap between clothing and human body: Effect of air gap orientation and body movement. *Int. J. Heat Mass Transf.* **108**: 271–291.

Vallerand, A. L., Savourey, G., Hanniquet, A. M., Bittel, J. H. (1992) How should body heat storage be determined in humans: By thermometry or calorimetry? *Eur. J. Appl. Physiol.* **65**(3): 286–294.

van Marken Lichtenbelt, W. D., Daanen, H. A. M., Wouters, L., Fronczek, R., Raymannn, R. J. E. M., Severens, N. M. W., van Someren, E. J. W. (2006) Evaluation of wireless determination of skin temperature using iButton. *Physiol. Behav.* **88**(4–5): 489–497.

Vehrs, P. R., George, J. D., Fellingham, G. W., Plowman, S. A., Allen, K. D. (2007) Submaximal treadmill exercise test to predict VO$_{2max}$ in fit adults. *Meas. Phys. Educ. Sci.* **11**(2): 61–72.

von Heimburg, E., Sandsund, M., Rangul, T. P., Reinertsen, R. E. (2017) Physiological and perceptual strain of firefighters during graded exercise to exhaustion at 40°C and 10°C. *Int. J. Occup. Saf. Ergon.*, doi: 10.1080/10803548.2017.1381468

Walker, A., Keene, T., Argus, C., Driller, M., Guy, J. H., Rattray, B. (2015) Immune and inflammatory responses of Australian firefighters after repeated exposures to the heat. *Ergonomics* **58**(12): 2032–2039.

Wang, F., Annaheim, S., Morrissey, M., Rossi, R. M. (2014) Real evaporative cooling efficiency of one-layer tight-fitting sportswear in a hot environment. *Scand. J. Med. Sci. Sports* **24**(3): e129–e139.

Wang, F., Gao, C. (2014) *Protective clothing: Managing thermal stress.* Cambridge: Woodhead Publishing.

Wang, F., Gao, C., Kuklane, K., Holmér, I. (2013) Effects of various protective clothing and thermal environments on heat strain of unacclimated men: The PHS (predicted heat strain) model revisited. *Ind. Health* **51**(3): 266–274.

Wang, F., Kuklane, K., Gao, C., Holmér, I. (2011) Can the PHS model (ISO7933) predict reasonable thermophysiological responses while wearing protective clothing in hot environments. *Physiol. Meas.* **32**(2): 239–249.

Wen, S., Petersen, S., McQueen, R., Batcheller, J. (2015) Modelling the physiological strain and physical burden of chemical protective coveralls. *Ergonomics* **58**(12): 2016–2031.

White, M. K., Hodous, T. K. (1987) Reduced work tolerance associated with wearing protective clothing and respirators. *Am. Ind. Hyg. Assoc. J.* **48**(4): 304–310.

White, M. K., Hodous, T. K. (1988) Physiological responses to the wearing of fire fighter's turnout gear with neoprene and GORE-TEX® barrier liners. *Am. Ind. Hyg. Assoc. J.* **49**(10): 523–530.

White, M. K., Hodous, T. K., Vercruyssen, M. (1991) Effects of thermal environment and chemical protective clothing on work tolerance, physiological responses, and subjective ratings. *Ergonomics* **34**(4): 445–457.

Wilkinson, D. M., Carter, J. M., Richmond, V. L., Blacker, S. D., Rayson, M. P. (2008) The effect of cool water ingestion on gastrointestinal pill temperature. *Med. Sci. Sports Exerc.* **40**(3): 523–528.

Winslow, C. E. A., Herrington, L. P., Gagge, A. (1936) A new method of partitional calorimetry. *Am. J. Physiol.* **116**(3): 641–655.

Winslow, C. E. A., Herrington, L. P., Gagge, A. (1937) Relations between atmospheric conditions, physiological reactions and sensations of pleasantness. *Am. J. Hyg.* **26**(1): 103–115.

Xu, X., Gonzalez, J. A., Santee, W. R., Blanchard, L. A., Hoyt, R. W. (2016) Heat strain imposed by personal protective ensembles: Quantitative analysis using a thermoregulation model. *Int. J. Biometeorol.* **60**(7): 1065–1074.

Yeoh, W. K., Lee, J. K. W., Lim, H. Y., Gan, C. W., Liang, W., Tan, K. K. (2017) Re-visiting the tympanic membrane vicinity as core temperature measurement site. *PLoS ONE* **12**(4): e0174120.

Young, A. J., Sawka, M. N., Epstein, Y., Decristofano, B., Pandolf, K. B. (1987) Cooling different body surfaces during upper and lower body exercise. *J. Appl. Physiol.* **63**(3): 1218–1223.

Zwiren, L. D., Freedson, P. S., Ward, A., Wilke, S., Rippe, J. M. (1991) Estimation of VO2max: A comparative analysis of five exercise tests. *Res. Q. Exerc. Sport.* **62**(1): 73–78.

9

3D Body Scanning Technology and Applications in Protective Clothing

Agnes Psikuta, Emel Mert, Simon Annaheim, and René M. Rossi

9.1 Introduction to Recent Development of 3D Body Scanning Technology

The first 3D whole body scanners appeared on the market already at the end of the previous century, and they were generally bulky, expensive, and with relatively low accuracy of a few millimeters (Daanen and van de Water 1998). The technology has improved in terms of scanner resolution and accuracy over the last decades. Besides the first basic technologies such as laser scanning, structured light projection, and stereophotogrammetry, new techniques, such as millimeter- and infrared-waves, emerged. The major improvement in all techniques included speed of scanning allowing capturing the body movement in some cases, color capturing, increased resolution, and accuracy to below 1 mm at simultaneously reduced cost. Second, many devices are coupled with dedicated software facilitating data handling, scan repair, and even deduction of the body dimensions (Daanen and Ter Haar 2013). This rapid development of the 3D scanning technologies continues to progress as shown by the launching in 2010 of the dedicated yearly international conference on 3D Body Scanning Technologies, the number of topics being addressed during this conference, and several new 3D scanners being launched every year.

The digitization of the human body surface has found successful application in various scientific and industrial fields, such as apparel industry for improvement of sizing charts, made-to-measure technologies and virtual try-on, in medicine for enhancement of plastic surgery outcomes, orthesis and prosthesis design, in animation industry for development of digital movies, and in fitness industry for tracking fat reduction and muscle build-up. In the field of protective clothing, this technology helped to select the right size of the clothing for individuals, reduce the number of sizes and stock of the clothing through anthropometric surveys, and optimize the clothing fit for males and females. On the other hand, a new opportunity for simulation of thermal effects in protective clothing emerged with a possibility to visualize and quantify the air layers within the ensemble that plays a crucial role for heat and mass transfer in clothing.

Before the development of the 3D scanning technology the determination of the size or the volume of the air layers required cumbersome and time-consuming methods, that is, vacuum suit method (Sullivan et al. 1987; Birnbaum and Crockford 1977; Crockford et al. 1972), photographic method (Kakitsuba 2004), or string method (McCullough et al. 1985). The 3D body scanning technique combined with 3D post-processing software offered a high-precision, non-invasive, and fast method to digitalize and analyze the spatial form of the dressed body, and hence, allowed investigating the distribution of the air layer thickness. To obtain the size of the air gap the 3D scans of the nude and dressed body were superimposed and the air gap thickness was calculated from either a selected number of points or from a discrete number of cross sections through the dressed body (Kim et al. 2002; Song 2007; Wang et al. 2006; Xu and Zhang 2009) or for the entire surface at high accuracy (Psikuta et al. 2012; Psikuta et al. 2015).

9.2 Clothing Air Gap and Contact Area

9.2.1 Importance of Air Gap and Contact Area for Heat and Mass Transfer in Clothing

The heat and mass transfer within the clothing system is a composition of a number of physical processes, such as sensible heat exchange (conduction, convection, and radiation), evaporation and condensation, sorption, as well as vapor and liquid water transfer (Whitaker 1998). In addition, factors associated with construction and use of a garment, such as air penetration and compression by wind, body posture and movement, and clothing fit, influence these processes significantly (Bouskill et al. 1998a; Holmer et al. 1999; Parsons et al. 1999). That is mainly due to the changing size and shape of the layers of air trapped between the skin and clothing, between clothing layers alone, and in the layer adjacent to the outer surface of clothing. The garment is a three-dimensional (3D) form created from the two-dimensional (2D) pattern on the flat fabric to cover complex geometry of the human body. This fact together with the fabric properties entails draping and sagging of the garment. Effectively, the thickness of the air layer is heterogeneous and varies over body parts and changes with body shape, posture, and movement. The heterogeneous thickness of the air layers within the clothing system influences the local heat and vapor exchange. Thus, the thermal, evaporative, and wicking properties of clothing depend not only on the properties of the fabric used for the garment but also on the magnitude and the temporal change of the contact area and air layer thickness.

The sensible heat loss through a stagnant air layer is transferred by conduction, radiation, and convection. The conductive heat loss rate through a stagnant air layer depends linearly on the distance between the human skin and the garment as well as the temperature gradient. The radiative heat transfer rate depends nonlinearly (to the power of four according to Stefan-Boltzmann law) on the temperature difference between human body and garment and the emissivity of both surfaces (Wissler and Havenith 2009). The convective heat transfer underneath the clothing can occur as forced convection by air volume displacement underneath the garment by wind compression or body movement (Danielsson 1993), and/or natural convection due to the buoyancy effect (Cengel and Ghajar 2007). Natural convection occurs only at a certain air gap thickness when the buoyancy force overcomes the shear and tensile forces in the fluid, which is estimated to be larger than about 8 mm by some authors from the textile research field (Spencer-Smith 1977; Bergman et al. 2011) and over 20 mm in engineering practice (Elsherbiny et al. 1982; Wakitani 1998; Shewen et al. 1996) and other clothing studies (Mert et al. 2015). Furthermore, when high-temperature gradients can occur, for example, in heat-protective clothing, the air gap thickness threshold for the natural convection will be lower. Such studies using the benchmark tests showed that the natural convection underneath the fabric system occurred between 6.4 and 13 mm of air gap thickness for temperature gradients of 400°C and 20°C, respectively (Cain and Farnworth 1986; Torvi et al. 1999).

The example in Figure 9.1 illustrates the significance of the air gap thickness for the sensible heat and vapor transfer in the one-layer clothing system (Figure 9.1a). The air trapped beneath garments and adjacent to the outer layer provides the bulk of both the thermal and evaporative resistances compared to the fabric [the specific thermal resistance of air is 39 m·K/W, and of a typical fabric of the same width as the air is 24 m·K/W (Lotens and Havenith 1991), the evaporative resistance increases by 2.3 m²·Pa/W per 1 mm of air layer thickness (Wissler and Havenith 2009; Psikuta et al. 2012)]. The thermal resistance increases inversely proportionally to the air gap thickness with the largest increase at low air gap values. The evaporative resistance of an air layer is directly proportional to its thickness and increases noticeably with only a small change in the air gap thickness (Figure 9.1b). The sensible heat transfer coefficient in the air gap changes nonlinearly with its increase being highly dependent on the air gap thickness in the range of 0 to about 15 mm and stabilizing at nearly constant value for air gaps larger than 30 mm. This trend is mainly due to change in conductive heat transfer coefficient with radiant and convective coefficients being nearly stable throughout the relevant range of air gap thickness (Figure 9.1c). Finally, Figure 9.1d shows an estimated share of each sensible heat transfer mechanism occurring in the air gap in the total sensible heat loss demonstrating the radiant heat transfer having the highest share of about 65% for air gaps larger than about 5 mm. The conductive pathway decreases inversely proportionally to have the smallest share for air gaps larger than about 12 mm, where the convective pathway joins in and stays on average at about 35% of the total heat loss.

The thermal conductivity is smaller for gases, such as stagnant air underneath the garment, while being higher for the solids such as fabrics, which can be in direct contact with the human body. Thus, the direct contact between the human body and the garment increases the total sensible heat loss owing to the effectively greater thermal conductivity of the porous fabric (Ismail 1988).

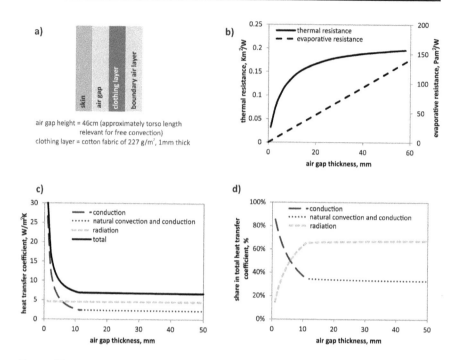

Figure 9.1

Thermal and evaporative resistances (b), sensible heat transfer coefficients (c), and estimated share of various sensible heat transfer mechanisms in total sensible heat loss (d) through a homogenous vertical air gap of 46 cm height at ambient temperature of 20°C and skin temperature of 34°C when wearing single layer garment made of a cotton fabric (227 g/m², 1 mm thick) (a).

Additionally, when the surfaces of the fabrics and/or the surface of the skin stay in contact, direct exchange of the liquid water through wicking can take place (Keiser et al. 2008; Umeno et al. 2001; Weder et al. 2008) and enhance additionally the heterogeneous heat and mass transfer (Umeno et al. 2001). Therefore, it is important to consider the spatial heterogeneity of the air gaps in mathematical clothing models and clothing design. The quantification of the contact area and air gap thickness (Psikuta et al. 2012; Mah and Song 2010d) has been enabled by the 3D scanning technique and advanced post-processing method. However, the thermal effect of vertical air layers with complex geometry was investigated only recently by Mert et al. (2015) and demonstrated that homogenous air gaps were more effective insulators than folded fabrics with the same average air gap. Additionally, although the presence of folded fabrics increased the sensible heat loss through the fabric, sizes of folds were not a crucial parameter for clothing insulation. Moreover, the sensible heat loss through the folded fabric increased with the increase in the contact area only when the contact area exceeded a certain threshold of about 42% in case of 30 mm fold thickness. On the other hand, existing mathematical clothing models assume a uniform air gap between the body and fabric layers or its full contact (Fan et al. 2000; Li and Holcombe 1998; Lotens and Havenith 1991). Such simplification facilitates the computation process but it disregards the non-uniform heat, vapor,

9. 3D Body Scanning Technology and Applications in Protective Clothing

and liquid water transfer, which depends on the presence of contact between surfaces (Umeno et al. 2001) and on the shape of the air layers trapped within clothing (Bouskill et al. 2002).

On the personal protective clothing side, air layers are of high relevance for the protection performance of the clothing (Song 2007; Mah and Song 2010d; Mah and Song 2010a). Therefore, the effect of the air gap thickness in personal protective clothing systems on its thermal protective performance has been widely investigated in various studies using bench-top tests on horizontal plane setup (Torvi et al. 1999; Lee et al. 2002; Fu et al. 2014), cylindrical system (Zhu et al. 2008; Zhu et al. 2009), flash fire manikin tests (Song 2007; Mah and Song 2010d; Mah and Song 2010a), and mathematical models (Ghazy and Bergstrom 2011, 2012; Ghazy 2014b). It has been shown that the predicted time to suffer from skin burns increases with the larger air gap thickness due to the decreased heat radiation and conduction due to the increased total thermal resistance of the clothing system (Fu et al. 2014). Based on data from flash fire manikin tests, it was found that the most severe burn injuries occur at body regions with smaller air gaps, and it was confirmed that the transferred energy to the skin decreases with the increase in the air gap thickness (Song 2007; Rossi et al. 2014). However, after a certain time the air gap thickness could vary due to the shrinkage of the fabric at high heat exposure, and hence, the time to observe skin burn decreases and the amount of the transferred energy increases (Ghazy 2014a; Song 2007). Also, increased free convection in some larger air gaps may play a more substantial role than depicted in Figure 9.1 due to possible higher temperature gradients between the body and the inner clothing layer. Consequently, the contribution of the air layer to thermal protective performance of the personal protective garment was higher with larger air gap thickness, whereas it was lower with smaller air gap thickness and the shrunken air layers at higher heat exposures.

9.2.2 Research Up-to-Date

Several studies have attempted to characterize the air layers in garments (Daanen et al. 2005; Havenith et al. 2010; Kakitsuba 2004; Kim et al. 2002; Lee et al. 2007; Song 2007; Wang et al. 2006; Xu and Zhang 2009; Zhang and Li 2010; Bouskill et al. 1998b). In these investigations only an average air gap thickness has been estimated for the whole garment from either the air volume trapped within the garment (Daanen et al. 2005; Havenith et al. 2010; Bouskill et al. 1998b; Lee et al. 2007) or the ratio between clothing and body surface area (Kakitsuba 2004). With the development of 3D scanning technologies (Daanen and Ter Haar 2013; Daanen and van de Water 1998), new opportunities for the contactless visualization and analysis of air layers emerged. The distribution of the air layer thickness was investigated using a 3D body scanning technique combined with a 3D post-processing software package that offered a high-precision, non-invasive, and fast method to digitalize and analyze the spatial form of the dressed body. To obtain the size of the air gap, the 3D scans of the nude and dressed body were superimposed and the air gap thickness was calculated from either a selected number of points or from a discrete number of cross sections through the dressed body (Song 2007; Wang et al. 2006; Xu and Zhang 2009; Zhang and Li 2010; Kim et al. 2002). The new 3D measurement technique was applied in apparel development for anthropometric measurements, automated body landmarking, and fit assessment (Choi and Ashdown 2011; Song and Ashdown 2012), in protective clothing research for burn risk prediction (Mah and Song 2010a; Song 2007) and modeling of thermal interaction in the human–clothing–environment system. However, none of the aforementioned methods allowed the systematic, accurate, and detailed evaluation of the local and average air gap thickness, nor did they address the issue of contact area in ensembles.

In the study by Psikuta et al. (2012) a method to accurately determine the air gap thickness and the contact area between clothing and the human body through an advanced analysis of 3D body scans of the nude and dressed body of a male manikin was introduced for the first time. This method allowed more accurate measurement of the air gap thickness and the contact area than other existing methods. Additionally, both parameters could be obtained for individual body parts or any number of body regions. Due to the high insulation capacity of air and the need for consideration of the fabric thickness for accurate determination of contact area between garment and skin, the thickness of the air gap for such applications should be investigated in the order of millimeters or tenth of millimeters; any inaccuracies higher than that can be critical. There are 3D whole body scanners available on the market that provide scanning accuracy of up to 5 mm according to the manufacturers (e.g., VITUS XXL of up to 1 mm, TC2 of up to 3 mm, or Size Stream 3D Body Scanner of up to 5 mm). Accuracy can be compromised when scanning dark or reflective surfaces or in adverse lighting conditions. Subsequent post-processing of 3D scans may also lead to further loss of scan accuracy. To achieve the relevant required accuracy of the newly developed method, including scanning, post-processing of scans, and determining distance between skin and garment surfaces, was extensively tested and proven to provide a high repeatability of maximum up to 1 mm depending on the scanner (Psikuta et al. 2015). Thus, the method can account for the thickness of fine fabrics of the same thickness magnitude as the scanner accuracy, which results in accurate determination of the garment contact area. Several further studies describing the local distribution of the air gap thickness and the contact area on upper and lower body have been conducted using this method (Bohnet 2013; Frackiewicz-Kaczmarek et al. 2015a, b; Mark 2013; Mert et al. 2016a; Mert et al. 2016b). They emphasized a regional systematic trend in the distribution of these two parameters in relation to the garment ease allowance (the difference between the body and the garment girths) at the corresponding body landmarks.

9.2.3 Definitions of Air Gap Thickness and Contact Area

The digital data of the 3D scanned body represents originally a cloud of points. The air gap thickness is determined in the post-processing phase as an average distance between points on the surface of nude and dressed manikin from two super-imposed scans (Figure 9.2). Such averages can be determined for the entire body, the entire garment, individual body parts, or any arbitrary body region even with a small resolution separately.

This method also provides an opportunity to determine the fraction of the surface area of the manikin that stayed in contact with the garment. To describe this parameter, the contact area (CA) between the garment and the skin was defined as a ratio coefficient:

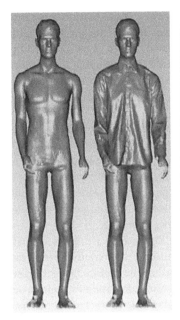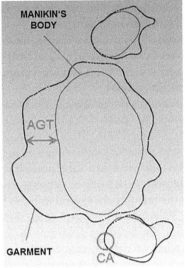

Figure 9.2

3D scans of the nude and dressed manikin in identical body position to be super-
imposed for determination of the air gap thickness (AGT) and the contact area
(CA) as well as a cross section through the aligned scans at waistline level indicating the sought parameters.

$$CA = \frac{A_{contact}}{A_{covered}}$$

where $A_{contact}$ is the skin area in direct contact with a garment and $A_{covered}$ is the
skin area covered by this garment. The surface area of the body in contact with
garment was determined based on the distance between the surfaces of the nude
and dressed manikin. Theoretically, if the distance between the surface of the
dressed manikin and the surface of the nude manikin is equal to the thickness
of the fabric, the air gap thickness beneath the garment is zero, and it can be
classified as contact between skin and garment. However, uncertainties of the
individual phases of the evaluation were in the order of magnitude of the fabric
thickness and had to be accounted for. These included the inaccuracy of the 3D
scans given by the manufacturer of the 3D scanner and the average alignment
error resulting from imperfect imposing of scans using uncovered body parts as
the reference shapes (Psikuta et al. 2012). These lumped measurement uncertainties are added to the fabric thickness of the garment and their sum defined as
the distance between the skin and the garment recognized as contact area. The
methodology of obtaining the exact accuracy for determination of the air gap
thickness and the contact area of this dedicated method for a given 3D scanner
was described in detail by Psikuta et al. (2015).

9.2.4 Methodology of 3D Scanning and Post-Processing

To ensure the best possible quality of the scans, the 3D scanner needs to be used
according to the guidelines of the producer with regards to operating conditions

including ambient temperature, light, frequency of calibration, operation time in one scanning session, etc. The manikin has to be set in the identical posture for nude and dressed situations to be scanned and to allow proper alignment of the scans. This can be achieved, for example, by equipping the manikin with additional locks against turning of body parts and an external construction supporting its arms and feet in a fixed position to ensure identical body positioning for all measurements.

The scanning protocol includes the scanning of the nude and dressed manikin with sample garments using the 3D scanner. To assess the reliability of the proposed technique, the manikin is redressed and scanned four to six times for each garment to allow for random changes in garment drape to analyze the repeatability of the measurements. After putting on a garment, all unusual positions of the garment, such as twisting or clinging, are removed and the garment is allowed to rest on the manikin for several minutes before scanning.

The 3D scans of nude and dressed manikins are post-processed in the next step using the dedicated software. The intensity of the required post-processing is dependent on the 3D scanner brand as well as the surface inspection software chosen for alignment of the manikin scans and calculation of their surface distance to obtain the required parameters such as the contact area and the air gap thickness. This step is challenging due to the complexity of 3D forms counterbalanced by available computing power, a high precision aligning of the 3D scans despite surface deficiencies, and repeatable slicing of the irregular shape of the dressed body into the body parts. Consequently, many manipulations of the 3D scans have to be done manually. Generally, the post-processing procedure consists of the following steps:

Step 1: cleaning of the 3D scan surface by removing scanning artefacts and closing surfaces with deficiencies;

Step 2: super-imposing of 3D scans of nude and dressed manikin in the 3D space using uncovered body parts as the reference shapes;

Step 3: slicing super-imposed manikins into regions, for which the parameters were sought, the division usually corresponds to the boundaries of body coverage for typical garments (neck, wrists, ankles) and the body shape change that induces a different draping pattern of the garments (chest, elbows and knees);

Step 4: computation of the contact area and the distribution of the air gap thickness for each region.

9.2.5 Model of Air Gap Thickness and Contact Area

Several studies have attempted to quantify the air layer distribution in a variety of garments (Bohnet 2013; Frackiewicz-Kaczmarek et al. 2015b; Mark 2013; Mert et al. 2016a; Mert et al. 2016b). They emphasized regional systematic trends in the distribution of air gap thickness and contact area in relation to the garment ease allowance (the difference between the body and the garment girths) at corresponding body landmarks. These trends were consistent and homogeneous between various garment fits and types. This fact together with the observed low variability between 3D scanning repetitions despite complete redressing of the manikin indicated a high potential for a precise correlation model. Such a model was developed based on the database of the garments available in the literature

(Psikuta et al. 2018). Several studies that used 3D body scanning techniques to determine the sought parameters came up from the same laboratory (although measured on two different manikins and using two different 3D scanners), and hence, they represented high consistency of methodology and output data. In total, the database of garments comprised 51 clothing pieces including 28 upper body garments, 22 lower body garments, and one coverall. The individual repetitions were gathered in a data pool over which an exploratory data analysis was done showing a linear dependence of the air gap thickness and the contact area on the ease allowance at the corresponding body landmarks. In the next step, the linear regression analysis was conducted, including the uncertainty evaluation. The exemplary analysis for two contrasting body regions, such as upper back with its small constant air gap thickness and large contact area and lumbus with its larger air gap thickness highly dependent on garment ease allowance and low contact area, is depicted in Figure 9.3. Table 9.1 shows the linear regression coefficients for individual body regions related to ease allowance according to Equations 9.1 and 9.2.

$$AGT = slope_{AGT} \cdot EA + intercept_{AGT} \tag{9.1}$$

$$CA = slope_{CA} \cdot EA + intercept_{CA} \tag{9.2}$$

where AGT is the air gap thickness in mm, CA is the contact area in % of the total region area, EA is the ease allowance of the garment at corresponding body region

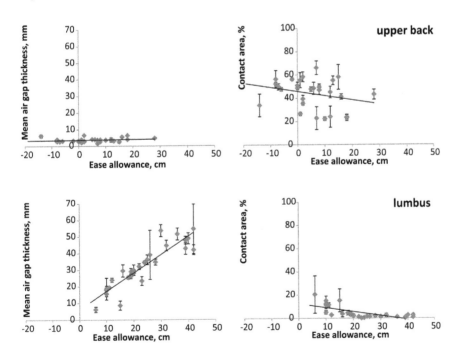

Figure 9.3

Air gap thickness and contact area with standard deviation in relation to the ease allowance at corresponding landmarks of the individual garments for upper back and lumbus including linear regression lines included in the model in Table 9.1.

Table 9.1 Coefficients of the Linear Regression Analysis for Air Gap Thickness and Contact Area Related to Ease Allowance for Individual Body Regions at Upper and Lower Body

	Body Region		Upper Body										Lower Body					
			Upper Chest	Lower Chest	Abdomen	Anterior Pelvis	Upper Back	Lower Back	Lumbus	Posterior Pelvis	Upper Arm	Lower Arm	Anterior Pelvis	Posterior Pelvis	Anterior Thigh	Posterior Thigh	Shin	Calf
Air gap thickness	Slope	mm/cm	0.1	0.2	0.6	0.7	0	0.3	1.1	1	0.9	1.1	0.3	0.2	1.1	1.4	1.8	0.7
	intercept	mm	4.4	5	0.2	10.1	3.5	8.1	7.2	11.6	2.9	5.6	4.1	4.1	-0.6	1.3	8.5	7.2
Contact area	Slope	%/cm	-0.6	-1.1	-0.6	-1.2	-0.4	-1.2	-0.3	-1	-1.7	-1	-1.5	-1.5	-4.6	-3.7	-1.5	-2.2
	intercept	%	33.8	45.3	23.9	23.5	45.5	34	12	20	34.9	18.9	29.3	38.3	59.6	44.1	17.9	36.3

9. 3D Body Scanning Technology and Applications in Protective Clothing

in cm, and slope and intercept are the linear equation coefficients for individual body regions as given in Table 9.1.

The dependence of air gap thickness on ease allowance is minimal at some body parts such as upper and lower chest and back as well as anterior and posterior pelvis approximating 0.1–0.3 mm/cm of ease allowance (Table 9.1). This means that regardless of the garment fit and the fabric used, the air gap thickness remains nearly constant at these body parts due to gravity force (upper chest and back) or snug adjustment to stay in place (pelvis) (Psikuta et al. 2012). At the remaining body parts, the air gap thickness increased proportionally with an increase in the ease allowance at the rate between 0.6 and 1.8 mm/cm of ease allowance (Table 9.1). Similar but reversed trends were observed for the correlation between contact area and ease allowance (negative slope, Table 9.1). That means that the contact area between the body and the garment decreases with the increase in the ease allowance. The uniformity of contact area at the upper chest and back through a variety of garments is due to gravitational force compelling the fabric to rest on the body, whereas the concavity of the body under protruding body region is responsible for low contact area at lumbus and abdomen.

Figure 9.4 shows the results of the error propagation study and demonstrates the influence of the standard error in prediction of the air gap thickness and contact area on garment thermal and evaporative resistance simulated using the model described by Mert et al. (2015). The dotted lines show the change in both parameters due to the predicted standard error of 1.2 and 6.7 mm for upper back and lumbus within the range of air gap thickness occurring at these body regions (2–8 and 6–55 mm, respectively; Figure 9.3). The effect of the standard error on thermal resistance is non-uniform throughout various air gap thicknesses, which is related to nonlinear radiative heat transfer. The evaporative resistance is uniformly affected by the prediction error throughout the different air gap thickness, which is related to the linear influence of the air gap thickness on water vapor diffusion between limiting clothing layers.

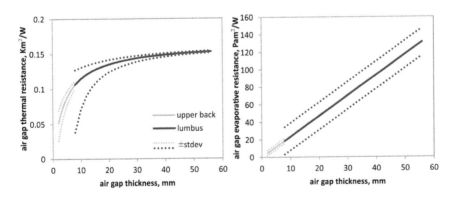

Figure 9.4

Thermal and evaporative resistances simulated using model described by Mert et al. (2015) at T_{air} = 20 °C, T_{skin} = 35°C, v_{air} = 0.2 m/s, for single fabric layer with R_{cf} = 0.05 m²·K/W and variable air gap thickness including change of these parameters due to standard error of air gap prediction in the model for upper back and lumbus.

9.3 Effect of Air Gap on Clothing Thermal Performance

To confirm the relevancy of the regression model for predicting thermal and evaporative properties of fabric-air-layer assemblies as well as thermophysiological response once applied as clothing characteristics on the human, further error propagation studies were carried out using fabric assembly model by Mert et al. (2015) and human thermoregulation model by Fiala and Havenith (2016). The scheme of the use of models at different complexity level and their inputs and outputs is shown in Figure 9.5.

The majority of clothing models reported in the literature assumes either a full contact between the garment and the body or a certain (arbitrary) homogenous air gap. The problem of unknown air gap was first addressed by Havenith et al. (2010) who used a tracking gas method to estimate total air volume trapped underneath the clothing. This information allowed the calculation of a realistic average air gap thickness of a given ensemble. Some detailed fabric models with integrated physiological model suggested using no air gap for tight clothing assuming that the entire garment is in full contact with the body or some arbitrary values for an average air gap thickness (e.g., Fan et al. 2000). However, in reality, the air gap thickness is much larger (Frackiewicz-Kaczmarek et al. 2015b; Mert et al. 2016a). For example, an ensemble consisting of tight shirt and tight jeans represents an average air gap thickness of 6.8 mm, and another ensemble consisting of loose shirt and loose jeans represents an average air gap thickness of 16.6 mm. Such a difference has a substantial impact on heat and mass transfer through the air gap and the entire clothing system. Second, even if realistic average air gap is used, there is still great variation of the air gap over body parts as demonstrated by the distinct examples in Figure 9.3 and variety of slopes in Table 9.1. In the case of the aforementioned exemplary ensembles the variability among body parts varied between 2.8 and 17.9 mm for the front thigh and lumbus, respectively, for tight-fitting ensemble (mean of 6.8 mm) and between 6.2 and 37.2 mm for the upper chest and lumbus, respectively, for loose-fitting ensemble (mean of 16.6 mm). The physiological effect of such variability can be quite significant especially when it

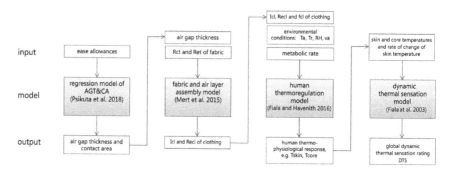

Figure 9.5

The scheme of simulation order using a cascade of various models (Psikuta et al. 2018; Mert et al. 2015; Fiala and Havenith 2016; Fiala et al. 2003) as well as required and resultant input and output parameters, where *Rct, Ret, Icl,* and *Recl* are thermal and evaporative resistances of the fabric and clothing at individual body regions, respectively, *fcl* is the local clothing area factor determined by 3D scanning, *Ta, Tr,* RH, and *va* are ambient and radiant temperatures, relative humidity, and air velocity, respectively.

occurs at large areas and/or physiologically significant body parts, such as, for example, trunk. This issue was discussed in detail by Psikuta et al. (2018).

Since both thermal and evaporative resistances are greatly affected by the magnitude of the air gap thickness, it is expected that this effect will be also pronounced in human thermal response when wearing various ensembles. Several options for calculation of the air gap thickness that researchers have available at the moment were evaluated to demonstrate the improvement of the simulation accuracy when using the presented air gap distribution model over practice and data available up to date. A frequent assumption found in literature is attributing 0 mm air gap thickness to tight-fitting clothing (e.g., Fan et al. 2000, Lotens et al. 1991). As seen in Table 9.1 (intercept), this assumption is nearly not represented at any body region even if ease allowance was equal to zero.

In the simulated example shown in Figure 9.6, the mean and rectal temperatures comparison between garment with a hypothetical 0 mm air gap and realistic tight- and loose-fitting ensembles are shown. It can be seen that the difference between the thermophysiological response for the assumed no-air-gap case is greatly different than that for the tight-fitting ensemble. The observed differences approximated in the range 0.5–2.3°C, 0.7–3.2°C, 1.2–5.8°C, 0.02–0.19°C, 0.0–1.4, and 0.03–0.27 for mean, upper back, and lumbus skin and body core temperatures, dynamic thermal clothing (DTS), and skin wetness, respectively. Such substantial differences are physiologically relevant and may lead to false estimation of the resultant thermophysiological state of the human body. Although the differences for mean skin and body core temperatures were rather small (within experimental error in human studies), the local skin temperature differences can be critical for local and overall thermal sensation and comfort prediction. Finally, the effect of the clothing fit was evident in the thermophysiological and

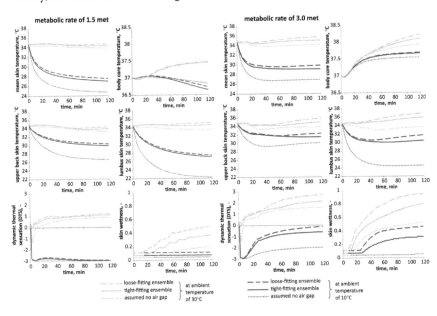

Figure 9.6

Mean skin, body core, upper back, and lumbus skin temperatures, dynamic thermal sensation, and skin wetness predicted for assumed no-air-gap case and an air gap distribution in tight- and loose-fitting ensembles for a combination of the environmental conditions (10 and 30°C) and activity levels (1.5 and 3 Mets).

sensational response showing the differences of up to 0.9°C, 0.9°C, 1.3°C, 0.13°C, 0.6, and 0.16 for mean, upper back, and lumbus skin and body core temperatures, DTS, and skin wetness, respectively, in the tested scenarios (Figure 9.6). This comparison proves that the dedicated design of the clothing can have an influence on the human thermophysiological response beyond the statistical error, and hence can be used to customize and to support the desired thermal clothing effects. These differences would inflate toward more extreme scenarios with higher metabolic rates, more water vapors impermeable clothing, and higher ambient temperature and the relative humidity in hot exposures, and low metabolic rates and low clothing thermal resistance in cold exposures.

The issue of ensemble fit was until now not treated as a parameter to model thermal behavior, and hence, protective performance of ensembles but rather as a resultant feature associated with wearing comfort and garment ergonomics. This series of studies demonstrated that besides ergonomic considerations, designing of the thermal processes in the clothing is possible to a great extent. Since the air gap provides the bulk of thermal and evaporative resistances to the clothing system, even small changes in its thickness may result in considerable change in heat loss as shown in Figure 9.1b and 9.1c. The human body is sensitive to the range of air gap thickness observed in typical casual and protective clothing as demonstrated in Figure 9.6. This sensitivity may be even more pronounced for more extreme scenarios or working conditions for which the capacity of the human body for dealing with heat or cold may be at the limit and the clothing performance may play a decisive role for the duration of exposure until failure and associated well-being and health risks. Furthermore, the quantification of the air gaps and related thermal processes in the clothing at the design stage provides a unique opportunity to influence these processes and consciously configure the ensemble for a given purpose to mitigate the adverse effects of the exposure.

9.4 Summary

This chapter presented the progress in assessment and simulation of the protective clothing performance, thanks to the emerged 3D body scanning technique and dedicated 3D data processing software. This technique was used in several studies allowing the development of an advanced model of air gap thickness able to predict the distribution of the air gap thickness and the contact area parameters locally and reliably. For the first time the available results on these parameters were summarized and further compiled into a statistical model predicting the air gap thickness and contact area for 14 body regions based on the ease allowances for the given body landmarks. Such a reliable and detailed model is crucial for mathematical models of heat and mass transfer in clothing to realistically predict thermal behavior of the clothing system and its possible impact on human thermal and perceptual response. It was demonstrated for several selected scenarios, including different clothing fit levels, that the up-to-date assumptions and methods for determination of the air gap thickness can produce a substantial error for all whole-body, mean, and local physiological parameters, and hence lead to false estimation of the resultant physiological state of the human body, thermal sensation, and comfort. Thus, the use of this detailed model shall contribute to the improvement of the simulation of human thermophysiological and perceptual response to thermal environments in occupational settings and enhance the clothing design for protective and functional apparel to balance the environmental and bodily influence on clothing performance. This tool and the developed approach shall help researchers improve human thermal comfort, health, productivity, safety, and overall sense of well-being when wearing clothing.

References

Bergman, T. L., Lavine, A. S., Incropera, F. P., Dewitt, D. P. (2011) *Fundamentals of Heat and Mass Transfer.* Wiley, New York Seventh Edition.

Birnbaum, R. R., Crockford, G. W. (1977) Quantitative-evaluation of clothing micro-environment air exchange. *Ergonomics* 20 (5):562–562.

Bohnet, A. (2013) Distribution of the air gap thickness and contact area in male and female clothing (BSc thesis). Albstadt-Sigmaringen University, Albstadt.

Bouskill, L., Livingston, R., Parsons, K., Withey, W. R. (1998a) The effect of external air speed on the clothing ventilation index. In: Hanson, M. A. (ed) *Contemporary Ergonomics.* Taylor & Francis Ltd, London, pp. 540–544.

Bouskill, L., Sheldon, N., Parsons, K., Withey, W. R. (1998b) The effect of clothing fit on the clothing ventilation index. In: Hanson, M. A. (ed) *Contemporary Ergonomics.* Taylor & Francis Ltd, London, pp. 510–514.

Bouskill, L. M., Havenith, G., Kuklane, K., Parsons, K. C., Withey, W. R. (2002) Relationship between clothing ventilation and thermal insulation. *Aihaj* 63 (3):262–268.

Cain, B., Farnworth, B. (1986) Two new techniques for determining the thermal radiative properties of thin fabrics. *Journal of Thermal Insulation* 9:301–322.

Cengel, Y. A., Ghajar, A. J. (2007) *Heat and mass transfer fundamentals and applications.* Boston, McGraw-Hill 4th edition.

Choi, S., Ashdown, S. P. (2011) 3D body scan analysis of dimensional change in lower body measurements for active body positions. *Textile Research Journal* 81 (1):81–93. doi:10.1177/0040517510377822

Crockford, G., Crowder, M., Prestidge, S. (1972) Trace gas technique for measuring clothing microclimate air exchange rates. *British Journal of Industrial Medicine* 29 (4):378.

Daanen, H., Hatcher, K., Havenith, G. (2005) Determination of clothing microclimate volume. In: Tochihara, Y., Ohnaka, T. (eds) *Environmental Ergonomics - The Ergonomics of Human Comfort, Health and Performance in the Thermal Environment,* vol 3. Elsevier Ergonomics Book Series. pp. 361–365.

Daanen, H. A. M., Ter Haar, F. B. (2013) 3D whole body scanners revisited. *Displays* 34 (4):270–275. doi:10.1016/j.displa.2013.08.011

Daanen, H. A. M., van de Water, G. J. (1998) Whole body scanners. *Displays* 19 (3):111–120. doi:10.1016/s0141-9382(98)00034-1

Danielsson, U. (1993) *Convection coefficients in clothing air layers* (PhD thesis). The Royal Institute of Technology, Stockholm.

Elsherbiny, S. M., Raithby, G. D., Hollands, K. G. T. (1982) Heat-transfer by natural-convection across vertical and inclined air layers. *Journal of Heat Transfer-Transactions of the Asme* 104 (1):96–102.

Fan, J. T., Luo, Z. X., Li, Y. (2000) Heat and moisture transfer with sorption and condensation in porous clothing assemblies and numerical simulation. *Int J Heat Mass Transf* 43 (16):2989–3000. doi:10.1016/s0017-9310(99)00235-5

Fiala, D., Havenith, G. (2016) Modelling Human Heat Transfer and Temperature Regulation. In: Gefen, A., Epstein, Y. (eds) *The Mechanobiology and Mechanophysiology of Military-Related Injuries*. Springer International Publishing, Cham, pp. 265–302. doi:10.1007/8415_2015_183

Fiala, D., Lomas, K., Stohrer, M. (2003) First Principles Modelling of Thermal Sensation Responses in Steady-State and Transient Conditions. *ASHRAE Transactions* 109:179–186.

Frackiewicz-Kaczmarek, J., Psikuta, A., Bueno, M. A., Rossi, R. M. (2015a) Air gap thickness and contact area in undershirts with various moisture contents: Influence of garment fit, fabric structure and fiber composition. *Textile Research Journal* 85 (20):2196–2207. doi:10.1177/0040517514551458

Frackiewicz-Kaczmarek, J., Psikuta, A., Bueno, M. A., Rossi, R. M. (2015b) Effect of garment properties on air gap thickness and the contact area distribution. *Textile Research Journal* 85 (18):1907–1918. doi:10.1177/0040517514559582

Fu, M., Weng, W. G., Yuan, H. Y. (2014) Effects of multiple air gaps on the thermal performance of firefighter protective clothing under low-level heat exposure. *Textile Research Journal* 84 (9):968–978. doi:10.1177/0040517513512403

Ghazy, A. (2014a) Influence of thermal shrinkage on protective clothing performance during fire exposure: Numerical investigation. *Mechanical Engineering Research* 4 (2).

Ghazy, A. (2014b) Numerical study of the air gap between fire-protective clothing and the skin. *Journal of Industrial Textiles* 44 (2):257–274. doi:10.1177/1528083713483784

Ghazy, A., Bergstrom, D. J. (2011) Influence of the air gap between protective clothing and skin on clothing performance during flash fire exposure. *Heat Mass Transfer* 47 (10):1275–1288. doi:10.1007/s00231-011-0791-y

Ghazy, A., Bergstrom, D. J. (2012) Numerical Simulation of Heat Transfer in Firefighters' Protective Clothing with Multiple Air Gaps during Flash Fire Exposure. *Numer Heat Tr a-Appl* 61 (8):569–593. doi:10.1080/10407782.2012.666932

Havenith, G., Zhang, P., Hatcher, K., Daanen, H. (2010) Comparison of two tracer gas dilution methods for the determination of clothing ventilation and of vapour resistance. *Ergonomics* 53 (4):548–558. doi:10.1080/00140130903528152

Holmer, I., Nilsson, H., Havenith, G., Parsons, K. (1999) Clothing convective heat exchange—Proposal for improved prediction in standards and models. *Ann Occup Hyg* 43 (5):329–337.

Ismail, M. I. (1988) Heat transfer through textile fabrics: Mathematical model. *Applied Mathematical Modelling* 12:434–440.

Kakitsuba, N. (2004) Investigation into clothing area factors for tight and loose fitting clothing in three different body positions. *Journal of the Human-Environmental System* 7 (2):75–81.

Keiser, C., Becker, C., Rossi, R. M. (2008) Moisture transport and absorption in multilayer protective clothing fabrics. *Textile Research Journal* 78 (7):604–613. doi:10.1177/0040517507081309

Kim, I. Y., Lee, C., Li, P., Corner, B. D., Paquette, S. (2002) Investigation of air gaps entrapped in protective clothing systems. *Fire and Materials* 26 (3):121–126. doi:10.1002/fam.790

Lee, C., Kim, I. Y., Wood, A. (2002) Investigation and correlation of manikin and bench-scale fire testing of clothing systems. *Fire and Materials* 26 (6):269–278. doi:10.1002/fam.808

Lee, Y., Hong, K., Hong, S. A. (2007) 3D quantification of microclimate volume in layered clothing for the prediction of clothing insulation. *Applied Ergonomics* 38 (3):349–355. doi:10.1016/j.apergo.2006.04.017

Li, Y., Holcombe, B. V. (1998) Mathematical simulation of heat and moisture transfer in a human-clothing-environment system. *Textile Research Journal* 68 (6):389–397. doi:10.1177/004051759806800601

Lotens, W. A., Havenith, G. (1991) Calcualtion of clothing insulation and vapour resistance. *Ergonomics* 34 (2):233–254. doi:10.1080/00140139108967309

Mah, T., Song, G. (2010a) Investigation of the Contribution of Garment Design to Thermal Protection. Part 2: Instrumented Female Mannequin Flash-fire Evaluation System. *Textile Research Journal* 80 (14):1473–1487. doi:10.1177/0040517509358796

Mah, T., Song, G. W. (2010d) Investigation of the Contribution of Garment Design to Thermal Protection. Part 1: Characterizing Air Gaps using Three-dimensional Body Scanning for Women's Protective Clothing. *Textile Research Journal* 80 (13):1317–1329. doi:10.1177/0040517509358795

Mark, A. (2013) *The impact of the individual layers in multi-layer clothing systems on the distribution of the air gap thickness and contact area* (MSc thesis). Albstadt-Sigmaringen University, Albstadt.

McCullough, E. A., Jones, B. W., Huck, J. (1985) A comprehensive data base for estimating clothing insulation. *ASHRAE* (2888):29–47.

Mert, E., Böhnisch, S., Psikuta, A., Bueno, M.-A., Rossi, R. M. (2016a) Contribution of garment fit and style to thermal comfort at the lower body. *Int J Biometeorol*:1–10. doi:10.1007/s00484-016-1258-0

Mert, E., Psikuta, A., Bueno, M.-A., Rossi, R. M. (2016b) The effect of body postures on the distribution of air gap thickness and contact area. *Int J Biometeorol*:1–13. doi:10.1007/s00484-016-1217-9

Mert, E., Psikuta, A., Bueno, M. A., Rossi, R. M. (2015) Effect of heterogenous and homogenous air gaps on dry heat loss through the garment. *Int J Biometeorol* 59 (11):1701–1710. doi:10.1007/s00484-015-0978-x

Parsons, K. C., Havenith, G., Holmer, I., Nilsson, H., Malchaire, J. (1999) The effects of wind and human movement on the heat and vapour transfer properties of clothing. *Ann Occup Hyg* 43 (5):347–352.

Psikuta, A., Frackiewicz-Kaczmarek, J., Frydrych, I., Rossi, R. (2012) Quantitative evaluation of air gap thickness and contact area between body and garment. *Textile Research Journal* 82 (14):1405–1413. doi:10.1177/0040517512436823

Psikuta, A., Frackiewicz-Kaczmarek, J., Mert, E., Bueno, M. A., Rossi, R. M. (2015) Validation of a novel 3D scanning method for determination of the air gap in clothing. *Measurement* 67:61–70. doi:10.1016/j.measurement.2015.02.024

Psikuta, A., Mert, E., Annaheim, S., Rossi, R. M. (2018) Local air gap thickness and contact area models for realistic simulation of hu-man thermophysiological and perceptual response. *Int J Biometeorol* 62:1121–1134. doi:10.1007/s00484-018-1515-5.

Rossi, R. M., Schmid, M., Camenzind, M. A. (2014) Thermal energy transfer through heat protective clothing during a flame engulfment test. *Textile Research Journal* 84 (13):1451–1460. doi:10.1177/0040517514521115

Shewen, E., Hollands, K. G. T., Raithby, G. D. (1996) Heat transfer by natural convection across a vertical air cavity of large aspect ratio. *Journal of Heat Transfer-Transactions of the Asme* 118 (4):993–995. doi:10.1115/1.2822603

Song, G. (2007) Clothing air gap layers and thermal protective performance in single layer garment. *Journal of Industrial Textiles* 36:193.

Song, H. K., Ashdown, S. P. (2012) Development of Automated Custom-Made Pants Driven by Body Shape. *Clothing and Textiles Research Journal* 30 (4):315–329. doi:10.1177/0887302x12462058

Spencer-Smith, J. L. (1977) The Physical Basis of Clothing Comfort, Part 2: Heat Transfer through Dry Clothing Assemblies. *Clothing Research Journal* 5 (1):3–17.

Sullivan, P. J., Mekjavic, I. B., Kakitsuba, N. (1987) Determination of clothing microenvironment volume. *Ergonomics* 30 (7):1043–1052. doi:10.1080/00140138708965994

Torvi, D. A., Dale, J. D., Faulkner, B. (1999) Influence of air gaps on bench-top test results of flame resistant fabrics. *Journal of Fire Protection Engineering* 10 (1):1–12.

Umeno, T., Hokoi, S., Takada, S. (2001) Prediction of skin and clothing temperatures under thermal transient considering moisture accumulation in clothing. *ASHRAE Transactions* 107 (Part1, 4418):71–81.

Wakitani, S. (1998) Flow patterns of natural convection in an air-filled vertical cavity. *Physics of Fluids* 10 (8):1924–1928. doi:10.1063/1.869708

Wang, Z., Newton, E., Ng, R., Zhang, W. (2006) Ease distribution in relation to the X-line style jacket. Part 1: Development of a mathematical model. *Journal of the Textile Institute* 97 (3):247–256. doi:10.1355/joti.2005.0239

Weder, M., Rossi, R. M., Chaigneau, C., Tillmann, B. (2008) Evaporative cooling and heat transfer in functional underwear. *International Journal of Clothing Science and Technology* 20 (1–2):68–78. doi:10.1108/09556220810850450

Whitaker, S. (1998) Coupled transport in multiphase systems: A theory of drying. *Advances in Heat Transfer* 31:1–104.

Wissler, E. H., Havenith, G. (2009) A simple theoretical model of heat and moisture transport in multi-layer garments in cool ambient air. *European Journal of Applied Physiology* 105 (5):797–808. doi:10.1007/s00421-008-0966-5

Xu, J. H., Zhang, W. B. (2009) The Vacant Distance Ease Relation between Body and Garment. Icic 2009: Second International Conference on Information and Computing Science, Vol 4, Proceedings:38–41. doi:10.1109/Icic.2009.318

Zhang, Z. H., Li, J. (2010) Relationship between Garment Fit and Thermal Comfort. Paper presented at the Textile Bioengineering and Informatics Symposium Proceedings, Vols 1–3.

Zhu, F., Zhang, W., Song, G. (2008) Heat transfer in a cylinder sheated by flame-resistant fabrics exposed to convective and radiant heat flux. *Fire Safety Journal* 43:401–409.

Zhu, F. L., Cheng, X. P., Zhang, W. Y. (2009) Estimation of Thermal Performance of Flame Resistant Clothing Fabrics Sheathing a Cylinder with New Skin Model. *Textile Research Journal* 79 (3):205–212. doi:10.1177/0040517508093591

10

Instrumented Flash Fire Manikin for Maximizing Protective Clothing Performance

Mengying Zhang and Guowen Song

10.1 History of Flash Fire Manikin System

In the 1940s, Baker and Smith designed the first flame manikin to perform a qualitative evaluation of a shirt's burn rate.[1] Leonard Colebrook and Vera Colebrook also used a similar manikin for a study published in 1949. Lit candles or alcohol lamps were utilized to create flame toward the bottom edge of the garment to test clothing flammability.[2] However, these initial flame manikins were used without the installation of sensors; thus, they could only be used to evaluate the flammability of clothing by empirical observation of the clothing burn rate and, therefore, could not be used for quantitative research.

The turning point in the development of combustion dummies occurred after World War II. Due to the emergence of high-speed aircraft and nuclear weapons, the U.S. Air Force's interest in reducing skin burns increased sharply, and improving flame-retardant garments emerged as the key strategy to solving this particular problem. Therefore, during the 1960s, the Air Force began to develop a flame manikin system for determining the thermal protective performance of the Air Force's protective clothing. To simulate the scenario of pilots flying over a large fire, they sprinkled aviation fuel on a water surface in a 7.62 × 6.10 m pit and ignited the gasoline to create a comparable fire zone. A manikin was then hoisted by a crane and controlled across a simulated fire for three seconds. To obtain the rise in temperature of human skin during thermal exposure, the flame manikin was covered with a leather material, along with temperature detecting paper and a melting point indicator, which were installed on the manikin's surface. The temperature detecting paper was used to measure the temperature-time data of the leather surface. Since the temperature of the leather material rises at precisely three times the rate of living human skin, the temperature rise curve of the skin surface can be predicted from the measured temperature rise curve of the leather surface, thereby accurately predicting skin burns based on the relationship between skin temperature and burns. This was the first time that skin burns had been used in a thermal protective performance evaluation of clothing. From this research, an index was developed as a guide to determine the percentage of skin burns, to better evaluate the thermal protective performance of clothing. This study achieved a major breakthrough in qualitative and quantitative analysis of the overall thermal protective performance of various types of clothing.[3]

To facilitate a more systematic, scientific, and accurate testing of the thermal protective performance of different clothing, the U.S. Air Force contracted with the Acurex Corporation to design and build an appropriate manikin that would allow a more systematic and scientific approach for garment evaluation protocols. Thus, a special type of high-temperature-resistant Teflon-coated epoxy-glass material was developed to build the manikin, and the thermal response of this material was similar to that of skin during extreme thermal exposure. There were 124 newly developed skin simulant sensors uniformly mounted on the manikin's surface area to obtain comprehensive temperature-time data.[4]

In addition, the manikin's data acquisition system was optimized so that it could continue recording the data; thus, the data acquisition system and burn analysis system were integrated. Moreover, a new computer code was developed to convert the temperature-time data measured by the sensors into surface heat flux data, and skin burns were then calculated based on the overall depth and severity of the skin burn.[4]

In the 1970s, some performance aspects were added by the E. I. DuPont company, improving the manikin developed by Acurex Corporation to include the following aspects: In terms of a simulated environment, the flame manikin was installed indoors. Initially, garments were ignited with a triangular tab of filter paper. Later, large industrial burners were used to create a flash fire to engulf the manikin, which mimicked the threat of a military or industrial fire. Since then, the Thermo-Man system allowed a simulation of a flash fire environment under strict laboratory-controlled conditions.[5] In terms of skin burn calculations, E. I. DuPont revised the calculation program to consider the phase change of water in skin tissue at 100°C, which refined the skin burn prediction to be more accurate. In terms of instrumentation, a computer-controlled data acquisition system controlled the experiment, recorded the data, carried out the calculations, and plotted the results.[6,7] Later, this flame manikin was registered as Thermo-Man.

Similar to Thermo-Man's composition and work application principle, the Tennessee Eastman Company built its own flame manikin system in 1979. The manikin was constructed using a fiberglass polyester material, and its surface was coated with a high-temperature epoxy resin. The difference between the thermal inertia of these materials and that of human skin was matched within 20%.[8]

The above models and improvements were all constructed for a male flame manikin system. At that time, women's garments were sometimes burned using male manikins in flammability experiments, but this could result in garment burning behavior and predicted skin injury results that were less accurate for female wearers, especially for areas with a distinctly different body contour, such as the upper torso. Because of this, the University of Minnesota began designing a female flame manikin in 1973 and completed the project in 1975. The female manikin was 1.7 m tall and made of polyester and fiberglass. The polyester fiberglass material was then coated with a mixture of ground asbestos, plasterboard, and building cement to maintain the thermal inertia of the manikin material close to that of human skin. There were 44 sensors embedded in the manikin, located mainly on the right torso. Some sites were specifically selected to represent body areas that frequently sustain severe injuries in clothing fires, due to what is known as heat entrapment.[1] The test flame was generated by igniting the bottom edge of the garment, so that the flame did not completely engulf the manikin and clothing; therefore, it did not actually simulate a flash fire environment.

The above-mentioned Acurex Corporation's flame manikin, DuPont's Thermo-Man, Tennessee Eastman's flame manikin, and the Minnesota Female flame manikin were early developments. After the Thermo-Man system was created by Acurex Corporation, the flame manikin system was built with a fixed pattern and, therefore, subsequent flame manikin systems were primarily designed or modified on the more traditional basis of Thermo-Man. These subsequent manikin systems include BTTG's PALPH and SOPHIE, Harry Burns at the University of Alberta, PyroMan at North Carolina State University, and a flame manikin built at Donghua University.

BTTG built RALPH (Research Aim Longer Protection against Heat), a male-shaped manikin, in 1989, followed by the female-shaped manikin SOPHIE (System Objective Protection against Heat In an Emergency) in 2006. RALPH and SOPHIE have a total of 135 and 132 sensors, respectively, distributed over the head, torso, legs, arms, and hands, which monitor the temperature on the surface of the manikin during a test. (The feet of the manikin are not sensored.)[9,10] Thus, one of the primary features of the BTTG™ system is that it allows for the testing of complete ensembles of fire-protective clothing, that is, clothing, headwear, and gloves.

The flame manikin Harry Burns, developed at the University of Alberta, was quite similar to that of the Thermo-Man system. Six burners were used to produce a diffusion flame with a heat flux ranging from 67 to 84 kW/m², while there are 110 heat sensors made up of an inorganic material, Colorceran. Colorceran is an inorganic mixture of calcium, aluminum, silicate with asbestos fibers and a binder, and the product kpc, or its square root, has similar value with that of human skin.[11] Due to space limitations, as well as the number of existing instruments to test the thermal protective performance of gloves and shoes, the hands and feet of the flame manikin were not mounted with sensors.

The flame manikin at North Carolina State University, PyroMan, is also similar to the Thermo-Man system. The manikin is fitted with 122 heat sensors distributed uniformly over the body, excluding the hands and feet.[12] This manikin was initially installed in a chamber and surrounded by eight propane gas burners to produce a flash fire with a heat flux index of 63–126 kW/m². Later, a new slug calorimeter heat flux transducer was developed and incorporated; the advantage of this kind of sensor is that the thickness and material properties of the copper disk can be easily measured and controlled. Additionally, these sensors are extremely durable and can tolerate repeated flash fire testing. A thermal protective performance report is then generated that includes the individual heat sensor responses, total accumulated heat received by the heat sensors, percentage of the manikin body receiving second-degree burns, the percentage receiving third-degree burns, the total predicted burn injury reflected by a percentage of the entire body, and a diagram depicting the burn intensity distribution over the entire body surface.[6,12,13]

The flame manikin at Donghua University was developed by Measurement Technology Northwest. This flame manikin features a fully articulated, non-degrading, ceramic-composite body formulated with integrated sensors, control electronics, a computational burn model, and a computer. This model represents a complete turn-key package, allowing the operator to characterize and evaluate the performance of garments or protective clothing ensembles in a simulated flash fire environment for controlled heat flux, flame distribution, and duration. One of the most notable features is that the manikin is fully articulated, with joints at the shoulders, elbows, hips, knees, and ankles.[14] Therefore, the manikin can simulate different postures of the human body. In addition, the system includes a traversing chain-driven track mechanism mounted to the ceiling, capable of moving a manikin through the flash fire, simulating movement through the fire. Similarly, the flame manikin at Worcester Polytechnic Institute also installed a traversing chain-driven track mechanism, which is capable of moving a manikin through the test area.[15]

Despite the massive development of flame manikin systems, all of the above are stationary manikins. However, the human body is constantly moving in a fire, that is, a firefighter attempting to contain, extinguish, or escape from the fire. Therefore, to more accurately simulate the human body's state in a fire field and more precisely evaluate the thermal protective performance of clothing, some research institutions have used a more simplified method to build dynamic limbs, such as DuPont's Thermo-Leg. Thermo-Leg can simulate a human's running movement, and the leg material and sensors are the same as the Thermo-Man. Eighteen heat sensors, similar to the Thermo-Man sensors, were distributed over the entire surface of the leg. As with Thermo-Man, flash fire conditions were achieved by using four large industrial propane torches mounted around the leg. The leg was immersed during the full cycle with an average heat flux of 84 kW/m².[6]

From the above discussion, it could be ascertained that the composition and testing principles of different flame manikin test systems are basically similar. A typical flame manikin system consists of a combustion chamber, flame manikin, gas supply system, flash fire simulation system (burner and control panel), and a software control system, as depicted in Figure 10.1. The software control system controls the entire process of the experiment. During the experiment, the flame manikin is placed in the combustion chamber, after which the burners inject a propane gas flame to generate a flash fire environment. The uniformly distributed temperature sensors on the manikin surface record the temperature-time profile both during and after thermal exposure.

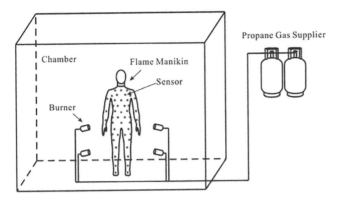

Figure 10.1

Schematic diagram of the flame manikin system.

These data are then calculated into heat flux data, which are then input to the skin burn analysis system as boundary conditions for the skin layer. The skin burn system then predicts the degree of the skin burn, based on the skin heat transfer model and Henriques's skin burn integral model.

The characteristics of different flame manikin systems are summarized in Table 10.1.

10.2 Combustion Chambers and Simulated Fire

DuPont designed a simulation system for imitating an indoor environment and improved it in the 1970s. Both the instrumented manikin and the exposure system are housed in the chamber and the flash environment can be created by igniting the propane gas with a large industrial burner. The chamber size should be sufficient to provide uniform flame engulfment of the manikin and allow for safe movement around the manikin for dressing without accidentally jarring and displacing the burners.[16]

The standard combustion chamber consists of multiple doors, windows, and heat-resistant walls, which are installed on one or both sides of the high-temperature-resistant glass, and through which the researchers can observe the phenomenon of the combustion process. The combustion chamber is equipped with a combustion manikin, a burner and a spray rotor, a gas transmission system, and a ventilation system. Several burners are divided into different groups and installed in the center area of the chamber, and distributed concentrically around the manikin. By adjusting the position of the burners, the flow of gas, the pressure of the pipes, the angle of the flame spray, etc., the simulated fire is capable of meeting the technical standards of the combustion dummy evaluation system.

The gas transmission system includes the fuel group, regulating valve, fuel pipeline, and computer control platform. Flame is generated by the cooperation of the burner and the gas transmission system, while the industrial propane gas is stored in the fuel tank. During the experiment, the gas enters the main

Table 10.1 Characteristics of Different Flame Manikin Systems

| Build Time | Research Institute | Flame Manikin System Characteristics | | | Importance |
		Exposure Environment	Manikin Materials	Sensors	
1960s	U.S. Navy[3]	Aviation fuel is spread on a water base in a pit and ignited	A store manikin is covered with leather	Temperature detector paper and melting point indicators	For the first time the skin burn index was used to evaluate the thermal protective performance of clothing
1970s	Acurex Corporation[4]	JP-4 fuel fires	0.64 cm thick high temperature Teflon-coated epoxy-glass material	124 thermocouple type burn damage sensors	Equipped the manikin with skin simulant heat flux transducers and developed a system for continuous data recording
1970s	Du Pont[5]	Eight large industrial propane gas burners	Flame resistant polyester resin reinforced with fiberglass	122 skin simulant sensors	Improved instrumentation and recording systems; mounted manikin in a laboratory to simulate flash fires using a series of gas burners fed with propane gas (Registered trademark Thermo-Man)
1975	University of Minnesota[1]	Use a series of gas flames to ignite the lower edges of garments on the mannequin	Polyester-fiberglass coated with a wet mixture of ground asbestos, plaster of Paris, and building cement.	44 skin simulant sensors made up of chromel-alumel thermocouples and Pyrex	The first adult female thermally instrumented mannequin
1979	Tennessee Eastman Company[8]	Unspecified	Polyester/fiberglass coated with epoxy resins	80 sensors made up of chromel-constantan thermocouples and Pyrex glass	Developed new skin simulant sensor
1989 (Male) 2006 (Female)	BTTG[10]	Six burners produce propane flames	Glass fiber and vinyl ester resin	RALPH and SOPHIE have a total of 135 and 132 sensors, respectively	Sensors were mounted at the hands and feet for the testing of complete ensembles of fire protective clothing, i.e., clothing, headwear, and gloves

(*Continued*)

Table 10.1 (Continued) Characteristics of Different Flame Manikin Systems

		Flame Manikin System Characteristics			
Build Time	Research Institute	Exposure Environment	Manikin Materials	Sensors	Importance
1992	University of Alberta[11]	Six burners produce propane diffusion flames, the average heat flux to be varied from about 67 kW/m² to 84 kW/m²	Flame-resistant polyester resin reinforced with fiberglass	110 skin simulant heat flux transducers	The flash fire exposure system was refined, so both the average heat flux and the duration of exposure can be varied; a new skin simulant sensor was developed
1992	North Carolina State University[12]	Eight industrial burners produce the flash fire	Flame resistant polyester resin reinforced with fiberglass	122 slug calorimeter heat flux sensors	Developed a new highly durable sensor to tolerate repeated testing
2011	Donghua University[14]	Six industrial burners produce the flash fire	Ceramic-composite body form	135 copper calorimeter heat flux sensors	Simulated different postures of the human body and crossing a flash fire

pipeline through the pressure reducing valve and is then directed to the burner through the pressure-stabilizing valve. In addition, a separate pipeline is available to relay the gas to the burner outlet. A computer control platform is used for the entire system operation, control, and detection signal processing. In the simulation experiment of the fire field, flames will completely engulf the manikin. However, the heat flow distribution and size of the combustion flame can be controlled by adjusting the distance between the burner and the manikin, the angle of the flame of the burner, the flow rate of the gas injection, and the pressure of the pipeline. This can satisfy the requirements the different ignition source demands.

The combustion chamber is equipped with an air supply system and exhaust system. The air supply system is primarily composed of a self-control air inlet and ventilation ducts, while the air exhaust system is mainly composed of a filtering device, a powerful exhaust fan, and ventilation ducts. The function of the ventilation system is to (1) provide sufficient oxygen necessary for the combustion experiment to ensure the complete combustion of the industrial propane gas; (2) allow for the rapid removal of the products of combustion and degradation after the test exposure; and (3) ensure a timely discharge of indoor heat to cool the dummy device and related equipment.

10.3 Thermal Sensors Employed in Manikin System

The design principle of the thermal sensor is to both simulate the thermal properties of human skin and obtain thermal information that reaches the surface of the skin, such as heat flux and cumulative thermal energy. The sensors utilized in flame manikin systems can be divided into two categories, namely, the copper slug sensor and the skin simulant sensor.

10.3.1 Copper Slug Sensor

The copper sensor is the most widely used type of sensor in the flame manikin system, and its prototype is derived from the copper disk heat flux sensor utilized in the TPP tester. At present the widely used copper slug sensor is mainly composed of a thermocouple, copper disk, copper ring, insulating material, and a shell. The emissivity of human skin is approximately 0.98, because the radiation properties of human skin are similar to the ceramic black painted sensor surface,[17] allowing almost all radiant heat to be absorbed by the skin. Therefore, to simulate the emissivity of human skin, the sensor surface is usually painted with a high-temperature-resistant black ceramic paint.[18]

In 1996, for the first time, Grimes et al. designed a copper slug sensor and applied it to the flame manikin system at North Carolina State University.[6,19] This sensor has numerous advantages, such as a fast response time and high precision, and has since been used by a number of other research institutions.[14,20] For the copper slug sensor, the thermocouple measures the temperature profile of the copper during exposure. To calculate the skin burn, it is necessary to convert the temperature-time data into heat flux data. Moreover, the thermal resistance of the copper disk is much smaller than the convective thermal resistance of the air; hence, the thermal resistance of the copper disk can be neglected. Therefore, in most cases, the radiative heat flux density can be calculated by employing the lumped heat capacity method. This method assumes that the temperature of the copper disk is spatially uniform and neglects the temperature gradient inside the copper. Additionally, based on the energy equation, the incident heat flux density can be calculated by applying the following formula:

$$q - h(T(t) - T_\infty) = \rho C_p C_L L \frac{dT(t)}{dt} + K_L(T(t) - T_0) \qquad (10.1)$$

where ρ, C_p, L, and C_L represent the density, specific heat capacity, thickness, and thickness deviation factor of the copper disk, respectively; K_L depicts an aggregate contact conductance between the copper disk and the supporting insulator; and $h(T(t) - T_\infty)$ is the heat loss due to natural convection.

In practice, the natural convection term is negligible; thus, the formula above can be simplified to the following:

$$q = \rho C_p K_m L \frac{dT(t)}{dt} + K_L(T(t) - T_0) \qquad (10.2)$$

where K_m and K_L constitute two correction factors obtained by the sensor calibration.

This kind of sensor displays good stability in repeated flash fire tests, and the heat flux measured by the sensor is certainly more accurate. However, since the thermal conductivity of the copper is much higher than that of human skin, the temperature rise rate of the copper is much faster than that of skin. To minimize the disturbances to the incident heat flux, the thickness of the copper disk must be adjusted, so as to force the temperature response of the copper disk to remain below the temperature response of human skin. Therefore, the thickness of the copper is a particularly critical parameter. According to Grimes et al.,[19] when the incident heat flux is 25 kW/m², the corresponding thickness of the copper disk is 1.524 mm. In these conditions, the temperature of the copper disk remains below both that of human skin and the skin simulant for the duration of the 15-s exposure. Therefore, the copper disk heat flux sensor is often utilized in short-burst, transient, high-heat, flux-flame manikin tests.

10.3.2 Skin Simulant Sensor

The copper slug sensor uses copper as a temperature-sensitive component, but the thermal properties of copper are diverse from those of skin. For example, compared with human skin, the thermal conductivity of copper is much higher, and the heat capacity is much lower. Those differences eventually lead to errors in the measurement of thermal information of the skin surface. That is why researchers also developed skin-simulating sensors during the same period. The sensors used in the DuPont Thermo-Man and University of Alberta flame manikins constitute typical skin simulant sensors.

The structure of a skin-simulating sensor is relatively simple and usually consists of only two parts, the skin simulation material and the thermocouple. Skin simulation materials need to possess thermal and optical properties that are a close representation of human skin; thus, thermocouples are used to acquire temperatures at or below the surface of a simulated material. Of all the skin simulant sensors at this stage, the incident energy of the skin surface remains a research focus; therefore, the skin simulant materials are actually supposed to simulate the thermal and optical properties of the skin surface.

For thermal performance, the skin simulant sensor could be considered as a semi-infinite flat plate; hence, the surface temperature can be expressed by the following formula:

$$T(0,t) - T_i = \frac{2q\sqrt{t}}{\sqrt{\pi k \rho c}} \tag{10.3}$$

where k and T_0 depict the thermal conductivity and initial temperature of the skin simulant material, respectively. The above-listed formula illustrates that the product kpc, the thermal inertia, or its square root, the thermal absorptivity, represent the grouping of thermal physical properties that should be similar to those of human skin.[11]

The University of Alberta used an inorganic material called Colorceran, which is made from a mixture of calcium, aluminum, silicate, and asbestos fibers in a certain ratio.[11] This kind of material not only exhibits a similar surface thermal inertia to the skin but also boasts a good flame-retardant and heat-resistant property in the long-lasting test. For better optical performance, the surface of the skin simulant sensor is often sprayed with a high temperature matte ceramic paint to simulate the emissivity of human skin.

Most skin simulant sensors are designed as solid cylinders, and the thermocouple is either attached to the surface of the skin simulant material or buried just below the surface. For example, the thermocouple of the Thermo-Man sensor is buried 0.127 mm below the surface. After the thermocouple has collected the temperature information of the simulated skin material's surface, it can be used to calculate the heat flux.

The calculation method for the heat flux of the skin simulant sensor differs from that of the copper slug sensor. Compared with copper, the thermal conductivity of skin simulant materials is low, and the corresponding thermal resistance is large. As a result, when heat incidents to the surface of the skin simulant sensor, a temperature gradient is generated. Thus, the lumped heat capacity method employed for calculating heat flux is no longer applicable. In this case, the Duhamel model or numerical inverse method is often used to calculate the temperature gradient in the skin simulant sensor. Furthermore, the skin surface temperature can be directly and quickly converted into a surface heat flux by using the Duhamel model; hence, it is widely used.

10.3.3 Comparison of Copper Slug Sensor and Skin Simulant Sensor

The copper slug sensor and skin analog sensor are both designed to calculate the incident energy on the skin's surface. The key difference between the two sensors is that, due to the larger thermal resistance of the skin simulant sensor, there is a temperature field in the skin simulant sensor; with the copper sensor, it is assumed that all the weight and temperature are concentrated on one point.

Copper slug sensors and skin simulant sensors present their own advantages and disadvantages. The test results are relatively stable and accurate due to the comparatively stable thermal conductivity and specific heat capacity of the copper during heat exposure. In addition, the response time of the copper sensor is very fast. However, the calculation of the heat flux is cumbersome, and its similarity to skin is often questioned. This, for skin simulant sensors, skin simulant materials can better replace the human skin for thermal response studies.

However, since the skin simulation material is usually produced from several different materials and is mixed in different proportions, the thermal conductivity of the skin simulant material makes it more difficult to attain an accurate determination or reading. Moreover, the properties of the skin simulant material often need to be calibrated or the parameter estimated by utilizing other sensors. In addition, the skin simulant sensor has some more practical problems, such as being difficult to clean.[21]

A summary of the sensors used in the flame manikin and related information is displayed in Table 10.2.

10.4 Standards

With the continuous improvement of flame manikin test methods, the relevant standards for evaluating thermal protective performance of clothing using this system have evolved as well. ASTM and ISO established the ASTM F1930-00 Standard Test Method for Evaluation of Flame Resistant Clothing for Protection Against Flash Fire Simulations Using an Instrumented Manikin, and ISO 13506-08 Protective Clothing Against Heat and Flame—Test Method for Complete Garments—Prediction of Burn Injury Using an Instrumented Manikin, respectively. Furthermore, test methods using the flame manikin system were homogenized in these standards. The ASTM F1930 standard has been developed to the 2015 version thus far. In different editions, the test and evaluation methods are basically similar, but with different parameters, such as skin thickness and thermal physical properties, while the ISO 13506 offers a standard basic reference to the ASTM F1930-00 standard test methods and parameters. The requirements for the flame manikin system in ASTM F 1930-15 and ISO 13506-08 are summarized in Table 10.3.

Table 10.2 Sensor Characteristics Used in Flame Manikin System

Sensor Type	Manikin	Characteristics	Structure Diagram
Copper slug sensor	NCSU Pyroman[19]	• The thermocouple closely adheres to the back side of the copper disk and measures the temperature-time profile	
	WPI flame manikin[15]	• Type T thermocouple measures the temperature-time profile of the copper disk	
Skin simulant sensor	Thermo-Man[18,22]	• Simulates skin • Measures heat flux data • Thermocouple is buried inside the skin simulant material	
	Alberta flame manikin[11,18,22]	• Skin simulant material is called Colorceran • Measures heat flux data • Thermocouple is attached to the surface of the skin simulation material	
	JSI flame manikin[23]	• Simulates skin • Measures heat flux data • Thermocouple is buried inside the skin simulant material	

In terms of testing and evaluation, the ASTM standard specifies indexes for using a certain percentage for skin burn injury and a diagram of the manikin showing the location and burn injury levels as second- and third-degree areas. However, the ISO standard also recommends the use of total energy transferred; therefore, the current evaluation indexes of the instrumented flame manikin test system include the three methods listed above.

10.5 Application and Models of Flash Fire Manikins

10.5.1 The Application of Flash Fire Manikins

Evaluation of thermal protective clothing has typically been performed through small-scale laboratory tests of flame resistance and thermal protective performance. These test methods may give useful information about the fabrics used in thermal protective clothing, but they do not necessarily predict how well

Table 10.3 Comparison of ASTMF 1930-15 and ISO 13506-08

	ASTM F1930-15[16]	ISO 13506-08[24]
Number of sensors	At least 100 (not including feet and hands)	At least 100 (not including feet and hands)
Chamber size	At least 2.1 by 2.1 by 2.4 m	At least 2.1 by 2.1 by 2.1 m
Manikin materials	Flame-resistant, thermally stable, nonmetallic materials that will not contribute fuel to the combustion process	Flame-resistant, thermally stable, such as ceramics or glass-reinforced vinyl ester resin; shell thickness should be at least 3 mm
Sensor-measuring capacity	$0–165 \text{ kW/m}^2$	$0–200 \text{ kW/m}^2$
Average exposure heat flux	$84 \text{ kW/m}^2 \pm 5\%$	$84 \text{ kW/m}^2 \pm 2.5\%$
Exposure duration	Up to 20 seconds	At least 8s (For single-layer garments: 5s or less. For ensembles: at least 5s.)
Data acquisition time	At least 60s	Up to 120s (For single-layer garments: 60s. For multilayer garments and ensembles: 120s.)
Skin Burn Injury Model	Henriques's Damage Integral Model	Henriques's Damage Integral Model

garments or garment assemblies will perform when exposed to a flash fire. The thermal protective performance of clothing is not only related to the properties of the fabrics but also affected by the pattern, style, fit, and local design features (such as the belts and the location of reflective strips) of clothing.[25,26] Flash fire manikin evaluation systems provide a more comprehensive understanding of the thermal protective performance of clothing system.

In 1992, using the Thermo-Man system, Behnke et al.[6] compared the thermal protective performance of single-layer garments made of Kevlar, Nomex, FR Cotton, and FR Wool, respectively. In general, second-degree burn injuries to 100% of the nude manikin occurred within about 1 s and third-degree burn injuries to 100% occurred in about 2 s of exposure. In contrast, manikins with a single layer garment predicted a second-degree burn in about 4 s and third-degree burn in about 5–6 s. These data indicated that garments composed of these four kinds of fabrics could provide a certain degree of thermal protection by reducing the amount of heat transmitted to the skin. By comparing these garments, it was found that FR Cotton garments ignited during the flash fire and continued to burn for 7 s after the initial flame exposure ended. As a result, a large amount of heat transferred to the sensors, resulting in more severe skin burns. In addition, the researchers studied the effect of garment design features by comparing the severity of the burn injury over the manikin torso. Since the front side of the garment contained several pockets, providing a double layer of fabric protection, the back area of the manikin received more burn damage than the chest.

There have been multiple studies conducted on the effect of fabric properties, clothing style and fit, the air gap layer between clothing and manikin, and thermal shrinkage on the thermal protective performance of clothing using flash fire instrumented manikins.

Fabric properties. In 1992, Dale et al.[11] investigated the thermal protective performance of six types of protective garments and found that performance increased with increased fabric weight. In 1995, Pawar[27] studied the effect of fabric weight and heat flux level on clothing's protective performance. In this study, Kevlar/PBI, Nomex IIIA, and FR Cotton garments were exposed to simulated flash fires varying in heat flux. At exposure energies lower than 6 cal/cm^2 in instrumented manikin tests, material weight was the dominating factor in protective performance. Above the energy level of 8 cal/cm^2, differences in the intrinsic heat-resistant materials became apparent. For example, at thermal energy levels less than 6 cal/cm^2, the FR cotton material demonstrated comparable protective properties, but for thermal energy levels beyond 8 cal/cm^2, FR cotton showed a sudden increase in predicted percentage of burn injury, while for aramid fabrics, the loss of protection was gradual. In 2005, Rossi et al.[28] studied the relationship between fabric flame propagation rate and clothing thermal protective performance, and found that fabrics with high flame propagation rates were the garments with the shortest times to pain and to burns on the manikin. For fabrics made of natural fibers, flame propagation rate was a good index to predict the potential hazard. However, for synthetics and blends of natural and synthetic fibers, there was an increased burn risk due to the combined flame spread and melting of synthetic fibers; therefore, indexes related to heat transfer to the skin had to be considered as well. In 2011, Jin et al.[29] compared the thermal protective performance of firefighter garments containing thermal barriers treated with aerogel with that of commercial garments. The second-degree burn injury of aerogel-treated garments was relatively lower. Therefore, garments using a thermal barrier treated with aerogels exhibited higher thermal protective performance than the commercial garments.

Garment style and fit. In 1998, Crown et al.[7] evaluated the effect of local design features on thermal protective performance of garments made of the inherently fire-resistant fabric meta-aramid and its blends. It was found that loose-fitting garments with controlled fullness and appropriate closures provided somewhat better protection than close-fitting garments, and the close-fitting cuff closures on sleeves and pant legs were more effective than zipper closures. Two-piece garments provided better protection than one-piece coveralls, mainly due to the overlap of upper and lower layers under the waist; this effect was weakened when the garments were worn over long thermal protective underwear. Their study also illustrated that a stand-up collar offered better protection for the neck than a convertible collar. In 2000, Prezant et al.[13] compared the skin burn injuries of PyroMan while wearing short and long underwear and found no significant differences between the two types. Mah and Song[30,31] compared the thermal protective performance provided by women's and men's garments for a female flash fire manikin. In general, no significant differences were found in total burn injuries between the women's and men's styles. However, there were significant differences in some local areas, such as lower back, seat, and right thigh regions, due to the presence of the belt in women's garment.

Air gap size and distribution. The above studies show that clothing style and fit affect the thermal protective performance of the garment by changing the size and distribution of air gap layers between the garment and manikin. Therefore, researchers further studied the effect of style and fit on thermal protective performance by quantifying the size and distribution of air gap layers. In 2010, Mah and Song[30,31] measured the air gaps between a female mannequin and protective coveralls using a three-dimensional body scanner. It was found that the air gap sizes were not evenly distributed over the mannequin, and there were a greater number of smaller air gaps over the mannequin than larger ones. Then the flash-fire instrumented female mannequin evaluation system was used to investigate the effect of air gap size and distribution on thermal protection. In general, as the size of the air gap increased, the time before burns occurred also increased, and the amount of energy absorbed by the sensors decreased. However, this did not mean that larger air gaps could provide greater thermal protection, since convection heat transfer might have initiated within the space, negating any added benefit of a larger air gap. Song also found that in the shoulder, knee, and upper back areas, the protective garment was closer to the body; in the waist and thigh, larger air gaps were present. The size of the protective garments, as well as fabric drapability and stiffness, also affected the distribution of air layers over the surface of the manikin. In addition, Song[32] investigated the effect of thermal shrinkage based on the manikin test and found that garment shrinkage during exposure could greatly reduce the air gap and potentially cause a significant decrease in the performance of thermal protective clothing. From this, a numerical model was employed to predict that a 7–8 mm air gap for a coverall would provide optimal thermal protection. Below this range, insulation would increase as the size of the air gap increased, but beyond this range, convection would occur, reducing the effectiveness of the thermal barrier.

Thermal shrinkage. The thermal shrinkage of thermal protective clothing can greatly affect its protective performance by changing the size and distribution of air gaps between clothing and human body. Therefore, systematic investigations on thermal shrinkage under exposure to flash fire were conducted. In 2013, Li et al.[33] investigated thermal shrinkage of firefighters' clothing using the Donghua flame manikin system. It was found that thermal shrinkage increased with increasing exposure time and garment size, and decreased with increasing fabric density. During the exposure, little shrinkage was observed by the video camera at the initial stage, while as the exposure time continued, the garment began to shrink rapidly, accompanied by a small amount of smoke. In addition, they found that thermal shrinkage was unevenly distributed over the manikin, and the most severe locations were the arms and legs. In 2015, Wang and Li[34] investigated the level of thermal protection retained by fire protective clothing after repeated exposures to flash fires. They found that for fabrics with severe thermal shrinkage, such as Nomex IIIA and polysulfonamide, thermal protective performance of fabrics increased after repeated exposure, while the thermal protection of garments decreased. They concluded that thermal shrinkage would result in an increase in the thickness of fabric and a decrease in the air gap size between clothing and manikin. The increase in the fabric thickness after repeated exposure would increase thermal protection, while the decrease in the air gap size due to shrinkage would accelerate heat transfer from garment to human body, resulting in a decrease in its thermal protection level.

Some researchers have tried to establish the relationship between small-scale tests and flash manikin tests due to the wide range of applications, low cost, and high repeatability of small-scale tests. In 1995, Pawar[27] investigated the correlation between TPP value (with or without an air gap) and the body burn percentage. It was found that these two values did not show any correlation when the small-scale test was in contact test configuration, but showed a strong correlation when the small-scale test was in spaced test configuration. A similar study was conducted by Wang et al.,[26] who also found that no significant correlation between a bench scale test in contact configuration and a flame manikin test in terms of TPP value and body burn percentage. They explained that the thermal shrinkage that was not included in the bench-scale test was a key factor contributing to the weak correlation. In 2002, Lee et al.[35] examined the correlation between bench-scale test results and manikin test results considering air gap measurements. It was found that areas on the manikin with zero and small air gaps showed good correlation with small-scale tests conducted with zero air gap at similar incident heat flux levels. In other words, if the percentage area with small air gaps of a clothing system was known and if the bench scale tests were calibrated to be similar to that of the local incident heat flux on the manikin, bench scale tests could be used to provide useful information on full-scale manikin tests.

In addition to research on the effect factors of thermal protection, some researchers also built new indexes to completely evaluate the thermal protective performance of clothing. In 2014, Rossi and Camenzind[36] proposed total transferred energy and energy transmission factor as improved characterization methods of the performance of the garments. The total transferred energy is the integration of the measured heat flux with time, and the energy transmission factor is the quotient between the transferred energy on the clothed manikin divided by the transferred energy registered by the nude manikin during calibration. The experimental results showed that the calculation of the total transferred energy as well as the transmission factor produces results with high repeatability and low coefficients of variation. They demonstrated that, although the interpretation of these two factor results was probably more complex than burn predictions for the end users, they provided more detailed information about local differences in thermal protection. More importantly, the risk of misinterpretation is reduced since the newly built indexes were based on physical values instead of empirical data. In 2006, Schmid et al.[37] found that underwear can melt in the case of extreme heat flux exposures, leading to severe skin injuries. Therefore, they defined the critical thermal resistance of the outer layer as the insulation needed to avoid material degradation during a flame engulfment exposure. Using critical thermal resistance, they found that for meltable garments with high thermal resistance, a higher thermal resistance of the protective layer was needed to prevent material damages.

10.5.2 Numerical Models of Flash Fire Manikin System

Numerical models for thermal protective clothing could supplement experimental studies, save costs, expand research scope, and obtain valuable information that is difficult to measure in physical experiments, such as the temperature distribution of clothing–air gap–skin systems. Through numerical simulation, the heat transfer mechanisms of protective garments exposed to extreme thermal environments can be further understood.

Song et al.[12,38] developed a numerical model to predict the performance of protective garments (Kevlar/PBI and Nomex IIIA) during flash fire exposure. The thermally induced thermophysical properties of the protective fabrics and the distributions of air gaps between garments and the manikin were considered in the model. A parameter estimation method was used to estimate heat-induced changes in fabric thermophysical properties. The air gaps between Nomex IIIA garments and the manikin were considered as temperature-dependent for the 4 s exposure because of thermal shrinkage. The heat transfer was assumed to be one-dimensional; no mass transfer occurred in fabric and air gap; fabric was considered as gray body for radiation; the radiative heat flux penetrated the fabric to a certain depth; and thermal-chemical reaction and degradation of fabric were neglected. Based on the above assumptions, the energy balance equation of fabric was described as follows:

$$\rho_{fab}(T)c_{p\,fab}(T)\frac{\partial T}{\partial t} = \frac{\partial}{\partial x}\left(k_{fab}(T)\frac{\partial T}{\partial x}\right) + \gamma q_{rad}exp(-\gamma x) \tag{10.4}$$

where ρ, c_p, k are the density, specific heat, and thermal conductivity of the fabric, respectively, and γ is the extinction coefficient (m^{-1}) of the fabric.

The thermal energy transfer by conduction/convection from fabric to human skin across the air gap was described as follows:

$$q_{air,cond/conv}\big|_{x=L_{fab}} = h_{air,gap}(T_{fab} - T_{skin}) \tag{10.5}$$

where $h_{air,gap}$ is the heat transfer coefficient of air due to conduction and natural convection in the air gap. T_{fab} and T_{skin} are the temperatures of the inside surface of the fabric and the human skin, respectively.

In addition to the fabric–air gap–skin system model, the fabric–air gap–underwear–skin configuration was also modeled by Song, and the heat transfer model was described as follows:

$$\rho_{underwear}c_{p\,underwear}\frac{\partial T}{\partial t} = k_{underwear}\frac{\partial^2 T}{\partial x^2} \tag{10.6}$$

where $\rho_{underwear}$, $c_{p\,underwear}$, $k_{underwear}$ are the density, specific heat, and thermal conductivity of the underwear, respectively. Pennes bio-heat transfer equation and Henriques burn integral model were used to calculate the temperature distribution of the three-layer skin and predict the degree of skin burns. The numerical model was validated using the Pyroman® flash fire manikin.

A parametric study was then conducted using the developed numerical model. For single-layer garments exposed to intense fire conditions, fabric thickness was the major factor influencing thermal protective performance, and fabric thermal conductivity and volumetric heat capacity were important parameters controlling heat transfer. Decreasing thermal conductivity and increasing volumetric heat capacity could effectively improve the protective performance of a garment. In addition, the fabric emissivity, the initial temperature of garments, and the ambient temperature were also crucial factors influencing predicted burn injury.

Other researchers also developed three-dimensional numerical models to simulate the transient heat transfer through a flame manikin in a combustion chamber, using the advanced technique of computational fluid dynamics. Jiang et al.[39] developed a numerical simulator composed of a three-dimensional CFD code, FrontFlow/Red, to simulate the combustive flow and heat transfer in flames, and a one-dimensional code, AFFECTION, to simulate radiation and conduction heat transfer through garments and human skin. The simulator was validated by a flame manikin experiment based on the ISO standard. Wang et al.[40]

conducted a three-dimensional transient CFD simulation of heat and mass transfer in a flame manikin test. Although the research object was a nude manikin and the study does not consider heat transfer in clothing, this study proved the feasibility of using Ansys Fluent to simulate the flash fire manikin system. Later, Tian et al.[41] developed a full-scale simulation of the flame chamber equipped with a manikin based on the method that Wang et al. proposed. In their study, both single-layer and multilayer clothing were simulated to investigate thermal protective performance.

10.6 Summary

Flash fire manikin first developed in the 1940s, and it was developing a system in the 1970s. In recent years, this system was accepted and became an important measurement tool in protective clothing engineering. Several research institutions have further contributed to the evaluation system, such as University of Minnesota, Tennessee Eastman Company, BTTG, University of Alberta, North Carolina State University, and EMPA. High-temperature-resistant materials have been applied to simulate the human body. Copper slug sensors and skin simulant sensors were improved to precisely capture the temperature-time profile to eventually predict the degree of skin burn injuries. Currently, the characterization indexes of thermal protective performance of clothing measured by flash fire manikin include predicted second-degree burn injury, third-degree burn injury, total predicted burn injury percentage, and its distribution among human body. Studies based on the flash fire manikin system contributed to further understanding on how textile materials and ergonomics design affect the overall clothing protective performance. The developed numerical model and the improved flash fire evaluation system revealed the fundamentals on heat and mass transfer between hazardous exposure and clothing system associated with wearing protective clothing and clothing performance engineering. The challenges for realistically predicting the performance provided by clothing relies on how close we can simulate human movement and physiological response. The challenges include simulation of the dynamic air gap developed between clothing and the human body, the localized physical compression, and moisture either from sweating or outside from firefighting. The current flash fire manikin system only simulates the intense fire exposure environment. However, many burn injuries that happened during exposure are reported in low and middle heat exposure level. With the advance of new technology, the next generation of the flash fire system will be to simulate human movement effect and physiological responses.

References

1. Norton, M. J. T., Kadolph, S. J., Johnson, R. F. et al. Design, construction, and use of Minnesota woman, a thermally instrumented mannequin. *Textile Research Journal*, 1985, 55(1): 5–12.
2. Colebrook, L., Colebrook, V. The prevention of burns and scalds: Review of 1000 cases. *The Lancet*, 1949, 254(6570): 181–188.
3. Stoll, A. M., Chianta, M. A. Heat transfer through fabrics as related to thermal injury. *Transactions of the New York Academy of Sciences*, 1971, 33(7 Series II): 649–670.

4. Elkins, W., Thompson, J. G. Instrumented thermal mannikin. ACUREX CORP/AEROTHERM MOUNTAIN VIEW CA, 1973.

5. Behnke, W. P., Geshury, A. J., Barker, R. L. Thermo-Man® and Thermo-Leg: Large Scale Test Methods for Evaluating Thermal Protective Performance/Performance of Protective Clothing: Fourth Volume. ASTM International, 1992.

6. Bercaw, J. R., Jordan, K. G., Moss, A. Z. Estimating injury from burning garments and development of concepts for flammability tests/Fire Standards and Safety. ASTM International, 1976.

7. Crown, E. M., Ackerman, M. Y., Dale, J. D. et al. Design and evaluation of thermal protective flightsuits. Part II: Instrumented mannequin evaluation. *Clothing and Textiles Research Journal*, 1998, 16(2): 79–87.

8. Trent, L. C., Resch, III, W. A., Coppari, L. A. et al. Design and construction of a thermally-instrumented mannequin for measuring the burn injury potential of wearing apparel. *Textile Research Journal*, 1979, 49(11): 639–647.

9. http://www.bttg.co.uk/testing/ppe/ppe-testing/.2017.11.06

10. Eaton, P., Healey, M. The development of a 'female' form manikin as part of a test facility to assess the fire protection afforded by personal protective equipment. Altrincham: BTTG Fire Technology Services, 2006.

11. Dale, J. D., Crown, E. M., Ackerman, M. Y. et al. Instrumented mannequin evaluation of thermal protective clothing/Performance of Protective Clothing: Fourth Volume. ASTM International, 1992.

12. Prezant, D. J., Barker, R. L., Bender, M. et al. Predicting the impact of a design change from modern to modified modern firefighting uniforms on burn injuries using manikin fire tests/Performance of Protective Clothing: Issues and Priorities for the 21st Century: Seventh Volume. ASTM International, 2000.

13. Song, G. Modeling thermal protection outfits for fire exposures. North Carolina State University, 2003.

14. Li, X., Lu, Y., Zhai, L. et al. Analyzing thermal shrinkage of fire-protective clothing exposed to flash fire. *Fire Technology*, 2015, 51(1): 195–211.

15. Sipe, J. E. Development of an instrumented dynamic mannequin test to rate the thermal protection provided by protective clothing. Worcester Polytechnic Institute, 2004.

16. ASTM F1930-15. Standard Test Method for Evaluation of Flame Resistant Clothing for Protection Against Fire Simulations Using an Instrumented Manikin.

17. Hardy, J. D. The radiation of heat from the human body. *Journal of Clinical Investigation*, 1934, 13(4): 615–620.

18. Mandal, S., Song, G. Thermal sensors for performance evaluation of protective clothing against heat and fire: A review. *Textile Research Journal*, 2015, 85(1): 101–112.

19. Grimes, R., Mulligan, J. C., Hamouda, H. et al. The design of a surface heat flux transducer for use in fabric thermal protection testing/Performance of Protective Clothing: Fifth Volume. ASTM International, 1996.

20. Ellison, A. D., Groch, T. M., Higgins, B. A. et al. Thermal manikin testing of fire fighter ensembles. BS Thesis, Worcester Polytechnic Institute, USA, 2006.

21. Torvi, D. A., Hadjisophocleus, G. V. Research in protective clothing for fire-fighters: State of the art and future directions. *Fire Technology*, 1999, 35(2): 111–130.
22. Barker, R. L., Hamouda, H., Shalev, I. et al. Review and Evaluation of Thermal Sensors for Use in Testing Firefighters Protective Clothing. Annual Report, 1999.
23. Juricic, D., Musizza, B., Gasperin, M. et al. Evaluation of fire protective garments by using instrumented mannequin and model-based estimation of burn injuries/Control & Automation, 2007. MED'07. Mediterranean Conference on. IEEE, 2007: 1–6.
24. ISO 13506-08. Protective clothing against heat and flame - Test method for complete garments - Prediction of burn injury using an instrumented manikin.
25. Mah, T., Song, G. Investigation of the contribution of garment design to thermal protection. Part 2: Instrumented female mannequin flash-fire evaluation system. *Textile Research Journal*, 2010, 80(14): 1473–1487.
26. Wang, M., Li, X., Li, J. Correlation of bench scale and manikin testing of fire protective clothing with thermal shrinkage effect considered. *Fibers and Polymers*, 2015, 16(6):1370–1377.
27. Pawar Mohan. Analyzing the thermal protective performance of single layer garment materials in bench scale and manikin tests. North Carolina State University, 1995.
28. Rossi, R. M., Bruggmann, G., Stämpfli, R. Comparison of flame spread of textiles and burn injury prediction with a manikin. *Fire & Materials*, 2010, 29(6):395–406.
29. Jin, L., Hong, K. A., Namb, H. D. et al. Effect of Thermal Barrier on Thermal Protective Performance of Firefighter Garments. *Journal of Fiber Bioengineering & Informatics*, 2011, 4(3):245–252.
30. Mah, T., Song, G. Investigation of the Contribution of Garment Design to Thermal Protection. Part 1: Characterizing Air Gaps using Three-dimensional Body Scanning for Women's Protective Clothing/Architectural Institute of Japan, 2010:199–200.
31. Mah, T., Song, G. Investigation of the Contribution of Garment Design to Thermal Protection. Part 1: Characterizing Air Gaps using Three-dimensional Body Scanning for Women's Protective Clothing. *European Physical Journal C*, 2010, 71(4):1607.
32. Song, G. Cothing Air Gap Layers and Thermal Protective Performance in Single Layer Garment. *Journal of Industrial Textiles*, 2007, 36(3):193–205.
33. Li, X., Lu, Y., Zhai, L. et al. Analyzing Thermal Shrinkage of Fire-Protective Clothing Exposed to Flash Fire. *Fire Technology*, 2013, 51(1):195–211.
34. Wang, M., Li, J. Thermal protection retention of fire protective clothing after repeated flash fire exposure. *Journal of Industrial Textiles*, 2016, 46.
35. Lee, C., Kim, I. Y., Wood, A. Investigation and correlation of manikin and bench-scale fire testing of clothing systems. *Fire & Materials*, 2002, 26(6):269–278.
36. Rossi, R. M., Camenzind, M. S. M. Thermal energy transfer through heat protective clothing during a flame engulfment test. *Textile Research Journal*, 2014, 84(13):101–110.

37. Schmid, M., Annaheim, S., Camenzind, M. et al. Determination of critical heat transfer for the prediction of materials damages during a flame engulfment test. *Fire & Materials*, 2016, 40(8):1036–1046.

38. Song, G. W., Barker, R. L., Hamouda, H. et al. Modeling the Thermal Protective Performance of Heat Resistant Garments in Flash Fire Exposures. *Textile Research Journal*, 2004, 74(12):1033–1040.

39. Jiang, Y. Y., Yanai, E., Nishimura, K. et al. An integrated numerical simulator for thermal performance assessments of firefighters' protective clothing. *Fire Safety Journal*, 2010, 45(5):314–326.

40. Wang, Y., Wang, Z., Zhang, X. et al. CFD simulation of naked flame manikin tests of fire proof garments. *Fire Safety Journal*, 2015, 71:187–193.

41. Miao, T., Wang, Z., Li, J. 3D numerical simulation of heat transfer through simplified protective clothing during fire exposure by CFD. *International Journal of Heat & Mass Transfer*, 2016, 93:314–321.

11

Smart Firefighting Clothing

Anna Dąbrowska

11.1 Introduction

Recent technological achievements tightly connected with development of miniaturized, energy-efficient, and low-cost electronic devices, as well as significant interest in internetworking of those devices with precisely programmed functionalities toward advanced services have led to the 4th Industrial Revolution (Industry 4.0) (Schwab, 2016). People want to have those services available and usable anytime and anywhere. This has generated a need for an "ambient intelligence" in which smart devices are embedded into everyday surroundings. In this field, clothing has a significant potential to be the most appropriate location for smart systems that guarantees their closeness and unobtrusiveness, as well as personalization of services (Cho et al., 2010).

The fusion of physical and cyber worlds has had an impact on all domains of human life and activities, including exploration of such areas as occupational safety and health (OSH). Until recently, most of the research in this field concerned development of smart safety systems intended to be used in rescue actions (i.e., for firefighters, mine rescuers, chemical rescuers) when direct hazards for health and life may occur and worker's protection can be guaranteed only by means of personal protective equipment (PPE). This research direction has led to a creation of a new kind of PPE that, thanks to embedded wearable electronics,

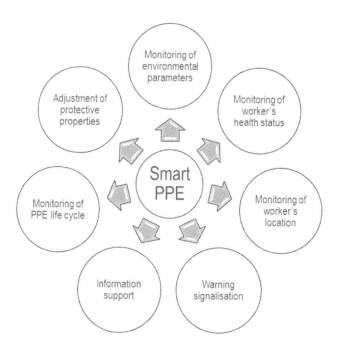

Figure 11.1

Additional functions that can be provided by means of smart PPE. (From Dąbrowska, A., et al., *A need for a new conceptual framework for occupational risk management in Smart Working Environment*, Warsaw, PEROSH, 2015.)

can monitor human-, environment- and work-related parameters, provide information support and warning signalization, as well as adjust its protective properties (Figure 11.1) (Podgórski et al., 2017; Dąbrowska et al., 2015). Clothes as our second skin have a particular potential to acquire an additional functionality as a personalized and flexible information platform (Park & Jayaraman, 2001).

Over the last decade, researchers have paid particular attention to smart protective clothing for firefighters, as they are exposed to one of the most hazardous and physically demanding working conditions. Depending on the rescue action, they

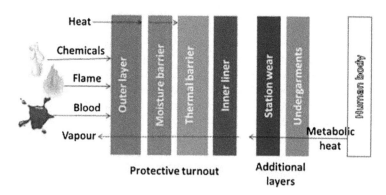

Figure 11.2

A schematic diagram of firefighter's protective clothing and function of each layer. (From Nayak, R., et al., *Fire Science Reviews*, 3(4), pp. 1–19, 2014.)

11. Smart Firefighting Clothing

may face situations in which they are directly exposed to fire, high temperature, atmosphere with hazardous contaminants, limited visibility, as well as thermal and physical strain (Coca et al., 2010; Pietrowski, 2012). Therefore, protective clothing for firefighters is a combination of several layers and each of them has a specific function to ensure safety and health of the user (Figure 11.2) (Nayak et al., 2014).

Networks of interconnected sensor motes embedded into the physical environment enable real-time monitoring and interaction with the environment and human beings, including firefighting application. They can expand basic protective functions of the firefighting clothing and provide the incident commander with a clear view on firefighters' status, optimization of their activities' schedule, and contribute to an increased awareness of the surrounding hazards (Sha et al., 2006). To achieve this goal, a Wireless Body Area Sensor Network (WBASN) integrated with the firefighting clothing can be used. WBASN is a network composed of a number of small sensor nodes or motes with predefined sensing capability. They combine a wide range of information technology, hardware, software, networking, and programming methodologies (Ghosh et al., 2013).

Firefighters are always at risk of cardiac failure resulting from the exposure to high temperature and radiant heat, as well as high level of physical exertion (Ghosh et al., 2013). Wireless Body Area Sensor Network can be used to monitor firefighters' surroundings and give feedback to the rescue operation station and fire chief as a basis for further decisions (Ghosh et al., 2013). Therefore, the aim of this chapter is to analyze state-of-the-art advancements in the field of smart firefighting clothing, including sensor and communication technologies, as well as wireless body area sensor networks and on that basis, to indicate future trends and challenges in this area.

11.2 Sensors for Firefighter's Health State Monitoring

Firefighting operations are often related to exposure to a combination of varied harmful factors such as high weight of protective clothing (about 40 kg), high level of physical effort, and high ambient temperature that can affect a firefighter's physical and mental ability (Gaura et al., 2009). In addition, according to the information provided by the National Fire Protection Association (NFPA) (Fahy et al., 2017), 38% of U.S. firefighters' deaths were caused by sudden cardiac disease, and another 38% by internal trauma, while only less than 15% were caused by crashes and 4% due to falls. Besides the fact that the number of firefighter deaths has significantly decreased over the last decade, analysis of the NFPA indicates that about 70 firefighters per year died on duty in the United States.

Taking into account the above, several research works have been already performed to analyze the causes of sudden cardiac fatal events among firefighters (Payne, 2015). According to Kales et al. (2003), the main reason is underlying coronary heart disease; however, within analysis performed by Smith et al. (2013) a combination of both off-duty and workplace factors was indicated. The authors stated that firefighters are exposed to strenuous physical activity, emotional stress, and environmental pollutants that can affect the cardiovascular system and as a consequence can lead to cardiac arrest. They indicated that sudden cardiac deaths more often occurred with duties related to fire suppression, and in firefighters with underlying atherosclerosis and/or structural heart disease.

Therefore, within recent years a particular interest on monitoring of firefighters' vital signs has been observed. In this field, body sensor network has been recognized as the most promising solution that can provide the complete information about the functioning of a firefighter's cardiovascular system. It is expected that BSNs together with self-learning function based on big data analytics will create personal profiles of the user and on that basis to provide early detection of anomalies (Potirakis et al., 2016; Kumar & Vigneswaran, 2016).

Prevalence of heart issues causes a significant focus on constant monitoring of electrocardiograms and other physiological parameters such as heart rate, blood pressure, or respiration (Gonzales et al., 2015; Kumar & Vigneswaran, 2016). For such long-term monitoring textiles are a preferred platform for sensors that have to be used in direct contact with skin (Shyamkumar et al., 2014). Therefore, many attempts have been made to develop unobtrusive textile-based body-monitoring sensors and electrodes that will not cause skin irritation (Catrysse et al., 2004; Cho et al., 2010). Shyamkumar et al. (2014) distinguished two approaches to fabrication of smart clothing: either through integration of sensors with the finished clothing (e.g., by stitching) or through introduction of smart materials during the fabrication process (e.g., weaving). However, currently those two approaches are often merged; sensors are manufactured during textile fabrication process and then integrated with clothing through the stitching process.

Textile electrodes (so-called "textrodes") for **electrocardiogram** (ECG) and **heart rate** monitoring were developed by Van Langenhove and Hetleer and integrated with a belt worn on a thorax (Van Langenhove & Hertleer, 2004). For this purpose, a knitted structure with stainless-steel fibers was used (Figure 11.3) (Catrysse et al., 2004). Developed textile electrodes were subjected to evaluation, which confirmed accuracy of obtained signals in comparison to conventional electrodes (Cho et al., 2010). Gonzales et al. (2015) developed a microfiber polyester shirt with seamlessly integrated dry textile ECG electrodes made of silver woven fabric. To ensure a good quality of the signals, electrodes were sewn onto elastic bands with a Velcro adjustment system.

Further approach regarding ECG electrodes concerns use of deposition and coating techniques such as sputtering, screen printing, electro-spinning, carbonizing, and evaporation deposition. Those conductive coatings provide higher conductivity; however, they also result in lower durability to washing cycles

Figure 11.3

A view of the ECG textrode. (From Catrysse, M., et al., *Sensors and Actuators, A*(114), pp. 302–311, 2004.)

11. Smart Firefighting Clothing

than electrodes with stainless steel. Ink-jet printing is also considered to be a promising technique for development of ECG electrodes on a textile substrate. This technique requires a formulation of conductive inks or printable capacitive structures on fabrics, which will ensure elasticity and utility parameters on a high level (e.g., abrasion resistance) (Shyamkumar et al., 2014; Rai et al., 2012).

A combination of ECG and **breathing rhythm** monitoring system on a textile substrate was proposed by Jourand et al. (2009). For this purpose, to provide ECG monitoring, three textile dry electrodes (Coosemans et al., 2006) were applied, and for respiration monitoring, two bi-axial accelerometers measure angular and radial displacement. Those sensors were integrated with a T-shirt and tested with human subjects' participation. Textile-based sensors for respiration monitoring were also made of stainless-steel yarn knitted in a belt (called "Respibelt") with an adjustable stretch worn on thorax (Catrysse et al., 2004). The measurements were performed on a basis of changes in circumference and length resulting from breathing, which cause variations in inductance plethysmography and resistance (Solaz et al., 2006). A deformation of the chest due to the breathing was also a basis for Ramos-Garcia et al. (2016) who developed a cover-stitched stretch sensor for breathing monitoring and integrated it with a commercially available shirt. Another solution was proposed by Brady et al. (2005) who developed a foam-based pressure sensor. This sensor on a basis of compression of the foam structure was able to measure chest expansion due to breathing. However, there are also known textile sensors for respiration rate measurement based on optical fibers, which can detect the wavelength-shift resulting from the strain (Cho et al., 2010) or by means of magnetometers (Solaz et al., 2006).

A cuffless noninvasive technique for **blood pressure** monitoring based on pulse transit time was proposed by Zhang et al. (2006). The authors developed a health shirt, which is able to record ECG and photoplethysmogram (PPG) and used them for blood pressure estimations. In this solution, ECG recording was performed on two wrists with another electrode placed on the forearm to improve signal quality, and PPG was captured from the fingertip. The health shirt was evaluated by tests with participation of 10 volunteers. A comparative analysis of the results of blood pressure estimated by means of the health shirt and by means of an automatic meter showed good coherence and potential for its use in daily life. A similar solution was developed by Atomi et al. (2017) who also proposed a wristwatch-type PPG sensor to provide a noninvasive method for continuous monitoring of blood pressure (Figure 11.4).

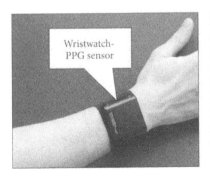

Figure 11.4

A view of the wristwatch-type PPG sensor. (From Atomi, K., et al., *Computational and Mathematical Methods in Medicine*, 1(1), pp. 1–10, 2017.)

In the case of firefighters, **temperature** is also a particularly important biosignal, as it can be a basis for identification of the firefighter overheating. To perform such a measurement, several methods can be used (Solaz et al., 2006), such as textile-embedded thermocouples (Van Langenhove & Hertleer, 2004) or thermistor-based sensors (Teller, 2004). Soukup et al. (2014) developed a temperature and humidity sensor based on specially prepared conductive yarns embroidered on the textile fabric. Husain et al. (2013) proposed a temperature sensing fabric with a double-layer knitted structure made of polyester (a basal yarn) with an embedded metallic wire (a sensing element) (Figure 11.5). As a metallic wire, nickel, tungsten, and copper wires in a bare and insulated form were applied, which are characterized by variation of electrical resistance due to the change in their temperature. Performed analysis indicated that insulated wires were unaffected by strain-dependent resistance errors, and all the samples could be used in a high humidity environment. This means that such fabric can be successfully used in underwear for firefighters and intended for measuring temperature in the undergarment microclimate.

Other biosignals that are frequently measured for medical purposes are galvanic skin response (GSR), electroencephalogram (EEG), electromyogram (EMG), sweat pH, and blood oxygen saturation (Cho et al., 2010; Jeong & Yoo, 2010). **Galvanic skin response** represents electrical conductivity between two points on the user's arm, and it is affected by the sweat from physical activity and by emotional stimuli (Solaz et al., 2006). **Electroencephalography** is a method for measuring electrical activities of the brain by using electrodes along the scalp skin, which is used in medical diagnosis as well as neurobiological research. Kumar and Thilagavathi (2014) developed textile EEG electrodes using a layered structure with conductive and nonconductive fabrics. Performed laboratory tests indicated that those electrodes can be used for high-quality recordings even with human beings with at least thin hair. Such electrodes may be useful in brain–computer interfaces

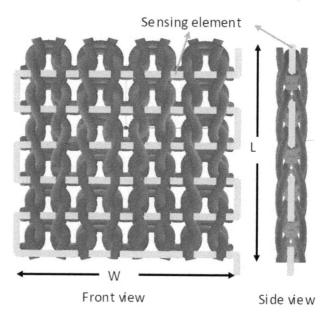

Figure 11.5

A scheme of the temperature sensing fabric. (From Husain, M. D., et al., *International Journal of Textile Science*, 2(4), pp. 105–112, 2013.)

11. Smart Firefighting Clothing

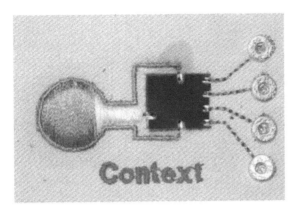

Figure 11.6

A view of the contactless EMG sensor. (From Taelman, J., et al., *Contactless EMG sensors for continuous monitoring of muscle activity to prevent musculoskeletal disorders*, Brussel, IEEE Benelux EMBS Symposium, 2006.)

or detection of drowsiness, and their textile structure will be more comfortable for the user. Another kind of sensor for biosignal monitoring, **electromyogram** sensors, enables monitoring of muscle activity to prevent musculoskeletal disorders. An exemplary solution of such contactless sensor developed within Context project is presented in Figure 11.6 (Taelman et al., 2006).

Recently, monitoring of sweat parameters has become a significant point of interest in biomedical applications as analysis of **sweat pH** and sweat rate may provide valuable physiological information regarding human performance (e.g., a need of rehydration). Coyle et al. (2009) developed textile-based sensors to perform such measurements with the use of superabsorbent material and pH-sensitive dye. In this solution, a change of color of the pH-sensitive dye is measured by a paired emitter–detector LED configuration.

To prevent hypoxemia during highly demanding activities, **blood oxygen saturation** (SpO$_2$) can be measured. For this purpose, Krehel et al. (2014) developed textile-based sensors enabling long-term photopleysmogram (PPG) monitoring. The authors embroidered optical fibers into textiles to provide "light-in light-out" properties. A comparative analysis with a commercially available system indicated a good correlation of obtained results.

11.3 Sensors for Monitoring of Environmental Parameters and Firefighter's Location

In the firefighting environment, numerous hazards can exist with a potential impact on health and safety, such as thermal factors, fire smoke, darkness and low visibility (and related localization issues), as well as physical hazards tightly connected with trips, falls, and crushing (Grant et al., 2015). Therefore, monitoring of those factors is almost as essential as vital signs to provide safety and health of the firefighters during their operations.

During firefighting operations, rescuers are exposed to a risk of **burn injuries** resulting from the contact with hot surfaces and flame, as well as to radiant heat and physiological thermal stress (Grant et al., 2015). That is why Mrugala et al. (2012) proposed a temperature sensor measurement system for firefighter gloves (Figure 11.7).

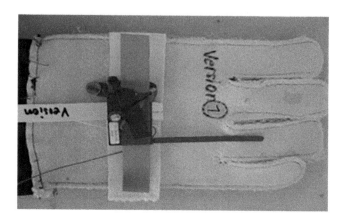

Figure 11.7

Firefighter glove with temperature sensor measurement system. (From Mrugala, D., et al., *Procedia Engineering*, 1(47), pp. 611–614, 2012.)

The designed smart gloves enable both convection and contact temperature measurement, as well as gesture recognition (haptic feedback).

Fire smoke, being **a mixture of solid and liquid aerosols, vapors, and gases** coming from the combustion process, is usually harmful or even toxic for the respiratory tract (Grant et al., 2015). An inhalation of this toxic smoke during operations is also a reason for a huge number of fatalities among firefighters. Carbon monoxide, hydrogen cyanide, and methane are primary toxic gases that require monitoring during fire operations to provide the firefighter with early alert and the possibility to take relevant protective measures (Ghosh et al., 2013). For this purpose, Azavedo et al. (2014) developed a wearable system for firefighters' smoke exposure monitoring that enables registration of exposure to carbon monoxide and nitrogen dioxide. Moreover, the authors also proposed a mobile application running in a smartphone for data processing to evaluate the level of exposure with regards to the acceptable threshold values and run the alarm if needed.

Location, orientation, and body movement measurements are the other crucial aspects in ensuring safety of the firefighting team, as well as managing their operations, especially in the case of high smoke density and darkness (Grant et al., 2015). Tracking firefighters' paths during fire operations gives a possibility for immediately pulling them back from service in the case of increased risk. Analysis of body movements, gestures, and positions is an important source of information for activity classification, removing noise from biosignals, and interpretation of physiological status (Troster, 2004). Navigation support can help firefighters to quickly leave a dangerous location, reduce time of the evacuation, and find other firefighters in distress (Nilsson et al., 2014). To achieve this goal, the following sensors and technologies are used: radio-frequency identification (RFID), accelerometers, gyroscopes, magnetometers, piezoelectric sensors, barometric sensors (with a use of a reference sensor at a known height), imaging sensors, thermal infrared cameras, Doppler radar, and global positioning system (GPS) (Cho et al., 2010; Jeong & Yoo, 2010).

Localization of firefighters in indoor environments is still a research problem (Schubert & Scholz, 2010). Monitoring of firefighters' mobility inside the critical area can be performed by means of sensor motes distributed at different locations and measuring the Received Signal Strength Indicator (RSSI) values received from those sensors (Ghosh et al., 2013). Nilsson et al. (2014) developed a real-time cooperative localization system called TOR (Tactical lOcatoR) based on dual foot-mounted inertial sensors and RF-based inter-agent ranging. In this system, Ultra-Wideband (UWB) transceivers are used with cooperative localization techniques. Each firefighter was equipped with inertial navigation system (INS) including triaxial accelerometer and gyroscope, which were mounted in both shoes to reduce the error growth. Developed system was subjected to evaluation in simulated real utility conditions. On the basis of the obtained results, the authors stated that the TOR system is able to provide position accuracy of about 2–3 m.

In terms of firefighters' safety, analysis of body movement can also provide important information. Iacono et al. (2006) monitored movements of legs and arms of the firefighter, as well as the firefighter's position (laying/standing), by means of a position sensor on the chest and four accelerators distributed in the clothing on each limb. On that basis, detection of firefighter's physical activity (e.g., running, walking, weaving the arms) as well as a lack of movement was possible.

Inertial sensors such as accelerometers and gyroscopes can be also used for detection of firefighters' **physical injuries** resulting from trips, falls, and crushing (Grant et al., 2015; Ghosh et al., 2013). Within many research studies, different positioning of fall detection sensors was applied (Figure 11.8). In the case of one sensor, the most common location was the user's waist. However, single sensors were also mounted on the wrist, head, neck, trunk, chest, back, shoulder, armpit, ear, thigh, and foot. Some research studies were also performed with multiple sensors to support fall detection algorithms or to identify the most appropriate location (Pannurat et al., 2014). To provide prevention functions, wearable devices for fall detection are often combined with vital signs monitoring (Ghasemzadeh et al., 2010), as well as protective measures such as a jacket-worn airbag (Fukaya & Uchida, 2008: Tamura et al., 2009) or a belt-worn airbag (Shi et al., 2009).

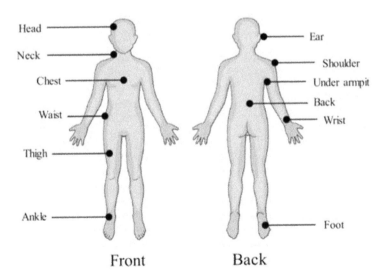

Figure 11.8

Positioning of fall detection sensors. (From Pannurat, N., et al., *Sensors*, 14(7), pp. 12900–12936, 2014.)

11.4 Communication in Smart Firefighting Clothing

Usually communication architecture of the Wireless Body Area Networks consists of three tiers of communications (Figure 11.9): Intra-BAN, Inter-BAN, and Beyond-BAN communications (Negra et al., 2016). Intra-BAN communications refer to radio communications of about 2 m around the human body, which can be divided into two subcategories: (i) communications between body sensors and (ii) communications between body sensors and the portable personal server device (PS). Inter-BAN communications refer to communications between the PS and one or more access points (APs). Beyond-BAN communications is intended for use in metropolitan areas. To bridge the two networks for inter-BAN and beyond-BAN communications, a gateway device (e.g., personal digital assistant or PDA) can be used for a wireless link between these networks (Chen et al., 2011).

There are several communication technologies and standards, such as Bluetooth, Bluetooth Low Energy (BLE), Zigbee and IEEE 802.15.4, IEEE 802.11 (Wi-Fi), IEEE 802.15.6, as well as other radio technologies, including UWB, ANT protocol, Zarlink technology, or Rubee active wireless protocol (Negra et al., 2016). Figure 11.10 describes the most suitable position for BAN in the power vs. data rate spectrum taking into account varied technologies.

Bluetooth technology (Adibi, 2015; Negra et al., 2016) was applied for a short-range communication with a high level of security. This technology guarantees that each device can communicate with seven other devices. Bluetooth devices operate in the 2.4 GHz ISM band (Industrial, Scientific and Medical band), the corresponding coverages range from 1 to 100 m, and maximum data rate is 3 Mbps. It is advantageous in the applications with varied data rates, network coverage, and power requirements. The most convenient application of the Bluetooth is short-term communication with high data rate (Negra et al., 2016).

In the case when less power consumption is possible, **Bluetooth Low Energy technology** (Alam & Hamida, 2015) is considered to be a better choice. This technology is appropriate for usually tiny devices, for example, intended for health monitoring. In this case, maximum data rate is limited to 1 Mbps; however, much quicker

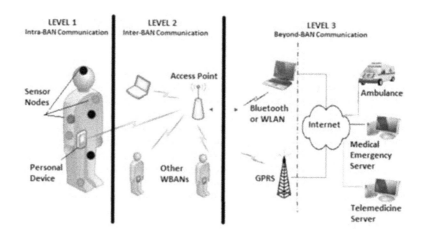

Figure 11.9

General architecture of Wireless Body Area Networks. (From Negra, R., et al., *Procedia. Computer Science*, 1(83), pp. 1274–1281, 2016.)

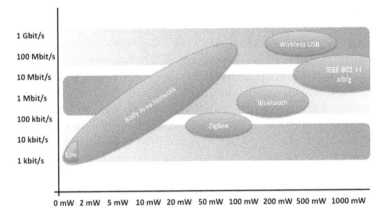

Figure 11.10

Data rate vs. power. (From Karulf, E., *Body Area Networks (BAN)*, available at: http://www.cse.wustl.edu/~jain/cse574-08/ftp/ban.pdf, 2008.)

synchronization of the devices can be obtained, important especially in critical applications requiring alarm generation and emergency response (Negra et al., 2016).

ZigBee (Adibi, 2015) is a technology intended for radio-frequency applications with a low data rate, long battery life, and secure networking. Security and privacy are guaranteed by means of its 128-bit security support for authentication. A basis for this technology is **IEEE 802.15.4** standard. ZigBee technology includes a sleep mode, so this feature highly contributes to a much longer battery life (Acampora et al., 2013). ZigBee-based devices operate in 868 MHz, 915 MHz, and 2.4 GHz, so in the case of WBAN applications, interferences with wireless local area network transmission can occur. Moreover, this technology is characterized by low data rate at about 250 kbps so it is not suitable for large-scale industrial applications, but can be useful in personal use (Negra et al., 2016).

IEEE 802.11 (Vallejos de Schatz et al., 2012) is a set of standards (802.11a, 802.11b, and 802.11c) for wireless local area network (WLAN). Wireless 802.11a uses the frequency band of 5 GHz but is not compatible with two other standards, while 802.11b and 802.11g transmit at 2.4 GHz and are compatible between them (Potirakis et al., 2016). A set of standards 802.11 enables large data transfer (e.g., video conferences, video streaming) due to a high level of data rate. Regarding WBAN, an important advantage is the fact that currently all mobile devices such as smartphones, tablets, and notebooks have integrated Wi-Fi; however, it is characterized by high energy consumption, which can limit application areas (Boulemtafes & Badache, 2016; Negra et al., 2016).

IEEE 802.15.6 (Astrin, 2012) is a standard that was specially designed for WBAN applications and it supports communications inside and around the human body. It uses several frequency bands from the Narrowband (400, 800, 900, 2.3, and 2.4 MHz) to the Ultra-Wideband (3.111.2 GHz) and the Human Body Communication (1050 MHz) (Chen et al., 2014; Negra et al., 2016). IEEE 802.15.6 standard is dedicated for applications with a wide range of data rates and low energy consumption. It has three security schemes, and the maximal data rate is 10 Mbps. This standard will not meet requirements for WBAN requiring high quality of audio and video transmissions (Alam & Hamida, 2015; Negra et al., 2016).

Low power consumption Bluetooth technology and Wi-Fi technology seem to be the best choices for the wireless communication between smart clothing and the outside world. Smart clothing is directly connected to the user's smartphone by Bluetooth. However, for special applications, another communication system is required that should be mounted in regions where smart clothing is used. Then smart clothing connects to the external network through short radio technology for indoor environments, such as low-power Wi-Fi. More energy is consumed for wireless communication to the smartphone, so adoption of the Bluetooth 4.0 is recommended to minimize energy consumption. Another option is Low-Power Wi-Fi standard (IEEE 802.11), which is characterized by low power consumption with frequency below 1 GHz, and improved coverage capability of Wi-Fi signal (Chen et al., 2016).

In firefighting application, the incident commander of a team should continuously receive information about the team members and their surroundings and on that basis decide on operation management. In such a case, Potirakis et al. (2016) proposed two kinds of wireless nodes ("commander" and "firefighter" nodes) to provide data transmission during the operation. In this system, results of measurements from each "firefighter" node are wirelessly transmitted to the central "commander" node and then to a personal computer.

11.5 Wireless Body Area Sensor Network within Smart Firefighting Clothing

Wireless Body Area Sensor Networks consist of wearable electronic devices that are able to collect, process, and communicate data gathered within a human body area. Besides the fact that the interest on this topic began several years ago, only the current technological progress allows for achieving satisfactory results in this field. The existing work on WBASNs is mainly focused on sensor node's energy saving, intra-BAN network design, implantable microsensors, physiological signal acquisition, etc. (Chen et al., 2016). Sangwan and Bhattacharya (2015) distinguished three kinds of WBASNs:

Managed WBASN in which decision on the data collected from one or more than one sensor node is taken by a third party where it is analyzed. Then, this third party decides on further data processing. Such a network is connected to other networks via Wi-Fi or GSM (Karulf, 2008).

Autonomous WBASN, which consists of actuators along with the sensor nodes that can cause relevant action on the human body according to the collected data without a need for waiting for third-party decisions (Karulf, 2008).

Intelligent WBASN, which is a combination of the two above-mentioned networks. In this WBASN simple decisions are taken by its own actuator nodes, and complex issues are sent to the third party. If within the presumed period of time there is no response, the WBSN makes a decision on its own (Sangwan & Bhattacharya, 2015).

A need for monitoring of firefighters' safety was noticed by Ghosh et al. (2013) who proposed a Wireless Body Area Sensor Network (WBASN) including gathering various physiological data and concentrations of potentially hazardous gases in a firefighter's vicinity. The developed WBASN also included generation of an alert in the case of exceeding the prescribed thresholds, as well as monitoring of firefighter's mobility inside the crisis area.

In a framework of the *i-Protect* project a new PPE system based on integration of sensors, nanomaterials, and ICT solutions with protective clothing was developed (Pietrowski, 2012). The main objective of the project was to develop an advanced PPE system guaranteeing active protection and information support for three groups of first responders: firefighters, chemical, and mine rescuers. Within the project, ergonomics of the design and its coherence with end-users' needs as well as predicted working conditions were taken into account. Moreover, research works on the development of microsensors for real-time monitoring of environmental parameters such as temperature, concentration of the selected toxic gases, and oxygen deficiency were performed. To ensure monitoring of the first responders' health status (i.e., body temperature and heart rate), optical fibers were provided and integrated with the underwear. One of the crucial goals of the project was also to ensure wireless communication between sensors and communication units as well as between each rescuer and the Rescue Command Centre.

In the *PROeTEX* project, the inner garment with integrated textile electrodes for ECG measurement, piezoresistive textile patch for respiration measurement, and temperature measurement for core temperature estimation were applied to improve safety of firefighters. In addition, this garment can be completed by a peripheral capillary oxygen saturation sensor in a form of a large electrode for the dehydration estimation. The main electronic system was located in the outer garment. It combines several sensors enabling location determination, activity and posture determination, as well as environmental parameters monitoring (temperature, heat flux, and gas concentration). Bluetooth communication module was used to link this sensor network with a long-distance communication module by means of a textile antenna. Moreover, acoustic and optical alarms were also implemented to the clothing (Berling, 2013; Voirin, 2015). A concept of the *PROeTEX* project is presented in Figure 11.11.

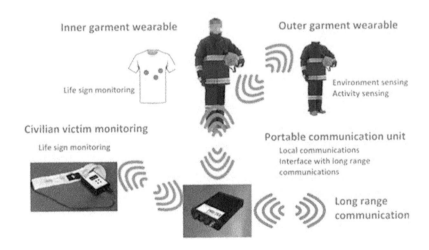

Figure 11.11

Scheme of the overall concept of the PROeTEX project. (From Voirin, G., Working garment integrating sensor applications developed within the PROeTEX project for firefighters, En: K. Kinder-Kurlanda & C. E. Niham, edits, *Ubiquitous Computing in the Workplace. What Ethical Issues? An Interdisciplinary Perspective*, Cham: Springer, pp. 25–36, 2015.)

Smart protective clothing aimed at improvement of safety of firefighters was also a goal of the *ProFiTEX* project (Berling, 2013; Hertleer et al., 2013). In this project, a strong emphasis was put on a tight cooperation with end-users at both stages of the designing and evaluation process. Regarding the design, the developed smart protective clothing was based on the results of the previously conducted *wearIT@work* project. It comprises a jacket for firefighters with embedded wearable electronics and a braided rope with beacons (called *Smart LifeLine*) for security issues, as well as data and energy transmission. In addition, firefighters were equipped with helmet-mounted display that indicated location of beacons, and as a result, the way back. A scheme of the concept of the ProFiTex project is shown in Figure 11.12.

Smart Personal Protective System for firefighters was also developed within the *Smart@Fire* project (Baxa & Soukup, 2016). The whole system includes several components such as heart rate module, integrated temperature sensors, alarm module, gas sensor modules, integrated illumination, Suit Control Unit (SCU) and Commander Control Unit (CCU), smart glove with integrated thermocouple, and IR temperature sensor and Arriana inertial localization system mounted in shoes (Figure 11.13). The firefighter suit set consists of a firefighter suit, sensor modules, acoustic alarm module, SCU, and the tool for sensor module demounting. WBASN is based on the Bluetooth technology version 4.0 (2.45 GHz). Moreover, wireless area network (WAN) was developed in a form of mesh network and signal repeaters were applied to prolong the communication distance (868 MHz). The developed Smart Personal Protective System enables environmental and physiological monitoring, IR thermal hotspot detection, localization, data visualization, and active illumination, according to the defined firefighters' needs.

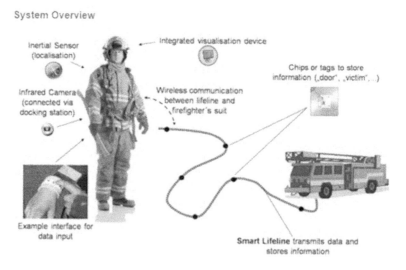

Figure 11.12

Scheme of the overall concept of the ProFiTex project. (From Kaufmann, H., *ProFiTex—Providing Fire Fighters with Technology for Excellent Work Safety*, available at: https://www.ims.tuwien.ac.at/projects/profitex, 2012.).

11. Smart Firefighting Clothing

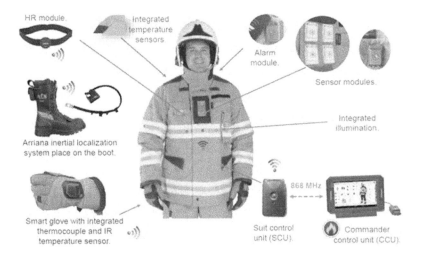

Figure 11.13

Scheme of the overall concept of Smart@fire project. (From Baxa, M., & Soukup, R., *Smart@Fire Final Conference: Smart Personal Protective System*. Brussels, University of West Bohemia, 2016.)

11.6 Challenges

Besides the fact that many research works have been already performed in the field of smart firefighting clothing, there are still several challenges that require further considerations to ensure market uptake. We can divide those challenges into the following categories:

- Physical technologies (e.g., textronics)
- Cyber technologies (e.g., Big Data analytics, cloud computing)
- Ergonomics (i.e., human factors)
- Ethics
- Standardization

Challenge 1: Physical Technologies

A literature review indicated that currently most solutions of smart firefighting clothing include separate electronic and textile parts. This means that such clothing still has plastic and/or metal components, which are stiff and can be uncomfortable during usage. Therefore, we should aim at development of **new integration techniques** and, as a consequence, further advancement in the **textronics** (i.e., textile electronics), which will guarantee unobtrusive monitoring functions with high accuracy. However, ensuring high level of washing and utility resistance, as well as resistance to harsh environment of textronic smart clothing, are key issues to be solved.

To achieve an increase in the market uptake of smart clothing, there is also a need for further developments in the field of **powering**, both in energy consumption management as well as in energy harvesting and wireless powering. Implementation of more wearables into the clothing, especially those that will exist as separate nodes, will require more batteries. This issue is extremely important taking into account the utility value of smart clothing (i.e., time needed for a battery exchange), as well as environmental aspects (i.e., amount of electrowaste). Therefore, we should aim at providing a maintenance-free smart clothing by means of limitation of additional activities that are required to provide operation of smart clothing.

Special attention should be also paid on the development of **actuation technologies**. Until now, the main interest was focused on providing monitoring functions in the clothing. However, such approach highly reduces capabilities of smart clothing, as it only provides information, without any reaction. That is why a key issue is to concentrate on actuators, based on nanoelectromechanical systems (NEMS), which will provide simultaneous, adequate, and customized response depending on the context.

Challenge 2: Cyber Technologies

Another future trend relates to **advanced algorithms** for cloud and pervasive computing, including predictive computations on a basis of Big Data analytics. A huge amount of data that can be collected from the sensor nodes in WBASN has a great potential for ensuring increased safety and health at work. A combination of physical smart components with predictive computations will lead to a creation of cyber-physical systems that can monitor human direct surroundings and on that basis provide adequate measures to increase safety and health as well as efficiency at work. However, this vision requires a change in paradigm concerning occupational safety and health (OSH) management from the periodic and collective approach to a real-time and personalized one (Podgórski et al., 2017).

Regarding challenge of the cyber technologies, **security** should be mentioned. It is particularly important in the case of smart firefighting clothing, as hacking the Wireless Body Sensor Network may result in wrong decisions during the rescue operations and exposure of firefighters to an increased risk of losing their health or even life. Hence, further developments in the field of encryption techniques to provide WSBAN reliability are required.

Challenge 3: Ergonomics

One of the main challenges that smart clothing designers will have to face is **human acceptance**, which is the result of such factors as usability and functionality (regarding wearables), comfort, and fashion (regarding clothing), as well as durability and safety (regarding integrated smart clothing) (Cho et al., 2010).

Protection and comfort are two key issues to be focused on within further research. Therefore, designing of smart clothing for **specific application including end-users' engagement** will be required. This issue is particularly important in the case of OSH-related applications, such as firefighting smart protective clothing.

Challenge 4: Ethics

Management of personal data, especially in the case of working environment, raises concerns related to **privacy**. Besides physiological monitoring, special attention needs to be paid to management of data coming from sensors that enable tracking of the user, including location sensors, voice, and picture recorder. Moreover, privacy not only of the user but also of the people around the user should be taken into account by means of background noise reduction or a use of fisheye lens onto the camera and recording with lower resolution (Ashbrook et al., 2010). In addition, clear rules should be elaborated, which will indicate how the personal data of the workers can be used, stored, analyzed, etc., taking into account ethics and privacy issues.

Challenge 5: Standardization

Application of smart technologies cannot provide an additional risk to the user, so **new testing methods and requirements**, as well as electronics that will guarantee its operation in harsh and complex environments, should be further developed. To reduce costs of such testing, as well as to evaluate performance of the smart protective clothing during its usage, further research on theoretical and simulation methods that do not require destructive testing is expected (Pan, 2011). Those activities should be supported by **standardization works** to work out a common approach in this field. Therefore, the European Commission has implemented a decision on standardization request regarding advanced garments and ensembles of garments that provide protection against heat and flame, with integrated smart textiles and nontextile elements for enhanced health, safety, and survival capabilities (mandate M/553 dated 6 of January 2017).

11.7 Conclusions

Smart firefighting protective clothing is a topic of significant interest from several years, what resulted in a huge number of sensors, as well as Wireless Body Sensor Networks intended for this application. However, it should be noticed that a market uptake of both kinds of clothing, as well as smart components indicated for integration with it, is still limited. Some issues being solved that may overcome this issue were described in Section 11.6. However, an important driving factor for implementation of smart firefighting clothing to the market may be also a hype on the Internet of Things technologies and fashion for a healthy lifestyle. This trend has caused that application of wearable technologies for safety and health purposes currently passes its revival.

A focus on providing smart clothing intended for firefighters results from the fact that their working environment is characterized by one of the most hazardous, unpredictable, and demanding environmental conditions. Therefore, a need for increasing safety of firefighters was placed on the first position. However, a need for improvement of occupational safety and health concerns also exists for other professional domains, including various types of industry, to which many of the solutions described in the chapter can be transferred.

What has changed in recent years is also a perspective from narrow approach focused on autonomous smart protective clothing (meaningless whether for firefighters or other groups) to wide approach related to smart working environment (SWE), in which various physical components (inter alia smart protective clothing or other kinds of personal protective equipment) of the working environment are linked together by means of cyber technologies creating cyberphysical system (CPS). Such CPS gives more possibilities in accident prevention and improvement of work safety in comparison to autonomous smart protective clothing. This results from the fact that data can be gathered and analyzed not only from the direct surrounding of the worker (as is the case of smart protective clothing) but also from other smart objects distributed in the working environment. Then, complex and real-time cloud computing, which also includes historical data, can allow predicting potentially dangerous situations before human exposure and prevent them.

11.8 Acknowledgments

The chapter has been based on the results of Phase IV of the National Programme "Safety and working conditions improvement," funded in the years 2017–2019 in the area of research and development works by the Ministry of Science and Higher Education/The National Centre for Research and Development. The Programme coordinator: Central Institute for Labour Protection—National Research Institute.

References

Acampora, G., Cook, D. J., Rashidi, P. & Vasilakos, A. V., 2013. A survey on ambient intelligence in healthcare. *Proceedings of the IEEE,* 101(12), pp. 2470–2494.

Adibi, S., 2015. *Mobile Health: A Technology Road Map (Vol. 5).* 1 ed. Cham: Springer.

Alam, M. M., & Hamida, E. B., 2015. Strategies for optimal MAC parameters tunning in IEEE 802.15.6 wearable wireless sensor networks. *Journal of Medical Systems,* 39(9), pp. 1–16.

Ashbrook, D., Lyons, K., Clawson, J., & Starner, T., 2010. Methods of Evaluation for Wearable Computing. En: G. Cho, ed. *Smart Clothing. Technology and Applications.* Boca-Raton: CRC Press, pp. 229–248.

Astrin, A., 2012. *802.15.6-2012 - IEEE Standard for Local and metropolitan area networks - Part 15.6: Wireless Body Area Networks.* Available at http://ieeex plore.ieee.org/document/6161600/

Atomi, K., Kawanaka, H., Bhuiyan, M. S., & Oguri, K., 2017. Cuffless blood pressure estimation based on data-oriented continuous health monitoring system. *Computational and Mathematical Methods in Medicine,* 1(1), pp. 1–10.

Azavedo, P. et al., 2014. A wearable system for firefighters smoke exposure monitoring. En: D. X. Viegas, ed. *Advances in forest fire.* Coimbra: Imprensa da Universidade de Coimbra, pp. 1312–1318.

Baxa, M., & Soukup, R., 2016. *Smart@Fire FInal Conference: Smart Personal Protective System.* Brussels, University of West Bohemia.

Berling, L., 2013. *Smart Textiles and Wearable Technology—A study of smart textiles in fashion and clothing,* Boras: Swedish School of Textiles, University of Boras.

Boulemtafes, A., & Badache, N., 2016. Design of Wearable Health Monitoring Systems: An Overview of Techniques and Technologies. En: A. A. Lazakidou, S. Zimeras, D. Iliopoulou & D. Koutsouris, edits. *mHealth Ecosystems and Social Networks in Healthcare.* Cham: Springer, pp. 79–94.

Brady, S., et al., 2005. *Garment-based monitoring of respiration rate using a foam pressure sensor.* Osaka, IEEE Computer Society.

Catrysse, M., et al., 2004. Towards the integration of textile sensors in wireless monitoring suit. *Sensors and Actuators,* A(114), pp. 302–311.

Chen, M., et al., 2011. Body Area Networks: A Survey. *Mobile Networks and Applications,* 1(16), pp. 171–193.

Chen, M., et al., 2016. Smart clothing: connecting human with clouds and big data for sustainable health monitoring. *Mobile Networks and Applications,* 21(5), pp. 825–845.

Chen, M., et al., 2014. A survey of recent developments in home M2M networks. *IEEE Communications Surveys & Tutorials,* 16(1), pp. 98–114.

Cho, G., Lee, S., & Cho, J., 2010. Review and reappraisal of smart clothing. En: G. Cho, ed. *Smart clothing. Technology and applications.* Boca Raton: CRC Press, pp. 1–35.

Cho, G., Lee, S., & Cho, J., 2010. Review and Reappraisal of Smart Clothing. En: G. Cho, ed. *Smart Clothing. Technology and Applications.* Boca Raton: CRC Press, pp. 1–35.

Coca, A., Williams, W. J., Roberge, R. J., & Powell, J. B., 2010. Effects of fire fighter protective ensembles on mobility and performance. *Applied Ergonomics,* 41(1), pp. 636–641.

Coosemans, J., Hermans, B., & Puers, R., 2006. Integrating wireless ECG monitoring in textiles. *Sensors and Actuators A: Physical,* 1(130–131), pp. 48–53.

Coyle, S., et al., 2009. *Textile sensors to measure sweat pH and sweat-rate during exercise.* London, Pervasive Health.

Dąbrowska, A., et al., 2015. *A need for a new conceptual framework for occupational risk management in Smart Working Environment.* Warsaw, PEROSH.

Fahy, R. F., LeBlanc, P. R., & Molis, J. L., 2017. *Firefighters Fatalities in the United States—2016.* Quincy, Massachusetts: NFPA.

Fukaya, K., & Uchida, M., 2008. Protection against impact with the ground using wearable airbags. *Industrial Health,* 46(1), pp. 59–65.

Gaura, E. I., Brusey, J., Kemp, J., & Thake, C. D., 2009. Increasing safety of bomb disposal missions: a body sensor network approach. *IEEE Transactions on Systems, Man and Cybernetics,* 39(6), pp. 621–636.

Ghasemzadeh, H., Jafari, R., & Prabhakaran, B., 2010. A body sensor network with electromyogram and inertial sensors: multimodal interpretation of muscular activities. *IEEE Transactions on Information Technology in Biomedicin,* 14(2), pp. 198–206.

Ghosh, S. K., Chakraborty, S., Jamthe, A., & Agrawal, D. P., 2013. *Comprehensive monitoring of firefighters by a wireless body area sensor network.* Bhopal, IEEE.

Gonzales, L., et al., 2015. *Textile sensor system for electrocardiogram monitoring.* Raleigh, IEEE Virtual Conference on Applications of Commercial Sensors (VCACS).

Grant, C., et al., 2015. *Research Roadmap for Smart Fire Fighting,* Gaithersburg: National Institute of Standards and Technology.

Hertleer, C., Odhiambo, S., & Van Langenhove, L., 2013. Protective clothing for firefighters and rescue workers. En: R. A. Chapman, ed. *Smart textiles for protection.* Cambridge: Woodhead Publishing Limited, pp. 338–363.

Husain, M. D., Atalay, O., & Kennon, R., 2013. Effect of Strain and Humidity on the Performance of Temperature Sensing Fabric. *International Journal of Textile Science,* 2(4), pp. 105–112.

Iacono, M., et al., 2006. *Monitoring fire-fighters operating in hostile environments with body-area wireless sensor networks.* Pisa, 5th Conference on Risk Assessment and Management in the Civil and Industrial Settlements.

Jeong, K. S., & Yoo, S. K., 2010. Electro-Textile Interfaces. Textile-Based Sensors and Actuators. En: G. Cho, ed. *Smart Clothing. Technology and Applications.* Boca-Raton: CRC Press, pp. 89–113.

Jourand, P., De Clercq, H., Corthout, R., & Puers, R., 2009. Textile integrated breathing and ECG monitoring system. *Procedia Chemistry*, 1(1), pp. 722–725.

Kales, S. N., Soteriades, E. S., Christoudios, S. G., & Christiani, D. C., 2003. Firefighters and on-duty deaths from coronary heart disease: a case control study. *Environmental Health*, 2(14).

Karulf, E., 2008. *Body Area Networks (BAN)*. Available at: http://www.cse.wustl.edu/~jain/cse574-08/ftp/ban.pdf

Kaufmann, H., 2012. *ProFiTex—Providing Fire Fighters with Technology for Excellent Work Safety*. Available at: https://www.ims.tuwien.ac.at/projects/profitex

Krehel, M., et al., 2014. Development of a luminous textile for reflective pulse oximetry measurements. *Biomedical Optics Express*, 5(8), pp. 2537–2547.

Kumar, L. A., & Vigneswaran, C., 2016. *Electronics in Textiles and Clothing. Design, Products and Applications*. 1 ed. Boca Raton: CRC Press.

Kumar, N. M., & Thilagavathi, G., 2014. Design and development of textile electrodes for EEG measurement using copper plated polyester fabrics. *Journal of Textile Apparel, Technology and Management*, 8(4), pp. 1–8.

Mrugala, D., Ziegler, F., Kostelnik, J., & Lang, W., 2012. Temperature sensor measurement system for firefighter gloves. *Procedia Engineering*, 1(47), pp. 611–614.

Nayak, R., Houshyar, S., & Padhye, R., 2014. Recent trends and future scope in the protection and comfort of fire-fighters' personal protective clothing. *Fire Science Reviews*, 3(4), pp. 1–19.

Negra, R., Jemili, I., & Belghith, A., 2016. Wireless Body Area Networks: Applications and technologies. *Procedia. Computer Science*, 1(83), pp. 1274–1281.

Nilsson, J.-O., et al., 2014. *Accurate Indoor Positioning of Firefighters using dual foot-mounted Inertial Sensors and Inter-agent Ranging*. IEEE, Monterey.

Pan, N., 2011. Developments in clothing protection technology. En: N. Pan & G. Sun, edits. *Functional textiles for improved performance, protection and health*. Cambridge: Woodhead Publishing Limited, pp. 269–290.

Pannurat, N., Thiemjarus, S., & Nantajeewarawat, E., 2014. Automatic Fall Monitoring: A Review. *Sensors*, 14(7), pp. 12900–12936.

Park, S., & Jayaraman, S., 2001. *Textiles and computing: Background and opportunities for convergence*. Atlanta, Proceeding ACN CASAS'01, pp. 186–187.

Payne, J. A., 2015. *"Sensing disaster": the use of wearable sensor technology to decrease firefighting line-of-duty deaths*, Monterey: Naval Postgraduate School.

Pieper, S., 2010. *On-site @ FDIC: Research illuminates the relationship between firefighting and cardiac strain*. Available at: http://my.firefighternation.com/profiles/blogs/onsite-fdic-research

Pietrowski, P., 2012. New PPE system development based on integration of sensors, nanomaterials and ICT solutions with protective clothing - i-Protect project approach. En: G. Bartkowiak, I. Frydrych & M. Pawłowa, edits. *Innovations in Clothing Technology & Measurement Techniques*. Warsaw: Technical University of Lodz Press, pp. 163–170.

Podgórski, D., et al., 2017. Towards a conceptual framework of OSH risk management in smart working environments based on smart PPE, ambient intelligence and the Internet of Things technologies. *International Journal of Occupational Safety and Ergonomics*, 23(1), pp. 1–20.

Potirakis, S. M., et al., 2016. Physiological parameters monitoring of fire-fighters by means of a wearable wireless sensor system. *IOP Conf. Series: Materials Science and Engineering,* 108(012011), pp. 1–9.

Rai, P., et al., 2012. Smart healthcare textile sensor system for unhindered-pervasive health monitoring. *Proc. SPIE,* 1(8344), pp. 1–10.

Ramos-Garcia, R. I., et al., 2016. *Analysis of a coverstitched stretch sensor for monitoring of breathing.* Nanjing, IEEE.

Sangwan, A., & Bhattacharya, P. P., 2015. Wireless Body Sensor Networks: A Review. *International Journal of Hybrid Information Technology,* 8(9), pp. 105–120.

Schubert, E., & Scholz, M., 2010. *Evaluation of wireless sensor technologies in a firefighting environment.* Kassel, IEEE.

Schwab, K., 2016. *The Fourth Industrial Revolution.* 1 ed. Geneva: World Economic Forum.

Sha, K., Shi, W., & Watkins, O., 2006. *Using wireless sensor networks for fire rescue applications: requirements and challenges.* East Lansing, IEEE.

Shi, G., et al., 2009. Mobile human airbag system for fall protection using MEMS sensors and embedded SVM classifier. *IEEE Sensors Journal,* 9(5), pp. 495–503.

Shyamkumar, P., et al., 2014. Wearable wireless cardiovasclar monitoring using textile-based nanosensor and nanomaterial systems. *Electronics,* 1(3), pp. 504–520.

Smith, D. L., Barr, D. A., & Kales, S. N., 2013. Extreme sacrifice: sudden cardiac death in the US fire service. *Extreme Physiology and Medicine,* 2(6).

Solaz, J.-S., et al., 2006. Intelligent textiles for medical and monitoring applications. En: H. R. Mattila, ed. *Intelligent textiles and clothing.* Cambridge: Woodhead Publishing Limited, pp. 369–398.

Soukup, R., Hamacek, A., Mracek, L., & Reboun, J., 2014. *Textile based temperature and humidity sensor elements for healthcare applications.* Dresden, IEEE.

Taelman, J., et al., 2006. *Contactless EMG sensors for continuous monitoring of muscle activity to prevent musculoskeletal disorders.* Brussel, IEEE Benelux EMBS Symposium.

Tamura, T., et al., 2009. A wearable airbag to prevent fall injuries. *IEEE Transactions on Information Technology in Biomedicine,* 13(6), pp. 910–914.

Teller, A., 2004. A platform for wearable physiological computing. *Interacting with computers,* 1(16), pp. 917–937.

Troster, G., 2004. The agenda of wearable healthcare. En: R. Haux & C. Kulikowski, edits. *IMIA yearbook of medical informatics 2005: Ubiquitous health care systems.* Stuttgart: Schattauer, pp. 125–138.

Vallejos de Schatz, C. H., Medeiros, H. P., Schneider, F. K., & Abatti, P. J., 2012. Wireless medical sensor networks: Design requirements and enabling technologies. *Telemedicine and e-Health,* 18(5), pp. 394–399.

Van Langenhove, L., & Hertleer, C., 2004. Smart clothing: a new life. *International Journal of Clothing Science and Technology,* 16(1/2), pp. 63–72.

Voirin, G., 2015. Working garment integrating sensor applications developed within the PROeTEX project for firefighters. En: K. Kinder-Kurlanda & C. E. Niham, edits. *Ubiquitous Computing in the Workplace. What Ethical Issues? An Interdisciplinary Perspective.* Cham: Springer, pp. 25–36.

Wang, H., et al., 2010. Resource-aware secure ecg healthcare monitoring through body sensor networks. *IEEE Wireless Communications,* 17(1), pp. 12–19.

Zhang, Y. T., et al., 2006. *A health-shirt using e-textile materials for the continous and cuffless monitoring of arterial blood pressure.* Boston, 3rd IEEE/EMBS International Summer School on Medical Devices and Biosensors.

Zhang, Z., Wang, H., Wang, C., & Fang, H., 2013. Interference mitigation for cyber-physical wireless body area network system using social networks. *IEEE Transactions on Emerging Topics in Computing,* 1(1), pp. 121–132.

12

Numerical Modeling for Heat and Moisture Transfer through Firefighting Protective Clothing

Yun Su and Guowen Song

12.1 Introduction

While extinguishing a fire or completing an emergency rescue, firefighters are commonly subjected to heat and flame, hot objects, hot liquids, and steam that can result in skin burn injuries and even death. Thermal protective clothing is an effective option to minimize skin burns. To better characterize the performance of thermal protective clothing, laboratory simulation methods have been employed to simulate various fire scenarios, such as flash fire and thermal radiation. For example, Thermal Protective Performance (TPP) tester and flame manikins can be used to evaluate the protective performance of fabric and clothing under flash fire [1]. The thermal protection of fabric and clothing under medium- and high-level thermal radiation can be examined by using Radiant Protective Performance (RPP) tester and radiant manikin [2]. In addition, a bench top tester (stored thermal energy test [SET]) was developed to more precisely evaluate the thermal protective performance of fabrics in low-level radiant heat exposure [3].

Although the experimental measurement of thermal protective performance effectively characterizes the performance of protective clothing, these tests are time-consuming and costly. Also, it is difficult to reuse tested fabric or clothing after heat exposure due to thermal shrinkage and thermal degradation, which can result in serious resource waste. With the rapid development of computer technology, numerical simulation as a nondestructive test has attracted the attention of researchers and become an important scientific research method. The complexity of heat and moisture transfer in clothing makes the analytical approach intractable, whereas computational modeling is both a quick and efficient alternative for analysis of heat and moisture transfer mechanisms. For one thing, the numerical model is a simulative platform to represent heat and moisture transfer in heat sources, protective clothing, air gaps, and skin tissue, as well as predict the required time to cause second- and third-degree skin burn injuries. For another, parametric studies can be carried out based on the heat and moisture transfer model, thus determining the effect of exposure conditions, fabric properties, and air gap size on thermal protective performance of clothing.

It is important to understand the mechanisms associated with heat and moisture transfer in firefighting protective clothing. In this chapter, model developments in evaluating heat and moisture transfer in firefighting protective clothing are discussed in detail. The human thermal response model is reviewed in terms of the thermal regulation model, skin bio-heat transfer model, and skin burn model. Finally, the application progress and developing directions of the numerical model are presented, aimed at establishing an improved numerical model and providing a better understanding of heat and moisture transfer behavior in firefighting protective clothing.

12.2 Heat and Moisture Transfer Mechanism

The heat and moisture transfer between human body, clothing, and environment is a dynamic process, as shown in Figure 12.1. Firefighting protective clothing is usually composed of an outer shell, a moisture barrier, and a thermal liner, which is required by NFPA 1971-15. Due to the specific geometry of the human body's shape and unique fabric properties, the air gap between a garment and the human body is unevenly distributed over the body when the firefighting protective clothing is dressed on the human body [4]. The air gap is also entrapped in between various layers of multilayer protective clothing. The air gap model between the clothing and human body, the skin bio-heat transfer model, and

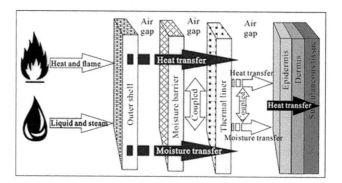

Figure 12.1

Diagram of heat and moisture transfer in "fire environment–multilayer protective clothing–human body."

Henriques' burn model are usually integrated into the clothing model to calculate skin temperature and the skin burn time.

The driving forces of heat and moisture transfer are primarily temperature difference, concentration difference, and pressure difference inside and outside of protective clothing. Thermal energy is usually transported through protective clothing by means of conduction, convection, and radiation. Convective heat transfer generally occurs on the outer surface of protective clothing only due to the small porous structure for these fabric systems [5]. Therefore, the heat transfer rate in multilayer protective clothing is determined by radiant and conductive heat transfer. Conduction is dependent on the percentage of fiber, air content, and moisture content within fabric; protective clothing's porous material is designed as such specifically so that thermal radiation can penetrate the protective clothing. This radiative heat transfer in protective clothing can be simulated based on a two-flux model, radiative transfer equation (RTE), and Beer's law. In addition, multilayer protective clothing can store thermal energy to affect the heat transfer rate during exposure.

Water within multilayer protective clothing can present as three phases (including bound water, liquid water, and water vapor). The moisture transfer process in protective clothing is quite complex. The primary methods of moisture transfer include molecular diffusion, water permeation, convective mass transfer, and capillary flow [6]. Heat and moisture transport in protective clothing are coupled, meaning that moisture within protective clothing can change the clothing's thermophysical properties, such as thermal conductivity and specific heat capacity, as well as its optical properties including radiant absorption coefficient and transmissivity [7,8]. Moreover, moisture absorption/desorption is affected during phase change. Therefore, fabric can absorb or release a large amount of energy, affecting the heat transfer process. The reverse is also true: heat transfer can change the existing status of moisture in a clothing system.

12.3 Numerical Models of Heat and Moisture Transfer in Protective Clothing

Numerical models of heat and moisture transfer in protective clothing can be divided into two categories: dry heat transfer models and wet heat transfer models. In certain fire scenarios, firefighters wear dry protective clothing to withstand heat and flame for short duration exposures. The impact of moisture transfer in protective clothing can be ignored in this situation. A dry heat transfer model is a simple model that represents heat transfer behavior in protective clothing. However, for most fire scenarios firefighters encounter, it is necessary to extinguish a fire and perform emergency rescue for long duration exposures. In these circumstances, protective clothing can be wetted by skin sweat, splashing water from a hose spray, and dew or rain [9]. Moisture transport in protective clothing exerts a significant effect on heat transfer. The coupled heat and moisture transfer model is a complex model that can determine the influence of moisture transfer on thermal protective performance.

12.3.1 Advances in Heat Transfer Model

Over the past decades, heat transfer models have been widely developed to study firefighting protective clothing. In the early 1970s, Morse et al. [10] simulated the thermal response of protective clothing-covered skin subjected to a JP-4 fuel fire, considering the effect of pyrolysis, ignition, and combustion of fabric on heat transfer. In the late twentieth century, Torvi [11] developed the classic heat transfer model for a single-layer fabric system, which mainly accounted for the simulation method of radiant heat transfer and proved that a one-dimensional model effectively simulated heat transfer in protective clothing. In recent years, more precise one-dimensional models have been established by other researchers. For example, a transient heat transfer model developed by NIST was employed to study the influence of the rate of radiant/convective heat flux on the TPP of fabrics [12]. Zhu et al. [13] used a new numerical model to investigate heat transfer in a cylinder sheathed by flame-resistant fabrics, which considered the effects of cylindrical geometry on heat transmission in fabrics. Concurrently, Mercer and Sidhu analyzed heat transport in "intelligent" protective clothing with embedded phase change material during radiant heat exposure [14].

The degradation reaction or pyrolysis of clothing can affect heat transfer through clothing exposed to heat and flame. Torvi [5] used apparent heat capacity to evaluate the impact of thermochemical reactions in the material. Zhu and Li developed a pyrolysis model of moist fabric that assumed a one-step chemical reaction [15]. This numerical model considered heat-induced changes in fabric's thermophysical properties (pyrolysis) and drying process.

A clothing numerical model developed by Song et al. [16] was employed to explain heat transfer behavior in a configuration that realistically simulated the shape of the human body. Models for air gaps in the fabric system were further improved by Ghazy [17–19], including multiple air gaps model of multilayer fabric system [18], and dynamical air gap models that considered body motion and thermal shrinkage of fabric [17,19]. A three-dimensional transient heat transfer in the flame manikin test of thermal protective clothing was proposed to predict the distribution of skin burns when exposed to flash fire, based on computational fluid dynamics (CFD) techniques [20,21]. The CFD techniques were also used to simulate the heat transfer in an air gap for investigating the effect of air gap orientations, heterogeneous air gap, and body movement [22,23]. Recently, Su et al. [24] developed a heat storage model that can be used to analyze heat storage during exposure and heat release without compression during cooling. A compressive heat transfer model in protective clothing was also developed for investigating the effects of applied pressure and applied temperature on skin burns [25].

12.3.1.1 Conduction

Conductive heat transfer in protective clothing is a dominant mode for transporting thermal energy owing to the small porous size within the fabric system. According to Fourier's law, one-dimensional heat transfer due to conduction is calculated by Equation 12.1.

$$-k_{fab}\frac{\partial T}{\partial x}=q_{cond} \tag{12.1}$$

where q_{cond} is the conductive heat flux, k_{fab} is the thermal conductivity of fabric system, and $\partial T/\partial x$ is the temperature gratitude along the x direction. The thermal conductivity of fabric systems is dependent on the volume percentage of dry fiber, air, and moisture content. It was considered that the thermal conductivity of fabric systems can vary with the increase in fabric's temperature under heat and flame [11].

The dead air within the interspaces of fabric is a good heat insulator since the thermal conductivity of air is far less than commonly flame-resistant fibers. A fabric's thermal conductivity (k_{fab}) was determined from the fiber to air fraction in dry fabric [11] as

$$k_{fab}(T) = V_{air}\% k_{air}(T) + (1 - V_{air}\%) k_{fiber}(T) \qquad (12.2)$$

where $V_{air}\%$ is the air content percentage contained in the fabric's pores, and k_{fiber} and k_{air} are the fiber thermal conductivity and the thermal conductivity of the air, respectively. The air content percentage ($V_{air}\%$) can be calculated using the density of fibers and air [11], which can be expressed as

$$V_{air}\% = \frac{\rho_{fiber} - \rho_{fab}}{\rho_{fiber} - \rho_{air}} \times 100\% \qquad (12.3)$$

where ρ_{fiber}, ρ_{fab}, and ρ_{air} are the fiber density, the fabric density, and the air density, respectively. The density of Nomex fibers is given by Futschik and Witte [26] as approximately 1443 kg/m³. The density of air at ambient conditions is 1.2 kg/m³ [27].

12.3.1.2 Convection

Convective heat transfer, while minimal in protective clothing, can occur in an air gap between protective clothing and a body's surface. In theory, conductive heat transfer presents a decrease with a rising of air gap size, while heat transfer driven by natural convection (due to density difference and temperature gradient) increases. The Rayleigh number (Ra) can be used to examine whether convective heat transfer will be significant, which is given as [28]

$$Ra = \frac{g\beta \Delta T L_{air}^3}{\alpha v} \qquad (12.4)$$

where g is the gravitational acceleration, β is the coefficient of thermal expansion, L_{air} is the air gap thickness, α is the thermal diffusivity of air, v is the dynamic viscosity of air, and ΔT is the temperature difference between the boundaries of the air gap. Heat transfer driven by steady natural convection in a horizontal air gap occurs when Ra is more than 1708 [11]. Transient natural convective heat transfer can occur for Ra greater than 5840 [11]. Torvi et al. [29] used flow visualization experiments to prove the initial of natural convection for an air gap of about 6–7 mm or above. However, there was a significant difference in heat transfer between horizontal and vertical air gap [23]. It was reported that a vertical air gap can present natural convection when Ra is more than 1000 [27].

In firefighting operations, the convective heat transfer in air gaps is driven not only by natural convection but also by forced convection due to body movement and thermal shrinkage of fabric [17,19]. This natural and forced convective heat transfer can both be calculated using Newton law of cooling, written as

$$q_{conv} = h(T_{fab} - T_{skin}) \qquad (12.5)$$

$$h = Nu\frac{k}{L_{air}} \tag{12.6}$$

where T_{fab} is the temperature of fabric's backside, T_{skin} is the temperature of skin's surface, and h is the convective heat transfer coefficient in an air gap that can be obtained by the Nusselt number (Nu). Buoyancy force resulting from density gradient of the air gap facilitates convective heat transfer. The orientation of an air gap has an important effect on buoyancy force. Nu for horizontal and vertical air gaps are respectively calculated by the following equations [28,30]:

$$Nu = \begin{cases} Nu \text{ for horizontal air gap:} \\ 1+1.44\left[1-\frac{1708}{Ra}\right]^{-}+\left[\left(\frac{Ra}{5830}\right)^{1/3}-1\right]^{-} \\ Nu \text{ for vertical air gap:} \\ \begin{cases} 1.0 & Ra \leq 1713 \\ 0.112Ra^{0.294} & Ra \geq 1713 \end{cases} \end{cases} \tag{12.7}$$

12.3.1.3 Radiation

Radiant heat transfer in porous materials is a complex process, involving absorbing, emitting, scattering, and transmitting [31]. Radiant heat transfer was not simulated in firefighting protective clothing in earlier studies. Based on guarded hot plate experiments in a vacuum, the importance of radiant heat transfer was questioned by Vershoor and Greebler [32] and other researchers until the 1950s. Since then, some experimental and numerical analysis methods have been employed to investigate the science of radiant heat transfer in porous media. In previous studies, some conduction models were used to approximately describe the principle of thermal radiation in porous media [33]. However, in scenarios for which heat transfer due to conduction was the main mode to transport thermal energy, these conduction models would not be suitable to simulate radiant heat transfer; in addition, these early models were not able to evaluate the effect of coupling radiation and conduction.

To deal with the above deficiencies, a two-flux model was proposed by Larkin to simulate radiant heat transfer in fibrous materials while assessing absorption and back-scattering [34]. At the end of the twentieth century, Tong et al. [35] proposed a spectral two-flux model of porous media to account for thermal radiation with absorbing, emitting, and scattering, which can be expressed as

$$\frac{dq_\lambda^+(x)}{dx} = -2\sigma_{a\lambda}q_\lambda^+(x) - 2\sigma_{s\lambda}b_\lambda q_\lambda^+(x) + 2\sigma_{a\lambda}e_{b\lambda}(T) + 2\sigma_{s\lambda}b_\lambda q_\lambda^-(x)$$

$$\frac{dq_\lambda^-(x)}{dx} = 2\sigma_{a\lambda}q_\lambda^-(x) + 2\sigma_{s\lambda}b_\lambda q_\lambda^-(x) - 2\sigma_{a\lambda}e_{b\lambda}(T) - 2\sigma_{s\lambda}b_\lambda q_\lambda^+(x) \tag{12.8}$$

where q_λ^+ and q_λ^- are the forward and backward radiant heat flux of micro-units, respectively, $\sigma_{a\lambda}$ and $\sigma_{s\lambda}$ are the average absorptivity and scattering coefficient, respectively, $e_{b\lambda}$ is the thermal radiation from blackbody emission, and b_λ is the back-scattering factor. There are two terms relating to scattering of radiation in Equation 12.8. The first is the second term on the right-hand side (Equations 12.8a and 12.8b), which represents the attenuation of energy by scattering, while the final term on the right-hand side represents the augmentation of thermal radiation owing to scattering.

To simplify the complexity of this solution for a two-flux radiant model, the effect of scattering on thermal radiation in fibrous materials can be ignored, as back-scattering and forward-scattering of radiation balance out in terms of fabric structure and one-dimensional heat transfer [36]. Therefore, the simple equations for radiant heat transfer are given as

$$\frac{dq_\lambda^+(x)}{dx} = -2\sigma_{a\lambda}q_\lambda^+(x) + 2\sigma_{a\lambda}e_{b\lambda}(T)$$

$$\frac{dq_\lambda^-(x)}{dx} = 2\sigma_{a\lambda}q_\lambda^-(x) - 2\sigma_{a\lambda}e_{b\lambda}(T)$$

(12.9)

Since its development, the two-flux model has been widely used by some researchers in modeling radiant heat transfer in fibrous insulation to investigate thermal and moisture performance under various experimental conditions [36]. Wan and Fan [37] employed a two-flux model to analyze radiant heat transfer in fibrous assemblies incorporating reflective interlayers. Fu et al. [8] investigated the influence of thermal radiation absorption due to moisture under low-level radiation exposure based on a similar two-flux model. Furthermore, the radiant transfer equation was used to simulate heat transfer in air gaps between different clothing layers [18] and between clothing and the human body [38]. The effect of self-emission and absorption of air gaps in fabric systems was studied in dynamic air gap models investigating body motion and thermal shrinkage of fabric [17,19].

Another meaningful radiative transfer model was used and validated experimentally by Torvi [11]. According to the structural model of fabric and the radiative properties, this study indicated that radiative transfer in the tested fabric conformed to Beer's law, as shown in Equation 12.10.

$$\frac{dq_\lambda(x)}{dx} = -\gamma q_\lambda(0)\exp(-\gamma x)$$

(12.10)

That is to say, thermal radiation through the fabric system can exponentially decay due to the fabric's absorption and scattering [39]. The extinction coefficient (γ) of fabric for thermal radiation was obtained by inputting absorption coefficient ($\sigma_{a\lambda}$) and scattering coefficient ($\sigma_{s\lambda}$) of the fabric [5]. Beer's law has been popularly used to analyze radiant heat transfer in thermal protection, such as thermal radiation in single-layer fabric exposed to 50/50 radiant/convective heat flux [12], thermal radiation in multilayer fabric systems under low heat intensity [40], thermal radiation in a cylindrical fabric model considering the effect of body geometry on heat transmission [13], and thermal radiation in a clothing model in a flash fire scenario [16]. According to the definition, Beer's law ignores scattering and emission source resulting from incident radiant heat in all directions, which aligns with the phenomenon that the decay of thermal radiation through a medium is far more than the emission of thermal radiation or the incident radiant heat (unless in a forward direction) [31]. Recently, an improved radiant heat transfer model considering the effect of emissions was derived and compared with the previous Beer's law model and experimental test [41]. The results, in contrast with previous research, showed that self-emission in multi-layer fabric system increases the rate of thermal energy transferred to human skin during heat exposure and ambience during cooling.

12.3.2 Development on Heat and Moisture Transfer Model

Moisture within porous fabric has three phases: solid (bound water), liquid, or vapor. The fiber and the absorbed water become the solid phase of the fabric, while the mixture of entrapped air and water vapor comprises the gas phase. It is usually supposed that the solid phase is stationary, and the swelling or shrinkage of fiber is ignored in most previous models. Early attempts to understand heat and moisture transfer through textile fabric were done by Henry [42] in the 1930s. Two parabolic partial differential equations were developed to simulate heat and moisture transfer behavior, and the coupled terms in these equations could explain sorption/desorption and latent heat transport of fiber. In the 1980s, Ogniewicz and Tien [43] first described water condensation in a heat and moisture transfer model. The results demonstrated that condensation rate and the resulting increase in heat transfer showed an increase with external humidity, temperature levels, and overall temperature differences. But the condensation was assumed to be in a pendulum state in this model.

Later, a more precise numerical model was established for analyzing moisture sorption and condensation. With regards to strongly hydrophilic fibers (such as wool and cotton), the moisture sorption process can be divided into two stages based on the adsorption kinetics of fiber [44]. In the first stage, the moisture sorption process is simulated on the basis of the Fickian diffusion with a constant diffusion coefficient. In the second stage, absorbed water can bring about the changes of fiber structure so that the rate of moisture absorption has a decrease with time [45]. The second-stage sorption of wool fibers follows an exponential relationship obtained by the experimental test results. Another commonly used heat and moisture transfer model was developed by Gibson [46] and based on continuous medium theory. The laws of energy conservation, mass conservation, and momentum conservation were employed to study heat and moisture transfer behavior that could reflect the distribution of temperature, concentration of each phase, and total pressure in fibrous material.

Chen [47] first described heat and moisture transfer in thermal protective clothing under radiant heat exposure based on a simple numerical model in 1959. It was assumed in this model that moisture transfer was driven by molecular diffusion, and the effect of phase change on heat transfer was ignored. In addition, the developed model was not effectively validated with the results measured by experimental method. Later, Prasad et al. [48] developed a heat and moisture transfer model in low heat intensity, which considered radiative heat transfer, phase change, and absorption/desorption of moisture in protective clothing. The numerical model was combined with the moisture sorption isotherm of fiber to calculate the distribution of temperature and moisture concentration in protective clothing. This model was further improved by Fu et al. [8] to investigate the effect of moisture within protective clothing on radiative heat transfer. The two-flux radiant heat transfer model was used to consider the effect of moisture on absorptivity and transmitivity of radiation. This improved model more precisely presented heat and moisture transfer behavior in firefighting protective clothing.

Vafai and Sozen [49] summarized and compared these classic heat and moisture transfer models. The results demonstrated that the heat and moisture transfer model developed by Gibson and Charmchi [46] was most suitable to simulate heat and moisture transfer in protective clothing for high-intensity heat exposures. However, this model did not address radiant heat transfer in clothing, in contrast to Torvi's model, which employed Beer's law to calculate the attenuation of thermal radiation in fibrous material. Thus, Chitrphiromsri and Kuznetsov [50] and Song et al. [51] combined Gibson and Charmchi's model with Torvi's model [5] to study moisture phase change, sorption/desorption, molecular diffusion, convection, and capillary liquid diffusion in thermal protective clothing while exposed to flame. The predictive results from this developed model indicated good agreement with experimental results for heat and flame. In addition, the developed numerical model was used by Chitrphiromsri et al. [52] to study "intelligent" multilayer firefighter protective clothing, which absorbed a significant amount of incident heat flux due to evaporation of the injected water during exposure, thus limiting temperature increase on the skin.

12.4 Modeling of Skin Heat Transfer and Burn Prediction

Firefighters in the act of suppressing fires usually encounter two thermal hazards, heat stress and skin burn injuries [53,54]. Heat stress results from metabolic heat produced by firefighters expending a large amount of energy. Metabolic heat energy can be decreased with thermal regulation of the body, for example, by increasing blood flow and heat transfer toward the environment [55]. However, because firefighters are required to wear personal protective equipment (PPE) and self-contained breathing apparatuses (SCBA) with large weight and low vapor permeability [56], it can be difficult to transport metabolic heat energy to the environment. Core temperature can be increased in this situation, thus leading to heat stress. Skin burn injuries are caused by ambient heat transfer (conduction, convection, and radiation), as the ambient temperature during heat exposure is far higher than the skin temperature. To evaluate skin burn injuries, researchers developed skin bio-heat transfer model that can simulate heat transfer in body tissues layers. The time to cause skin burns is obtained by calculating skin burn models, for example, Stoll criterion and Henriques' burn integral model.

12.4.1 Skin Bio-Heat Transfer Model

According to the depth of skin burn, burn injuries can be classified as first degree, second degree, third degree, and fourth degree based on the impact on the epidermis, dermis, and subcutaneous tissue. The heat transfer in fat, muscle, and core is usually ignored in skin burn predictive models, while blood perfusion and the metabolism can be considered by inserting two internal source functions. Most skin bio-heat transfer models are derived from Fourier's law. The assumption is that there is only one-dimensional heat transfer along the depth of skin and the thermal properties of each skin layer are constant. Also, the blood temperature is equal to the body's core temperature, and the flow rate of partial blood remains stable. The commonly used bio-heat transfer model in the prediction of skin burn is the Pennes' bio-heat transfer model, given as [57]

$$(\rho c_p)_{skin} \frac{\partial T}{\partial t} = \frac{\partial}{\partial x}\left(k_{skin}\frac{\partial T}{\partial x}\right) + w_b(\rho c_p)_b(T_b - T) + G_m \qquad (12.11)$$

where ρ_{skin} and $(c_p)_{skin}$ are the density and specific heat of each layer skin tissue, respectively, ρ_b and $(c_p)_b$ are the density and specific heat of blood, respectively, w_b is the rate of blood perfusion, T_b is the blood temperature, and G_m is the metabolic heat production. The mathematical form of perfusion heat source results from the assumption that tissue is supplied with a large number of capillary blood vessels uniformly distributed throughout its volume (a soft tissue model). The metabolic heat source can be measured by a temperature-dependent function, but it is usually treated as a constant value. On the basis of the Pennes' bio-heat transfer model, Torvi [58] investigated the effect of blood flow and metabolic heat on skin heat transfer. The resulting skin bio-heat transfer model in ISO 13506 originated from Pennes' bio-heat transfer model, while a simple skin heat transfer model was used in ASTM F1930 and ASTM F2731. The amendment of thermal properties of skin layers was used in this latter model to simulate the effect of blood flow and metabolic heat on skin heat transfer.

However, the Pennes' bio-heat transfer model is not suitable to describe an unstable heat transfer process, especially when the human body is subjected to short duration and high-intensity heat exposure. This is because Fourier's thermal conduction equation is developed based on the assumption that the rate of heat transfer in a medium is infinite [59]; that is, any local temperature disturbance causes an instantaneous perturbation in temperature at each point in the medium. To deal with the limitations of Fourier's law, non-Fourier heat conduction model was studied extensively in recent decades. The thermal wave model of bio-heat transfer (TWMBT) was developed by Liu [60] for finite speed heat transfer, given as

$$\left(\tau w_b(\rho c_p)_b + (\rho c_p)_{skin}\right)\frac{\partial T}{\partial t} + \tau(\rho c_p)_{skin}\frac{\partial^2 T}{\partial t^2} = \frac{\partial}{\partial x}\left(k_{skin}\frac{\partial T}{\partial x}\right) + w_b(\rho c_p)_b(T_b - T) + G_m$$

$$(12.12)$$

where τ is the thermal relaxation time. The predicted results from TWMBT were compared with the results obtained from Pennes' bio-heat transfer model when the skin was exposed to constant-temperature or constant heat flux surface. The results demonstrated significant difference between results from the TWMBT and Pennes' bio-heat transfer model for higher-level heat exposure, indicating that the TWMBT model could be used to predict the time to cause skin burns for short duration and high-level heat exposures. Zhu and Zhang [61] compared the TWMBT model and Pennes' bio-heat transfer model under radiant heat exposure, and recommended the TWMBT model to calculate second-degree burn times for higher radiant heat exposure. Another non-Fourier bio-heat transfer model is the Dual Phase Lag (DPL) model developed by Tzou [62]. This model considered the effect of microstructural intersection for nonhomogeneous structures on heat transfer.

$$\left(\tau_q w_b (\rho c_p)_b + (\rho c_p)_{skin}\right)\frac{\partial T}{\partial t} + \tau(\rho c_p)_{skin}\frac{\partial^2 T}{\partial t^2}$$
$$= \frac{\partial}{\partial x}\left(k_{skin}\frac{\partial T}{\partial x}\right) + \tau_t k_{skin}\frac{\partial^3 T}{\partial t \partial x^2} + w_b(\rho c_p)_b(T_b - T) + G_m \tag{12.13}$$

The DPL model of bio-heat transfer was employed to predict skin burn time under flame and radiant heat exposure. Results were compared with the results from the TWMBT, Pennes' model, and Stoll's criterion, as shown in Figure 12.2 [63]. A similar trend between the DPL model and the TWMBT was observed, presenting a significant difference with Pennes' model and Stoll's criterion.

In addition, most of the skin bio-heat transfer models used to determine thermal protection do not evaluate the effect of moisture diffusion and evaporation on skin heat transfer. Shen et al. [64] developed a heat and moisture transfer model for skin tissues. This model assumed that moisture diffusion and evaporation existed in three-layer skin tissues. However, although moisture diffusion and evaporation had an impact on heat transfer in three-layer skin tissues, blood perfusion was found to have no effect on the epidermis layer and moisture evaporation occurred only on the skin surface [65]. An improved heat and moisture model was proposed by Fu et al. [66] to cope with the above drawbacks. It was found that the predictive error between the skin heat flux obtained from this heat and moisture transfer model and experimental tests was 4.91%, which was far less than that between the results obtained from the heat transfer model and experimental results (11.70%)[67].

12.4.2 Skin Burn Prediction Model

The skin bio-heat transfer model can be used to determine the temperature distribution in skin layers exposed to various thermal environments. The required times to cause second- and third-degree skin burns are calculated based on

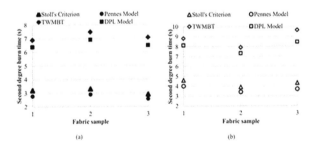

Figure 12.2

Comparison of second-degree burn time using Stoll's criterion, Pennes' model, TWMBT, and DPL model: (a) 80 kW/m² radiant heat exposure; (b) 80 kW/m² flame exposure. (From P. Talukdar, et al., *International Journal of Heat and Mass Transfer*, 78, 1068–1079, 2014.)

Henriques' burn integral model [68]. The calculated temperature at skin surface below 80 μm (or 200 μm) is used in Henriques' burn integral given as

$$\Omega = \int_0^t P \exp\left(-\frac{\Delta E}{RT}\right) dt \qquad (12.14)$$

where Ω is a quantitative measure of burn damage, R is the universal gas constant (8.31 J/mol °C), P and ΔE are the pre-exponential term and the activation energy of the skin, respectively, T is the absolute temperature at the basal layer or at any depth in the dermis, and t is the total time for which T is above 317.15 K. When Ω reaches a value of 1 at the epidermis–dermis interface and the dermis–subcutaneous tissue interface, the corresponding times are treated as second- and third-degree burn time, respectively. Full-scaled manikin tests, such as ISO 13506-2008, ASTM F1930-13, employ this method to assess second- or third-degree burn injuries in different body locations. Henriques' burn integral is also used in bench top tests including ASTM F2731-11, which can predict the minimum exposure time of firefighters during heat exposure and cooling phases. Theoretically, Henriques' burn integral model is suitable for evaluating skin burn injuries of different degrees under steady and transient heat exposure, but there are actually some limitations since these parameters in Henriques' burn integral model are different for different heat exposures, for example, pre-exponential and activation energy [69]. These values were suggested by Weaver and Stoll [70] for the basal layer and by Takata [71] for the dermal base.

Another widely used predictive model is Stoll's criterion, which presents the relationship between heat flux and tolerance time to second-degree burns under constant heat exposure levels (see Figure 12.3). Stoll criterion was developed by Stoll and Chianta [72] based on a series of animal tests. However, for more convenient predictions of burn time using Stoll's criterion, most researchers employed

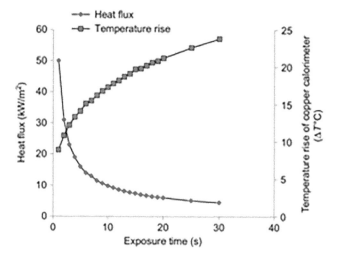

Figure 12.3

A relationship plot between exposure times, temperature increases, and heat flux based on Stoll's criterion. (From S. M. Guowen Song, Rene M. Rossi, *Thermal Protective Clothing for Firefighters*, Matthew Deans, UK, 2016.)

the relationship between temperature rise and time to second-degree burns to evaluate thermal protective performance [73]. This is because temperature changes on the skin's surface are obtained easily comparing with the variation in heat flux or thermal energy, for example, copper slug sensor. Because different sensors have different formulas between heat flux and temperature, the relationship between temperature rise and time to second-degree burns was not used for new sensors.

Recently, ASTM standard recommended measuring the relationship between accumulated energy on the skin's surface and second-degree burn time to predict skin burn time. Compared to the damage integral model, the criterion developed by Stoll is for a limited heat flux level (4–48 kW/m²), not suitable for a higher heat flux level [5]. Also, Stoll's criterion is suggested for rectangular heat pulses that are similar to constant heat exposure on unclothed human skin. Thus, Stoll's criterion is not suitable to evaluate transient heat exposure and third-degree burn injury. Therefore, Henriques' burn integral model is more flexible for various heat shape profiles. The difference between Stoll's criterion and Henriques' burn integral model was compared by Zhai and Li [74] when the protective materials were exposed to three levels of heat intensity. The results demonstrated that there was a discrepancy between the simulated burn curve and the Stoll curve at the beginning of a material test, so it was suggested to calibrate the damage integral model before application, especially for the new bio-heat transfer models.

12.5 Conclusions and Future Trends

Many heat and moisture transfer models have been developed for firefighting protective clothing in recent decades. By changing the boundary conditions of the external environment, the heat and moisture transfer model can be employed to simulate different exposure conditions, such as flash fire and high- and low-level thermal radiation. These heat and moisture transfer models can be used to develop a simulation platform, which is helpful for clothing designers and researchers in the engineering design process to identify the requirements of thermal protective clothing and determine the design scheme without making real prototypes, and thus reduce the design cost. The heat and moisture transfer model in firefighting protective clothing requires further development from the following three aspects.

12.5.1 Inverse Problem Designs of the Coupled Heat and Moisture Transfer Model

The objective of the development of heat and moisture transfer models is to study the mechanism associated with the heat and moisture transfer in clothing systems, and provide improvements on clothing comfort and functions. The optimal designs for comfort and function are based on the inverse solution of the numerical model. In other words, the optimal parameters for fabric and clothing, such as fabric thickness, density, thermophysical properties, and clothing air gap, are determined according to experimental conditions and human comfort requirements.

The inverse problems of textile material design on heat and moisture transfer properties are important and indispensable in applications, but there is little research exploring inverse problems of the heat and moisture transfer model in firefighting protective clothing. In recent years, Xu et al. [76,77] conducted some inverse designs of textiles based on the coupled heat and moisture transfer model under low-temperature conditions. The inverse heat and moisture transfer problems of thickness for single-layer and double-layer fabric in low-temperature conditions were studied by direct search-based optimization algorithms, such as the Golden Section method, particle swarm optimization (PSO), and the simulated annealing (SA) method [78–80]. The inverse problem of thermal conductivity determination under low temperature was solved based on the Fibonacci search method and the Genetic Algorithm (GA) [81].

12.5.2 Application of Computational Fluid Dynamics (CFD)

Multidimensional heat and moisture transfer has a great impact on heat transportation and moisture distribution in porous fabric. One-dimensional heat and moisture transfer along the thickness direction was mainly considered in previous heat and moisture transfer models. Recently, three-dimensional heat and moisture transfer model in the clothed body based on CFD has achieved rapid growth, especially in the thermal protective field [82–86]. Computational thermal manikins were developed to accurately simulate the irregular shape of a clothed body and analyze the air flow, radiant and convective heat transfer between various environments [83], and naked or clothed flame manikins under flash fire conditions [84,85]. In addition, the CFD method was employed to study the effect of air gap orientation and dynamic air gap on heat transfer through the air gap between firefighting protective clothing and the body's surface [22]. In the future, CFD should be used to provide a framework for modeling heat and moisture transfer in firefighting protective clothing, including radiative and convective heat transfer, diffusion and convection of liquid water and gas, capillary transport of liquids, sorption phenomena, and phase changes.

12.5.3 Integration of Heat and Moisture Transfer Model and Thermophysiological Model

The heat and moisture transfer model is usually integrated with a skin burn model to predict the time to produce skin burn injuries. There are few heat and moisture transfer models regarding heat stress imposed on firefighters. Firefighters are subject to a great deal of heat stress at fire sites [75]. Therefore, it is critical to integrate heat and moisture transfer models and thermal regulation models into the development process for firefighting protective clothing to determine body and core temperature distribution, especially for low-intensity thermal exposure. Due to the differences in thermophysiological characteristics throughout the body, multinode and multisegment thermal regulation models should be developed for presenting the thermal response of different body parts. To consider the effect of the body's geometrical shape in clothing, different air gap sizes should be evaluated for heat and moisture transfer models. Heat stress can be obtained based on these coupled models for exposure to various thermal environments. The relationship between heat stress and skin burn time can be analyzed for the same heat exposures, which will provide some effective instruction to simultaneously protect against skin burn injuries and heat stress.

References

1. G. Song, Modeling thermal protection outfits for fire exposures (2003).
2. K. L. Watson, From Radiant Protective Performance to RadManTM: The Role of Clothing Materials in Protecting against Radiant Heat Exposures in Wildland Forest Fires (2014).
3. R. L. Barker, A. S. Deaton, K. A. Ross, Heat Transmission and Thermal Energy Storage in Firefighter Turnout Suit Materials, *Fire Technology*, 47(3) (2011) 549–563.
4. G. Song, Clothing air gap layers and thermal protective performance in single layer garment, *Journal of Industrial Textiles*, 36(3) (2007) 193–205.
5. D. A. Torvi, *Heat Transfer In Thin Fibrous Materials Under High Heat Flux Conditions*, University of Alberta, 1997.
6. B. Das, M. D. Araujo, V. K. Kothari, R. Fangueiro, A. Das, Modeling and Simulation of Moisture Transmission through Fibrous Structures Part I: Water Vapour Transmission, *Journal of Fiber Bioengineering and Informatics*, 5(4) (2012) 341–358.
7. P. Łapka, P. Furmański, T. Wisniewski, Numerical modelling of transient heat and moisture transport in protective clothing, *Journal of Physics*: Conference Series, IOP Publishing, 2016, pp. 012–014.
8. M. Fu, M. Q. Yuan, W. G. Weng, Modeling of heat and moisture transfer within firefighter protective clothing with the moisture absorption of thermal radiation, *International Journal of Thermal Sciences*, 96 (2015) 201–210.
9. Y. Su, J. Li, Development of a test device to characterize thermal protective performance of fabrics against hot steam and thermal radiation, *Measurement Science & Technology*, 27(12) (2016) 125904.
10. H. L. Morse, J. G. Thompson, K. J. Clark, K. A. Green, C. B. Moyer, Analysis of the thermal response of protective fabrics, DTIC Document, 1973.
11. D. A. Torvi, J. D. Dale, Heat transfer in thin fibrous materials under high heat flux, *Fire Technology*, 35(3) (1999) 210–231.
12. S. Kukuck, K. Prasad, Thermal Performance of Fire Fighters' Protective Clothing. 3. Simulating a TPP Test for Single-Layered Fabrics, National Institute of Standards and Technology, Gaithersburg, MD (2003).
13. F. L. Zhu, W. Y. Zhang, G. W. Song, Heat transfer in a cylinder sheathed by flame-resistant fabrics exposed to convective and radiant heat flux, *Fire Safety Journal*, 43(6) (2008) 401–409.
14. G. N. Mercer, H. S. Sidhu, Mathematical modelling of the effect of fire exposure on a new type of protective clothing, *Australian & New Zealand Industrial & Applied Mathematics Journal* (2007).
15. X. Li, Y. Wang, Y. Lu, Effects of body postures on clothing air gap in protective clothing, *Journal of Fiber Bioengineering and Informatics*, 4(3) (2011) 277–283.
16. G. Song, R. L. Barker, H. Hamouda, A. V. Kuznetsov, P. Chitrphiromsri, R. V. Grimes, Modeling the Thermal Protective Performance of Heat Resistant Garments in Flash Fire Exposures, *Textile Research Journal*, 74(12) (2004) 1033–1040.
17. A. Ghazy, Influence of Thermal Shrinkage on Protective Clothing Performance during Fire Exposure: Numerical Investigation, *Mechanical Engineering Research*, 4(2) (2014).

18. A. Ghazy, D. J. Bergstrom, Numerical Simulation of Heat Transfer in Firefighters' Protective Clothing with Multiple Air Gaps during Flash Fire Exposure, *Numerical Heat Transfer, Part A: Applications*, 61(8) (2012) 569–593.

19. A. Ghazy, D. J. Bergstrom, Numerical simulation of the influence of fabric's motion on protective clothing performance during flash fire exposure, *Heat and Mass Transfer*, 49(6) (2013) 775–788.

20. Y. Y. Wang, Z. L. Wang, X. Zhang, M. Wang, J. Li, CFD simulation of naked flame manikin tests of fire proof garments, *Fire Safety Journal*, 71 (2015) 187–193.

21. Y. Y. Jiang, E. Yanai, K. Nishimura, H. L. Zhang, N. Abe, M. Shinohara, K. Wakatsuki, An integrated numerical simulator for thermal performance assessments of firefighters' protective clothing, *Fire Safety Journal*, 45(5) (2010) 314–326.

22. P. Talukdar, A. Das, R. Alagirusamy, Numerical modeling of heat transfer and fluid motion in air gap between clothing and human body: Effect of air gap orientation and body movement, *International Journal of Heat and Mass Transfer*, 108 (2017) 271–291.

23. Udayraj, P. Talukdar, A. Das, R. Alagirusamy, Numerical investigation of the effect of air gap orientations and heterogeneous air gap in thermal protective clothing on skin burn, *International Journal of Thermal Sciences*, 121 (2017) 313–321.

24. Y. Su, J. He, J. Li, Modeling the transmitted and stored energy in multilayer protective clothing under low-level radiant exposure, *Applied Thermal Engineering*, 93 (2016) 1295–1303.

25. Y. Su, J. He, J. Li, A model of heat transfer in firefighting protective clothing during compression after radiant heat exposure, *Journal of Industrial Textiles* (2016) 1528083716644289.

26. M. W. Futschik, L. C. Witte, Effective Thermal Conductivity of Fibrous Materials, ASME-PUBLICATIONS-HTD, 271 (1994) 123–123.

27. T. L. Bergman, F. P. Incropera, *Fundamentals of heat and mass transfer*, 7th ed., Wiley, Hoboken, NJ, 2011.

28. K. Hollands, G. D. Raithby, L. Konicek, Correlation equations for free convection heat transfer in horizontal layers of air and water, *International Journal of Heat & Mass Transfer*, 18(7) (1975) 879–884.

29. D. A. Torvi, J. Douglas Dale, B. Faulkner, Influence of Air Gaps On Bench-Top Test Results of Flame Resistant Fabrics, *Journal of Fire Protection Engineering*, 10(1) (1999) 1–12.

30. S. Ostrach, Natural convection in enclosures, *Journal of Heat Transfer*, 110(4b) (1988) 1175–1190.

31. M. F. Modest, *Radiative heat transfer*, 2nd ed., Academic Press, Amsterdam; Boston, 2003.

32. J. Verschoor, P. Greebler, Heat transfer by gas conduction and radiation in fibrous insulation (1952).

33. E. F. M. V. D. Held, The contribution of radiation to the conduction of heat, *Applied Scientific Research*, 4(3) (1953) 77–99.

34. B. K. Larkin, S. W. Churchill, Heat transfer by radiation through porous insulations, *Aiche Journal*, 5(4) (1959) 467–474.

35. T. W. Tong, C. L. Tien, T. W. Tong, C. L. Tien, Thermal radiation in fibrous insulations, *Journal of Building Physics*, 4(1) (1980) 27–44.

36. B. Farnworth, Mechanisms of Heat Flow Through Clothing Insulation, *Textile Research Journal*, 53(53) (1983) 717–725.

37. X. Wan, J. Fan, Heat transfer through fibrous assemblies incorporating reflective interlayers, *International Journal of Heat and Mass Transfer*, 55(25) (2012) 8032–8037.

38. A. Ghazy, D. J. Bergstrom, Numerical Simulation of Transient Heat Transfer in a Protective Clothing System during a Flash Fire Exposure, *Numerical Heat Transfer, Part A: Applications*, 58(9) (2010) 702–724.

39. F. P. Incropera, F. P. Incropera, *Fundamentals of heat and mass transfer*, 6th ed., John Wiley, Hoboken, NJ, 2007.

40. Y. Su, J. He, J. Li, Modeling the transmitted and stored energy in multi-layer protective clothing under low-level radiant exposure, *Applied Thermal Engineering*, 93 (2015) 1295–1303.

41. Y. Su, J. He, J. Li, An improved model to analyze radiative heat transfer in flame-resistant fabrics exposed to low-level radiation, *Textile Research Journal*, (2016) 0040517516660892.

42. P. S. H. Henry, Diffusion in Absorbing Media, *Proceedings of the Royal Society*, A Mathematical Physical & Engineering Sciences, 171(945) (1939) 215–241.

43. Y. Ogniewicz, C. L. Tien, Analysis of condensation in porous insulation, *International Journal of Heat & Mass Transfer*, 24(3) (1981) 421–429.

44. S. Mandal, G. Song, Thermal sensors for performance evaluation of protective clothing against heat and fire: a review, *Textile Research Journal*, 85(1) (2015) 101–112.

45. Y. Lu, J. Li, X. Li, G. Song, The effect of air gaps in moist protective clothing on protection from heat and flame, *Journal of Fire Sciences*, 31(2) (2013) 99–111.

46. P. W. Gibson, M. Charmchi, Modeling convection/diffusion processes in porous textiles with inclusion of humidity-dependent air permeability, *Int Commun Heat Mass*, 24(5) (1997) 709–724.

47. N. Y. Chen, *Transient heat and moisture transfer to skin through thermally-irradiated cloth*, Massachusetts Institute of Technology, 1959.

48. K. Prasad, W. H. Twilley, J. R. Lawson, Thermal performance of fire fighters' protective clothing: numerical study of transient heat and water vapor transfer, US Department of Commerce, Technology Administration, National Institute of Standards and Technology, 2002.

49. K. Vafai, M. Sozen, A Comparative analysis of multiphase transport models in porous media, *Annual Review of Heat Transfer* (1990).

50. P. Chitrphiromsri, A. V. Kuznetsov, Modeling heat and moisture transport in firefighter protective clothing during flash fire exposure, *Heat and Mass Transfer*, 41(3) (2005) 206–215.

51. G. Song, P. Chitrphiromsri, D. Ding, Numerical simulations of heat and moisture transport in thermal protective clothing under flash fire conditions, *Int J Occup Saf Ergo*, 14(1) (2008) 89–106.

52. P. Chitrphiromsri, *Modeling of Thermal Performance of Firefighter Protective Clothing during the Intense Heat Exposure* (2005).

53. J. Malchaire, A. Piette, B. Kampmann, P. Mehnert, H. Gebhardt, G. Havenith, E. Den Hartog, I. Holmer, K. Parsons, G. Alfano, Development and validation of the predicted heat strain model, *Annals of Occupational Hygiene*, 45(2) (2001) 123–135.

54. R. Rossi, Fire fighting and its influence on the body, *Ergonomics*, 46(10) (2003) 1017–1033.

55. N. Sugimoto, S. Sakurada, O. Shido, Changes in ambient temperature at the onset of thermoregulatory responses in exercise-trained rats, *International Journal of Biometeorology*, 43(4) (2000) 169–171.

56. J.-H. Kim, W. J. Williams, A. Coca, M. Yokota, Application of thermo-regulatory modeling to predict core and skin temperatures in firefighters, *International Journal of Industrial Ergonomics*, 43(1) (2013) 115–120.

57. H. H. Pennes, Analysis of tissue and arterial blood temperatures in the resting human forearm, *Journal of Applied Physiology*, 1(2) (1948) 93–122.

58. D. A. Torvi, J. D. Dale, A finite element model of skin subjected to a flash fire, *J Biomech Eng*, 116(3) (1994) 250–255.

59. P. Talukdar, A. Das, R. Alagirusamy, Heat and mass transfer through thermal protective clothing–A review, *International Journal of Thermal Sciences*, 106 (2016) 32–56.

60. J. Liu, X. Chen, L. X. Xu, New thermal wave aspects on burn evaluation of skin subjected to instantaneous heating, *Ieee T Bio-Med Eng*, 46(4) (1999) 420–428.

61. F. Zhu, W. Zhang, G. Song, Thermal performance assessment of heat resistant fabrics based on a new thermal wave model of skin heat transfer, *Int J Occup Saf Ergo*, 12(1) (2006) 43–51.

62. D. Tzou, A unified field approach for heat conduction from macro-to micro-scales, *Journal of Heat Transfer*, 117(1) (1995) 8–16.

63. P. Talukdar, R. Alagirusamy, A. Das, Heat transfer analysis and second degree burn prediction in human skin exposed to flame and radiant heat using dual phase lag phenomenon, *International Journal of Heat and Mass Transfer*, 78 (2014) 1068–1079.

64. W. S. Shen, J. Zhang, F. Q. Yang, Skin thermal injury prediction with strain energy, *Int J Nonlin Sci Num*, 6(3) (2005) 317–328.

65. D. J. Maitland, D. C. Eder, R. A. London, M. E. Glinsky, B. A. Soltz, Dynamic simulations of tissue welding, *P Soc Photo-Opt Ins*, 2671 (1996) 234–242.

66. M. Fu, W. Weng, H. Yuan, Numerical Simulation of the Effects of Blood Perfusion, Water Diffusion, and Vaporization on the Skin Temperature and Burn Injuries, *Numerical Heat Transfer, Part A: Applications*, 65(12) (2014) 1187–1203.

67. S. C. Jiang, N. Ma, H. J. Li, X. X. Zhang, Effects of thermal properties and geometrical dimensions on skin burn injuries, *Burns: Journal of the International Society for Burn Injuries*, 28(8) (2002) 713–717.

68. F. C. Henriques, Jr., Studies of thermal injury; the predictability and the significance of thermally induced rate processes leading to irreversible epidermal injury, *Archives of Pathology*, 43(5) (1947) 489–502.

69. L.-N. Zhai, J. Li, Prediction methods of skin burn for performance evaluation of thermal protective clothing, *Burns: Journal of the International Society for Burn Injuries* (2015).

70. J. A. Weaver, A. M. Stoll, Mathematical model of skin exposed to thermal radiation, *Aerospace Medicine*, 40(1) (1969) 24–30.

71 A. Takata, Development of criterion for skin burns, *Aerospace Medicine*, 45(6) (1974) 634–637.

72 A. M. Stoll, M. A. Chianta, A method and rating system for evaluation of thermal protection, DTIC Document, 1968.

73. Y. Su, Y. Wang, J. Li, Evaluation method for thermal protection of firefighters' clothing in high-temperature and high-humidity condition: A review, *International Journal of Clothing Science and Technology*, 28(4) (2016) 429–448.

74. L.-N. Zhai, J. Li, Correlation and difference between stoll criterion and damage integral model for burn evaluation of thermal protective clothing, *Fire Safety Journal*, 86 (2016) 120–125.

75. S. M. Guowen Song, Rene M. Rossi., *Thermal Protective Clothing for Firefighters*, Matthew Deans, UK, 2016.

76. D. Xu, Inverse problems of textile material design based on clothing heat-moisture comfort, *Applicable Analysis*, 93(11) (2014) 2426–2439.

77. D. Xu, Y. Chen, X. Zhou, Type design for bilayer textile materials under low temperature: Modeling, numerical algorithm and simulation, *International Journal of Heat and Mass Transfer*, 60 (2013) 582–590.

78. D. Xu, J. Cheng, X. Zhou, An inverse problem of thickness design for single layer textile material under low temperature, *Journal of Math-for-Industry*, 2(2010B-4) (2010) 139–146.

79. D. Xu, J. Cheng, Y. Chen, M. Ge, An inverse problem of thickness design for bilayer textile materials under low temperature, in: Journal of physics: Conference series, IOP Publishing, 2011, pp. 012–018.

80. D. Xu, M. Ge, Thickness determination in textile material design: dynamic modeling and numerical algorithms, *Inverse Problems*, 28(3) (2012) 035011.

81. Y. Yu, D. Xu, On the inverse problem of thermal conductivity determination in nonlinear heat and moisture transfer model within textiles, *Applied Mathematics and Computation*, 264 (2015) 284–299.

82. Y. Yan Jiang, E. Yanai, K. Nishimura, H. Zhang, N. Abe, M. Shinohara, K. Wakatsuki, An integrated numerical simulator for thermal performance assessments of firefighters' protective clothing, *Fire Safety Journal*, 45(5) (2010) 314–326.

83. Z. Wang, J. Li, M. Tian, CFD Simulation of Flame Manikin Test for Fire Proof Garment during Flash Fire Exposures and Cooling Phase, in: 2015 International Conference on Education, Management, Information and Medicine, Atlantis Press, 2015.

84. M. Tian, Z. Wang, J. Li, 3D numerical simulation of heat transfer through simplified protective clothing during fire exposure by CFD, *International Journal of Heat and Mass Transfer*, 93 (2016) 314–321.

85. J. Li, M. Tian, Personal thermal protection simulation under diverse wind speeds based on life-size manikin exposed to flash fire, *Applied Thermal Engineering*, 103 (2016) 1381–1389.

86. M. Tian, J. Li, Simulating the thermal response of the flame manikin with different materials exposed to flash fire by CFD, *Fire and Materials* (2016).

Index

W

Weber–Fechner law, 78
Wet Bulb Globe Temperature (WBGT), 72
Wi-Fi technology, 316, 318
Wireless area network (WAN), 320
Wireless Body Area Sensor Networks (WBASN) (smart firefighting clothing), 316, 318–320
    autonomous WBASN, 318
    intelligent WBASN, 318
    i-Protect project, 319
    managed WBASN, 318
    PPE systems, 319

    PROeTEX project, 319
    ProFiTEX project, 320
    Smart@Fire project, 320
    Smart LifeLine, 320
    Smart Personal Protective System, 320
Wireless local area network (WLAN), 317

Y

Yarn types and properties, 38

Z

Zarlink technology, 316
ZigBee, 316, 317
Zylon, 8

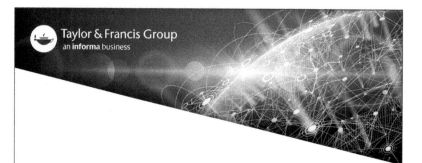